SMOKE AND MIRRORS:
GLOBALIZED TERRORISM AND THE
ILLUSION OF MULTILATERAL SECURITY

FRANK P. HARVEY

Smoke and Mirrors

Globalized Terrorism
and the Illusion of Multilateral Security

UNIVERSITY OF TORONTO PRESS
Toronto Buffalo London

© University of Toronto Press Incorporated 2004
Toronto Buffalo London
Printed in Canada

ISBN 0-8020-8948-8

Printed on acid-free paper

National Library of Canada Cataloguing in Publication

Harvey, Frank P. (Frank Paul), 1962–
Smoke and mirrors : global terrorism and the illusion of
multilateral security / Frank P. Harvey.

Includes bibliographical references and index.
ISBN 0-8020-8948-8

1. United States – Foreign relations – 2001– .   2. National security –
United States.   3. Terrorism – Political aspects – United States.
4. Security, International.   5. Canada – Military policy.   I. Title.

E895.H37 2004   327.73'009'0511   C2004-900722-X

This book has been published with the help of a grant from the Canadian
Federation for the Humanities and Social Sciences, through the Aid to
Scholarly Publications Programme, using funds provided by the Social
Sciences and Humanities Research Council of Canada.

University of Toronto Press acknowledges the financial assistance to its
publishing program of the Canada Council for the Arts and the Ontario
Arts Council.

University of Toronto Press acknowledges the financial support for its
publishing activities of the Government of Canada through the Book
Publishing Industry Development Program (BPIDP).

To Anupa, Kalli G, and Jay

Contents

Acknowledgments

I am grateful to the Social Science and Humanities Research Council of Canada (grant # 410-99-1235), the International Security Research and Outreach Program (ISROP, Arms Control and Disarmament Division, Department of Foreign Affairs and International Trade, Ottawa, Canada), the Security and Defence Forum (SDF, Directorate of Public Policy, Department of National Defence, Ottawa, Canada), and the Centre for Foreign Policy Studies (CFPS, Dalhousie University, Halifax, Canada) for research and financial support to complete various parts of this book. I would like to thank Graham Walker, Marcia Seitz Ehler, and Jay Nathwani for their excellent research assistance.

Portions of chapters 1 and 4 through 8 are derived from my writing in 'Addicted to Security: Globalized Terrorism and the Inevitability of American Unilateralism,' *International Journal* 59, no. 1 (Winter 2003–2004): 27–58; 'Dispelling the Myth of Multilateral Security After 11 September and the Implications for Canada,' in David Carment, Fen Osler Hampson and Norman Hilmer, eds., *Canada among Nations 2003: Coping with the American Colossus* (London: Oxford University Press, 2003), 200–218; 'The Future of Strategic Stability and Nuclear Deterrence,' *International Journal* 58, no. 2 (Spring 2003): 321–46; 'National Missile Defence Revisited, Again,' *International Journal* 56, no. 3 (Spring 2001): 347–60; 'Proliferation, Rogue-State Threats and National Missile Defence: Assessing European Concerns and Interests,' *Canadian Military Journal* 1, no. 4 (Winter 2000–2001): 69–78; and 'The International Politics of National Missile Defense: A Response to the Critics,' *International Journal* 55, no. 4 (Autumn 2000): 545–66.

Abbreviations

ABM	anti–ballistic missile
BMD	ballistic missile defence
CBRN	chemical, biological, radiological, and nuclear
CSIS	Canadian Security Intelligence Service
DFAIT	Department of Foreign Affairs and International Trade
IAEA	International Atomic Energy Agency
ICBM	inter-continental ballistic missile
ICC	International Criminal Court
ICT	information and communication technology
MAD	mutually assured destruction
MIC	military industrial complex
MIRV	multiple independently targeted re-entry vehicles
MOU	memorandum of understanding
NACD	non-proliferation, arms control, and disarmament
NATO	North Atlantic Treaty Organization
NPR	Nuclear Posture Review
NPT	Non-proliferation Treaty
R&D	research and development
START	Strategic Arms Reduction Treaties
TMD	theatre missile defence
UNSC	United Nations Security Council
USNSS	United States National Security Strategy
WMD	weapons of mass destruction

SMOKE AND MIRRORS:
GLOBALIZED TERRORISM AND THE
ILLUSION OF MULTILATERAL SECURITY

Introduction

Transformation and Complexity: Predicting Global Security after 9/11

Measured by their impact on international alliances, foreign and security policies of major powers, American military priorities, long-term defence plans, fluctuations in global financial and economic markets, and even patterns of military, economic and development assistance, the terrorist attacks on the World Trade Center and Pentagon (9/11/ 2001) and subsequent wars in Afghanistan (2001) and Iraq (2003) combined to produce tectonic shifts in world politics and international relations. Consider, for example, the impact of 9/11 on U.S. alliance priorities: much less emphasis is being placed on military-security cooperation with major allies in the context of, say, NATO, and much more energy is being directed towards cooperative agreements and strategic alliances with smaller regional powers to address a new set of security threats emerging from the nexus of terrorism and proliferation. Afghanistan, Uzbekistan, and Pakistan have arguably become more central to U.S. security interests than France and Germany, particularly in light of attempts by the latter two to quash both NATO and UN Security Council support for the U.S.-U.K.–led intervention in Iraq in March 2003, and similar efforts to delay passage of reconstruction resolutions after the war.

Cooperation with like-minded democracies throughout the cold war, ostensibly to protect Western values and liberal institutions, is being replaced with bilateral and multilateral coalitions of convenience with former Eastern European powers and Islamic nations in Asia and the Middle East. The fight for democratic institutions and ideologies is now secondary to the primordial obligation to protect 'American' lives and to establish 'homeland' security. It was not uncommon during the cold war for the U.S. government to ally with authoritarian regimes in the crusade to contain communism, but the post-9/11 application of this strategy to the war on terror will be even more ruthless in its demands, more unilateral in scope, more self-serving in its objectives, and far more selfish in expecting compliance from allies. American leaders are convinced that their state is becoming increasingly vulnerable to asymmetric threats from terrorists determined to acquire and use weapons of mass destruction. For officials in Washington, this is the one and only relevant lesson of 9/11, and the wars in Afghanistan and Iraq are but the latest illustrations of how that lesson is transforming American foreign and security policies and priorities.

This transformation did not begin on 9/11 – the international system was already experiencing important changes that were altering the dominant nuclear rivalry between the U.S. and Russia and shifting the focus towards terrorism and related threats from new and aspiring nuclear powers. But 9/11 served to warp this transformation forward at light speed. The attacks gave Washington a compelling reason (and justification) to accelerate the trend towards American unilateralism, particularly as it relates to the battle against nuclear, biological, and chemical weapons proliferation; the imperative to deploy a ballistic missile defence; the demise of the Anti-Ballistic Missile (ABM) Treaty; changes to the logic and practice of strategic stability; the diminishing relevance of multilateral arms control and disarmament; and the utility of pre-emption and preventive war as the cornerstone of American grand strategy. What we are witnessing today is the emergence of a more assertive form of U.S. unilateralism that will have a profound impact on the foreign and security policies of the United States and its allies for decades – a view that directly contradicts the conventional wisdom that multilateralism will emerge as the only viable option for U.S. security in an age of globalized terrorism and proliferation. This book challenges that wisdom and the policy prescriptions that it produces.

The system-wide turbulence created by the terrorist attacks is expected to rival in significance the collapse of the Soviet Union in 1989 and the end of the cold war. Indeed, the power, influence, and reach of the United States will virtually guarantee that the ripple effects of America's response to 9/11 will be global in scope and sustained in their impact. The 'post–cold war' decade is over – we are now in the midst of a 'post–11 September' era in which the fundamental weaknesses of (impediments to) multilateral approaches to security are being exposed.

Major events such as the end of the cold war, the terrorist attacks on 9/11, or the U.S.-U.K. war in Iraq in 2003 provide excellent opportunities to re-evaluate the strengths and weaknesses of competing theories of international relations, the accuracy of different predictions about the future derived from these theories, and the utility of alternative policy recommendations derived from these predictions. The errors that plagued optimistic expectations about the emergence of global stability after the cold war, for example, were based, in part, on questionable assumptions about the appeal of democratic values, and the emerging power and influence of international institutions and global liberal regimes. These institutions, according to some writers, were expected to impose sufficient discipline on state actors to prevent repeated defections from collective action and ultimately control international violence.[1] It was apparent to many others, however, that the decline of bipolar strategic rivalry was not sufficient to provide multilateral institutions with the coercive power they needed to compel states to cooperate.[2] Both optimists and pessimists agreed that the system had undergone fundamental change, but their predictions were often mutually exclusive.

Some predictions about the disintegration of former alliances after the cold war proved to be quite accurate,[3] while other predictions regarding shared perceptions of America's relatively benign hegemony and, by implication, the prevailing absence of 'power balancing' were far less impressive. John Ikenberry's explanation for why the international system would not experience the kind of balancing that realists expect in response to American hegemony seems to have been a tad premature, especially in light of events surrounding the 2003 war in Iraq.[4] As Ikenberry explains, the absence of balancing is not unexpected; it is odd only because the prevailing realist theories are based on flawed assumptions about the nature of international politics and the controlling influence of system structure: 'Realists' theories miss the institutional foundation of Western political order – a logic of order in which the binding and constraining effects of institutions and democratic polities reduce the incentives of western states to engage in strategic rivalry or balance against American hegemony.'[5] But if the actions of France, Germany, Russia, and China in the period leading up to and following the 2003 Iraq war do not constitute a well-coordinated effort to balance American hegemony, what does?

Other contributions to Ikenberry's edited volume, *America Unrivaled*, provide other, somewhat dated, explanations for (and predictions about) the lack of balancing in the system, most of which suffer from similar theoretical flaws. For instance, William C. Wohlforth predicted that 'conditions to make counter-hegemonic alliances ... are not only absent from the current unipolar system, but they are unlikely to be present for a very long time.' States are not likely to take on the challenge of counterbalancing the U.S. for fear of inviting 'focussed enmity.'[6] Apparently, several major powers are prepared to compete with Washington and risk the consequences of American enmity. Stephen Walt concludes that balance-of-threat theory is perhaps the most compelling explanation for the absence of balancing: 'The principle sources of threat explain why balancing behaviour has been muted thus far. The United States is by far the world's most powerful state, but it does not pose a significant threat to the vital interests of most of the other major powers.'[7] Obviously, U.S. actions in Iraq did pose a significant enough threat to the vital interests of France and Russia. Josef Joffe explains the lack of balancing in these terms: 'In explaining why balancing has not kicked in against the United States, antirealist theory would claim that the system is not destiny; structure qua distribution of power cannot explain behaviour, at least not in our day and age ... Rules and norms, democratic culture and economic interaction have dethroned "structure" as the ultimate arbiter of international politics.'[8] In the case of Iraq, however, democratic culture and economic interaction have not dethroned 'structure' as the ultimate arbiter – in fact, economic interactions between Iraq and both France and Russia can account for their decision to counterbalance U.S. efforts in the region, as predicted by structural theory. Similarly, democratic culture (and public support) in both France and the United States explains why both of these democracies took diametrically opposed positions on the war and why balancing occurred. Some might argue that the Iraq war was not really a case of balancing, since no state attempted to provide military support to the Iraqi regime. But I suspect none of these authors would have imagined, even in the months leading to the second gulf war, that France and Germany would be so vigorous in their effort to build a coalition against any U.S. foreign policy initiative, let alone one that was so obviously important to Washington.

   Perhaps the clearest illustration of premature closure of inquiry is Charles A. Kupchan's prediction that America's unipolar moment 'will not last ... The rise of other powers and America's waning and unilateralist internationalism will combine to make America's unipolar moment a fleeting one.'[9] The rise of Europe and 'America's diminishing appetite for global engagement' will compel the United States to 'fortify the homeland and rein in overseas commitments in an attempt to cordon itself off from such threats.'[10] Far from turning inward, the 2003 Iraq war clearly demonstrates that the U.S. is prepared to increase its overseas commitments by engaging in the most sweeping and expansive intervention and reconstruction projects in decades, far surpassing similar commitments in Bosnia, Kosovo, and Afghanistan. Moreover, the split(s) in Europe resulting from divisions over the Iraq war, in which sixteen of nineteen NATO members ultimately supported the U.S. position, does not bode well for Kupchan's predictions regarding the emergence of a sufficiently unified and powerful Europe to replace U.S. hegemony.

If predictions about the absence of balancing, the demise of unipolarity, and the emergence of multilateralism are wrong, then, by extension, the explanations put forward to account for these predictions are equally suspect.

Objectives and Overview

The purpose of this book is to explain the U.S. response to emerging threats of terrorism and proliferation in the aftermath of 9/11, and to offer what I regard as a slightly more accurate set of predictions about the future of American foreign and security policy – predictions that point to the deficiencies of multilateral approaches to security and corresponding pressures pushing successive U.S. governments to maintain an unwavering commitment to unilateral options in the decades ahead. The evidence is derived, in large measure, from two crucial case studies – the war in Iraq (chapter 3), and ballistic missile defence (chapters 4 and 5). The objective is to apply the evidence compiled from these cases to provide a more comprehensive basis on which to evaluate unilateral and multilateral approaches to security after 9/11. Among the central questions to be addressed are the following: What does 9/11 imply about the future of terrorism, or the pace and direction of proliferation? What does Washington's response suggest about its preference for multilateral versus unilateral responses, or its dependence on international institutions, organizations, and regimes? What impact will the attacks have on arms control and disarmament, strategic stability, or the doctrine of mutually assured destruction and pre-emption? What does

9/11 tell us about globalization, or the mutually reinforcing relationship between globalization and terrorism? How will these trends affect European and Canadian interests and responses?

The book is expected to make several important contributions to the literature by uncovering complex and interdependent linkages between globalism and security by introducing a new framework of analysis for comparing and evaluating unilateralism and multilateralism in a contemporary setting, and by addressing the implications of these trends for American and Canadian priorities.

Outline

Chapter 1 explores the security challenges confronting American officials in an era of globalization. In the aftermath of the 9/11 attacks and subsequent wars in Afghanistan and Iraq, the paramount challenge facing the policy and academic communities is to clarify these emerging threats and, more important, to assess the implications for designing effective security policy. Debate between proponents of multilateralism and unilateralism will be evaluated in this context.

After reviewing key impediments to contemporary research on globalism and security, and following a detailed description of the multiple positive, negative, and interactive effects on terrorism, the chapter highlights the many changes that continue to influence the dominant nuclear rivalry between the United States and Russia. Among the questions at the centre of the debate over ballistic missile defence (BMD) is whether BMD represents a destabilizing shift in a nuclear doctrine that maintained strategic stability throughout the cold war (i.e., nuclear deterrence; mutually assured destruction; the anti-ballistic missile treaty) or, alternatively, whether BMD is now an essential part of U.S. security against new and emerging threats.

The evidence outlined in the chapter suggests that the threshold for escalation of U.S.-Russia military security crises is significantly higher today than it has ever been. Relations between the longest-enduring nuclear rivals in history are becoming stronger and more cooperative in the aftermath of 9/11 and, consistent with expectations derived from an ever-expanding web of interdependencies and mutual vulnerabilities, there are powerful incentives to ensure that U.S.-Russia crises are prevented from escalating to the point of seriously jeopardizing this relationship. While all of this was true prior to 9/11, it is now etched in stone.

Chapter 2 addresses what appears to be the dominant critique of American foreign policy after 9/11, namely, Washington's fixation with maintaining independent, unilateral control over its own security despite the fact that successful unilateralism will, according to the conventional wisdom, become increasingly difficult in an era of globalization. But what appears on the surface to be an irrational reaction to the challenges of terrorism and the spread of weapons of mass destruction (WMD) is, in fact, a rational response derived from a comparative assessment of the costs, benefits, and risks of relying exclusively on multilateral alternatives. After clarifying the conceptual confusion in the literature on unilateralism and multilateralism, the chapter goes on to introduce a more systematic framework for establishing the presence/absence and successes/failures of unilateral and multilateral strategies. This discussion will set the stage for the two case studies – the 2003 Iraq war (chapter 3) and ballistic missile defence (chapters 4 and 5).

The case studies will inform the book's primary thesis (summarized in chapter 6): in the context of asymmetric security threats associated with globalized terrorism, American unilateralism is not only inevitable but, in many cases, preferable to multilateral alternatives. Multilateralism has never achieved the kind of success in the security field that would warrant giving its proponents the moral or intellectual authority to dismiss unilateral alternatives. The evidence from the case studies demonstrates that when compared with unilateral measures, multilateral approaches do not offer a better return in global, regional, or national security with fewer costs and risks, because the impediments to effective and timely, multilateral solutions are simply too great. By implication, there is no logical, empirical, legal, moral, or policy-relevant foundation for embracing multilateral approaches at the exclusion of all other options. While collective action may provide the best hope for achieving certain objectives, it is not the case that the global desire for peace and security is best served through existing multilateral mechanisms, particularly the United Nations and associated organizations and regimes. As the United States becomes increasingly threatened by terrorism after 9/11, the emerging security reality will create enormous pressures on American officials to respond with immediate unilateral initiatives. Many of these responses will be far from perfect, but they will offer a better return for Washington's security investment with fewer risks and a higher probability of at least some measure of success.

The imperatives driving American unilateralism, in turn, will have a direct impact on Europe and Canada, the implications of which are covered in chapters 7 and 8. The former revisits the question of multilateralism in Iraq through the prism of Canada's response to the crisis. The purpose is to highlight key deficiencies with Prime Minister Jean Chrétien's principled position against the war and to uncover some of the problems with Ottawa's almost religious commitment to contemporary multilateralism. Chapter 8 expands on the case by addressing the medium- to long-term implications of Canada's dishonest multilateralism and the prevailing 'weak state' approach to foreign and security policy that privileges 'distinction' over 'security' – Canada's default strategy when dealing with American initiatives. In a post-9/11 era of U.S. hegemony and unilateralism, however, this default strategy is becoming increasingly costly and counterproductive. The book concludes with a few observations about the moral foundations of multilateralism and raises important questions about the extent to which ongoing dependence on failed institutions is morally justifiable in an age of globalized terrorism. Unless Canadian officials establish a new grand strategy for dealing with the U.S. and the international community Canada (and Canadian interests) will continue to fade.

Chapter 1

Linking Globalism, Terrorism, and Proliferation

Globalism has emerged as the most important force in the international system.[1] It affects every aspect of global economic, political, diplomatic, and military behaviour and continues to influence major debates in the fields of international relations, foreign policy, international trade, development, global finance, immigration, governance and democratization, and global security (to name a few).[2] Globalism typically refers to one (or more) of five trends outlined in Table 1.1.

Thomas Friedman's analogy of 'three democratizations' (finance, technology, and information) in *The Lexus and the Olive Tree* is perhaps the most straightforward portrayal of globalization. Consider, for example, the system-transforming impact of global access to personal computers, wireless modems, cellphones, pagers, digitization and miniaturization, satellite television, expanding Internet access, and the scope and pace of international travel. Lael Brainard aptly sums up the effects this way: 'At its root it is about the planet getting smaller and the free flow of people, goods, capital, and ideas across borders. Globalization has been sped up by technology (computers, jets, etc.) and it makes traditional notions of sovereign nation-state powers more and more irrelevant.'[3] These trends create 'networks of interdependence at multicontinental distances with multiple ... linkages among nation states, markets and individuals.'[4] Although many scholars describe these system-wide pressures in different ways, they are virtually unanimous in the view that these forces are ubiquitous, powerful, inevitable, and irreversible.

The literature surrounding globalization typically encompasses research on 'what it is' and 'what it does.' The answers to these two questions often produce additional divisions in the field, many of which are reviewed by Steve Smith and John Baylis.[5] With respect to what it is,

TABLE 1.1
Defining Globalism

(a) **Death of geography:** Geographic boundaries and territorial borders/barriers are becoming increasingly insignificant, porous, and permeable (soft); state control over domestic economic, social, and cultural affairs is diminishing as state sovereignty (i.e., the capacity to protect and promote national interests and values) evaporates.

(b) **Death of distance (space and time):** Distances between countries (and cultures) are decreasing as information, communication, and transportation (ICT) technology continues to improve. Advancements in ICT also increase the efficiency of financial, trade, and military activities (i.e., the time required to perform these activities is declining, in some cases at exponential rates).

(c) **Sensitivity**[a]: As the planet shrinks, both small and large states are becoming more sensitive to economic, political, and military crises that occur in any part of the world; relatively minor political, economic, and military events are having a larger impact on more states and regions in the system.

(d) **Vulnerability (ripple effects):** States are becoming more susceptible to the negative consequences of these crises, and these consequences are interlinked and mutually reinforcing.

(e) **Death of independence:** All these pressures (a–d) combine to diminish the capacity of national governments (in both large and small states) to maintain independent control over their own defence and security.[a]

---

[a] Robert Keohane and Joseph Nye, *Power and Interdependence*, 3d ed. (New York: Longman, 2000). See also Keohane and Nye, 'Power and Interdependence in the Information Age,' *Foreign Affairs* 77, no. 5 (September–October 1998).

most observers concur that globalization has generated profound changes that transcend economic relations between and among states, global culture (homogeneity), and the substance and speed of communications. It represents, in other words, the collapse of time and space, and its effects range from the interpersonal to the international. Sceptics claim that the changes are not particularly new or unique but cyclical; the current international system simply represents the latest ebb in the typical ebb and flow of interdependence characteristic of similar changes in the past. With respect to debates over what globalization does, the evidence is equally inconclusive. The views range from claims that it provides the best hope for development and enhanced global governance, to mutually exclusive predictions that these same forces increase the power and independence of non-state actors and multinational corporations in ways that ultimately undermine the prospects for economic prosperity, global governance, and international security. As

Smith and Baylis note, 'there seems to be a paradox at the heart of the globalization thesis.'[6] In fact, there are many paradoxes.

SECTION I: Globalism, Security, and Terrorism: Complexity and Confusion

Globalism and Security

Notwithstanding the importance of globalization, the systemic and widespread transformations these forces continue to produce, and the large amount of research and writing on the subject, the precise nature of the link between globalization and security remains unclear and somewhat confusing. There are at least five explanations for this confusion.

First, the literature has tended to focus almost exclusively on the economic dimensions of globalization. Comparatively less attention has been devoted to the important question of how these pressures affect international, regional, or domestic security.[7] Second, the literature tends to treat the globalism-security nexus as unidirectional, a problem rarely acknowledged by those working on the subject. Globalism, however, is both a cause and an effect (a force and an outcome, an independent and dependent variable). Although these forces are typically cited as creating new and emerging security threats, the responses to these new threats, in turn, have a direct impact on globalization. There is no question, concludes Lael Brainard, 'that the drive to increase the security of our borders runs counter to the very forces that propel globalization. At least initially, it will be impossible to address the demands for greater security and scrutiny of cross-border movements – of goods, people, information, financial capital, even mail – without some setbacks to the drumbeat of faster, cheaper, and less bureaucratic that has driven economic activity.'[8] In other words, the 9/11 terrorist attacks have created pressures to reverse the nature, pace, and direction of globalization – an important issue to be discussed in more detail below.

Third, scholars who choose to overlook the first problem (by treating the globalism-security relationship as unidirectional) are faced with the additional challenge of tracking the implications for security across different levels of analysis. More specifically, globalism is likely to have varying degrees of influence over international security, regional security, domestic security, individual/human security, bilateral and/or multilateral security, and so on. These effects are likely to be interconnected and mutually reinforcing, making it very difficult to be definitive about the precise impact of globalism on security, the extent to which these effects are uniformly positive or negative, or whether the effects occur simultaneously or sequentially.

Fourth, in addition to exploring linkages across different levels of analysis, researchers must also disentangle the effects across potentially hundreds of different 'security-related' issue areas, each sufficiently complex to justify its own separate research program – e.g., terrorism; proliferation (separable into nuclear, chemical, and/or biological variants); international conflict and war; intra-state (ethnic) violence; transnational crime; drug trafficking; environmental security; humanitarian intervention; peacekeeping. The effects of globalization, both positive and negative, are likely to vary across these different dimensions, simultaneously enhancing and detracting from security. The fact that these effects are often mutually exclusive, interactive, multidimensional, and multidirectional is rarely acknowledged in the literature, and when it is acknowledged the implications of this complexity are almost never adequately addressed.

Fifth, further confounding the search for definitive answers to questions about globalism's impact on security is the fact that each of the above-mentioned security-related issue areas encompasses dozens of distinct components. With respect to terrorism, for example, one could explore the relationship between globalism and terrorist communication, the use of enhanced computer encryption by terrorists, terrorist financing, recruitment and networking, counter-terrorism, fundamentalism, cyber-crime, threats to critical infrastructure, the proliferation of weapons of mass destruction (WMD) to terrorist groups, border security, and immigration and refugee policy. Again, the literature rarely devotes enough attention to variations across (and interlinkages among) these different components of terrorism. Nor does it explain how they relate to one another or how these relationships both enhance and detract from our capacity to address emerging security threats.

Clarifying the precise relationship between globalism and security is perhaps one of the most challenging puzzles confronting those responsible for crafting contemporary foreign and security policy. Pushing the puzzle analogy a bit further, it is as if we are confronted not with one but several puzzles, each with an unknown number of pieces, each piece with images on both sides, and all without a clear sense of the final picture.

Globalism and Terrorism

Few dimensions of security provide a clearer illustration of the paradoxes and challenges facing the policy community than the issue of
global terrorism. For many, like Joseph Nye, the globalism-terrorism
relationship is obvious, and ominous: 'September 11th was a terrible
symptom of the deeper changes that were already occurring in the
world. Technology has been diffusing power away from governments,
and empowering individuals and groups to play roles in world politics
– including wreaking massive destruction – which were once reserved
to governments. Privatization has been increasing, and terrorism is the
privatization of war. Globalization is shrinking distance, and events in
faraway places, like Afghanistan, can have a great impact on American
lives.'[9] And according to Brainard, 'The technologies touted as the
handmaiden of globalization – the Internet, global financial networks,
and commercial aviation – proved their moral neutrality by enabling
terrorists to wreak havoc on a heretofore unimagined scale. And
America's openness and huge footprint in the international system
make us both more vulnerable and more attractive to terrorist attacks.'[10]

Most predictions about what to expect from globalism are not encouraging.[11] Religious and political fundamentalism are expected to
increase, because extremist ideas thrive in environments in which people
search for ways to create order in an otherwise complex, confusing, and
disruptive world. Advances in communication technology and access
to incredible amounts of information on the Internet will also make it
much easier for groups to share ideologies and grievances from abroad,
facilitating the transfer of fundamentalism to other regions, states, and
communities.[12] Yet, logically, fundamentalism could also decline as a
result of globalization. The instantaneous transmission of images and
values associated with Western democracy, justice, and the rule of law
could create a voice for balance and moderation that offers an alternative to extremist ideologies and hatreds. On the other hand, transmission of these same Western images could highlight the economic
inequalities that promote the divisions and associated hatreds at the
root of terrorism.

Thomas Homer-Dixon's work on 'complex terrorism' provides
perhaps the best overview to date of the paradoxical implications of
globalization. As he explains, 'Modern societies face a cruel paradox:
fast-paced technological and economic innovations may deliver un-

rivalled prosperity, but they also render rich nations vulnerable to crippling, unanticipated attacks. By relying on intricate networks and concentrating vital assets in small geographic clusters, advanced Western nations only amplify the destructive power of terrorists – and the psychological and financial damage they can inflict.'[13] Homer-Dixon describes three products of globalization that increase overall vulnerability to terrorist attacks: (1) the ease with which terrorists can now gain access to weapons of mass destruction,[14] (2) the growing complexity and interconnectedness of modern societies,[15] and (3) the geographic concentration of wealth, human capital, and communication linkages.[16]

Globalization simultaneously facilitates recruitment of terrorists, the expansion of informal terrorist networks, and the coordination, administration, financing, planning, and execution of terrorist attacks.[17] The Internet, global positioning satellites (GPS), and mobile communications have enabled terrorist groups to develop a more impressive command, control, communications, and intelligence capacities that rival that of many states.[18] The Internet is the most obvious tool through which terrorists can now share volumes of secure (or publicly accessible) information on weapons, tactics, communications, global contacts/networks, fundraising, critical infrastructure targets, and cyberattacks.[19] With respect to the latter, the Pakistani Hackerz Club illustrates the dangers of an emerging 'e-Jihad' in which computer-literate recruits to Hamas and Hezbollah now throw 'virtual electronic stones.'[20] The fact that close to 95 per cent of U.S. military traffic is transferred through civilian computer systems, producing more than twenty-two thousand cyber attacks (typically, computer viruses) in 1999 alone, is not encouraging.[21] Some studies put the number of attacks at a few hundred thousand a year, with only 10 per cent of penetrations being detected.[22] Over time these trends are expected to increase the number of targets, the probability of terrorist attacks, and the level of destruction if and when these attacks occur.[23]

On the other hand, the very global forces and technology terrorists exploit to fund, plan, and carry out attacks can also be used to enhance preventive and pre-emptive measures by increasing the efficiency, credibility, and effectiveness of global intelligence and response networks. These same technologies, for example, improve inter- and intra-agency (as well as interstate) intelligence coordination and facilitate the monitoring and tracking of terrorist activities, communications and financial transactions.[24] Although computers, cellphones, and the Internet are being used more frequently by terrorists, the frequency of use has made terrorists more vulnerable to counter-terrorist investigators.[25] New U.S. intelligence tools developed by the CIA and FBI (e.g., Threat Matrix) have already improved effectiveness of counter-terrorist operations. The Threat Matrix, explains Ann Scott Tyson, was developed after 11 September by the U.S. Counter-Terrorism Center.[26] It is an information- and intelligence-gathering device that compiles data on terrorist activity each day and transmits the information to the president and senior officials in the National Security Council and departments of Defense and State. The information is compiled from a variety of sources and includes 'signals intelligence (such as intercepted phone, e-mail, and radio communications), GPS and satellite imagery, and reports from human sources (such as recruited agents and detained terrorist suspects).' Experts within the FBI and CIA counter-terrorist units select those threats they consider credible and log each by the nature of the threat and its source. In addition to the matrix, there are several new computer and mathematical tools currently being used to help predict terrorist attacks and to decode encrypted messages.[27]

The debate over whether any of this technology will improve counter-terrorist efforts remains unresolved. According to experts with extensive experience within the CIA like Ruel Marc Gerecht, effective counter-terrorism is a myth, primarily because what is actually required cannot be achieved through information and communication technology alone:

> Westerners cannot visit the cinder-block, mud-brick side of the Muslim world – whence bin Ladin's foot soldiers mostly come – without announcing who they are. No case officer stationed in Pakistan can penetrate either the Afghan communities in Peshawar or the Northwest Frontier's numerous religious schools, which feed manpower and ideas to bin Ladin and the Taliban, and seriously expect to gather useful information about radical Islamic terrorism – let alone recruit foreign agents ... An officer who tries to go native, pretending to be a true-believing radical Muslim searching for brothers in the cause, will make a fool of himself quickly.[28]

Despite these obvious barriers to success, however, there are other opportunities derived from emerging technologies that can prevent certain types of terrorist attacks. Improved attacker identification software, for example, will allow authorities to pinpoint the precise source of cyber attacks in shorter periods of time. Like conventional deterrence, the prerequisites for successfully deterring a cyber attack include effective and immediate identification of the source, followed by a credible threat of retaliation and the capability to inflict unacceptable costs on the attacker.[29] Improved forensic capabilities to process clear, precise, and immediate targeting information is one of many strategies currently being developed by American counter-terrorist and critical infrastructure units.

Terrorism and Globalism

The discussion thus far has focused exclusively on the globalism-terrorism connection. But the relationship is not unidirectional – terrorist attacks and the security measures they generate can also reverse the pace and direction of globalization. Brainard's excellent study of the terrorism-globalism linkage highlights some of these countervailing pressures. 'When a sense of safety previously taken for granted is profoundly undermined, there is a natural tendency to pull up the drawbridges and pull back from the world. And when jobs and economic security are put at risk, there is a tendency to look towards protectionist solutions ... [terrorism] puts at risk many of the gains globalization has brought, but it likewise may present opportunities to smooth some of the rough edges of the globalization associated with American policies of the past decade. To skip to the punch line, the future trajectory of globalization is not preordained; it lies largely in our own hands.'[30]

Some of the effects of 9/11 on globalization are likely to be short-lived. For example, we can expect an increase in the use of email and fax as replacements for potentially deadly regular mail brought on by the anthrax attacks in the United States in 2001–2.[31] We can also expect to see an increase in the use of teleconferencing, conference calls, and Internet chat lines to replace face-to-face interaction, thus avoiding the added security risks that come with air travel and threats from diseases such as SARS. Other effects of 9/11 are likely to be perceived as either positive or negative, depending on where you sit. Security measures will create added pressures to modify the technology and infrastructure required for processing international economic, financial, and trade transactions – creating a 'members' club' that raises barriers to global economic integration for poorer nations and, by implication, reverses some of the positive aspects of economic globalization.[32] Consider current efforts by Canada and the United States to establish more efficient cross-border economic and trade transactions using integrated enforcement mechanisms. The standards yet to be negotiated for cross-border trade could conceivably raise barriers to U.S.-Mexico trade.

The central point is this: there is nothing about the process (or logic) of globalization that privileges positive or negative effects, precludes one outcome over another, or establishes globalism or terrorism as an independent (cause) or dependent (effect) variable. Both sets of hypotheses and associated predictions are logically sound, and each is equally defensible with reference to specific anecdotal evidence. Globalization exhibits both positive and negative effects and does not favour one set of consequences over another. These effects are not predetermined and are likely to play out in distinct ways for different individuals, groups, and states under different sets of circumstances. The answer to the question of whether globalism will increase or decrease the aggregate level of terrorism is yes. The same level of complexity and confusion applies to the relationship between globalism and proliferation.

SECTION II: Globalism and Proliferation: Strategic Stability in Transition

Before the terrorist attacks on 9/11 handed George W. Bush the prime directive of his administration (and subsequent ones), Washington and Moscow were in the midst of negotiations over a new, post–cold war nuclear relationship. The central issue at stake was how to revise the prerequisites for nuclear deterrence and global strategic stability without jeopardizing U.S. efforts to address new and emerging threats from WMD proliferation and terrorism.

Strategic stability is a catch-all expression used by scholars and practitioners to describe a set of interrelated concepts (such as mutually assured destruction), theories (for example, nuclear deterrence), policies (massive retaliation; flexible response; no first use), and treaties (Anti-Ballistic Missile Treaty), all designed during the cold war for one purpose – to stabilize the longest nuclear rivalry in history to prevent a nuclear exchange between the United States and Russia. The key was to balance strategic forces so that each side could survive a pre-emptive nuclear attack with a sufficiently large stockpile of ballistic missiles to launch a retaliatory strike. The logic was (and remains) elegant and persuasive: as long as the retaliatory (second) strike threatened enough devastation, there would be no rational reason to launch first.

Policy makers throughout the cold war were preoccupied with three central questions: What deters? How much is enough? And what if deterrence fails? The appeal of nuclear deterrence theory was its simple (and impeccable) logic, which provided straightforward answers to the core questions and guidelines for how to achieve deterrence stability. To work well, according to the theory, the balance of strategic forces had to promise *crisis stability* so that neither side would perceive an advantage in escalating violence in a crisis; *arms race stability* to minimize incentives to build more weapons; and *survivability* to maximize second-strike potential and *mutual vulnerability*.[33]

Although the perfect balance of air-, land-, and sea-launched strategic missiles was never entirely clear, there was one principle to which both sides adhered – nationwide ballistic missile defence (BMD) systems were to be prohibited. In the context of a highly charged and competitive cold war environment, national defence systems would be provocative, destabilizing, and exceedingly dangerous. They would undermine crisis stability by increasing pressure in a conflict to pre-empt so as to overwhelm the opponent's defences, jeopardize mutual vulnerability by making an opponent's second strike less threatening (and a first strike less costly and more rational), and create strong incentives for vertical proliferation.

The George W. Bush administration's decision to withdraw from the Anti Ballistic Missile (ABM) Treaty and to accelerate the testing and deployment of a limited, layered BMD system obviously raises important questions about the future of strategic stability and the evolution of nuclear deterrence. Given longstanding commitments to the non-proliferation, arms controls, and disarmament regime (NACD), itself founded on principles, theories, and doctrines developed throughout the cold war, this pullout from the treaty raises equally important questions for those who remain exclusively committed to multilateral arms control.

How significant is this decision? Does it indicate a fundamental shift in the U.S.–Russia nuclear doctrine? Is the shift permanent, especially in the aftermath of the attacks on New York and Washington on 11 September 2001? What are the implications for the future of strategic stability? Are the concepts (mutual assured destruction, or MAD), theories (deterrence), policies, and treaties still relevant? Do we need a more complex approach to strategic stability and arms control that acknowledges emerging threats of terrorism and proliferation of WMD to new and aspiring nuclear powers? If so, what would a future-oriented approach to deterrence and arms control encompass? How effective are unilateral and multilateral approaches to arms control in the context of this transformation?

Strategic Stability: Continuity and Change

Although the cold war officially ended well over a decade ago, we are only now experiencing the effects of a transition from one nuclear environment to another.[34] This fact was acknowledged by President George W. Bush in the 2002 U.S. National Security Strategy: 'It has taken almost a decade for us to comprehend the true nature of this new threat.'[35] Adjustments in nuclear strategies have been relatively slow because transitions, by definition, encompass both continuity and change, with features of the old and new nuclear environments interacting simultaneously. This explains why it is so difficult to resolve policy debates about the future of strategic stability – both sides are right and wrong about some things, and both sides can produce evidence to support some of their core arguments. The result is a collection of mutually exclusive conclusions that bipolar strategic stability is relevant and irrelevant; nuclear deterrence theory is valid and invalid; MAD is appropriate and inappropriate; the ABM Treaty is essential and obsolete; and BMD is stabilizing and destabilizing.

The position one takes in each debate depends almost exclusively on perceptions of change, particularly with respect to the dominant nuclear rivalry. And perceptions of change in turn depend on whether one focuses on numbers or on relationships. Those who claim that very little has changed since the end of the cold war tend to focus on numbers of nuclear weapons and the current balance of air-, land-, and sea-launched nuclear forces in the American and Russian arsenals. Those who believe the system has undergone fundamental and permanent change are more likely to focus on the constantly improving relationship between the U.S. and Russia and deteriorating relations between the U.S. and new or emerging proliferators. The relevant question is whether numbers determine the health and stability of a nuclear relationship or whether the health of a nuclear relationship determines the relevance and stability of numbers.

Continuity in Numbers

Proponents of the continuity thesis argue that as long as there are large numbers of nuclear weapons, and as long as abolition is excluded as a

serious policy option, the U.S. and Russia will maintain military sufficiency to render enemy nuclear forces ineffective and to extend deterrence to allies. Since nuclear deterrence stability is a property of the
balance in nuclear weapons, cold war or no cold war, the number of
existing weapons will remain important to strategic decisions made by
both sides. By extension, mutual deterrence and traditional approaches
to bipolar strategic stability will (and should) continue to be a defining
characteristic of the international system.

If one focuses on numbers and assumes that the nuclear balance has a
logic and force of its own, the level of conflict and cooperation between
nuclear rivals will always depend on that logical imperative. In other
words, the health and stability of a nuclear rivalry is a function of that
balance.[36] If the balance is not protected, the political relationship deteriorates; it is the numerical ratio in strategic forces that matters. The
arms race, says Leon Fuerth, is 'really about existential, and therefore
potentially irrational, fear. That is why nuclear capabilities are so much
more important as drivers in the psychological equation of war and
peace than are statements of intention. Capability endures; intentions
do not.'[37] If this is true, the thousands of Russian nuclear weapons and
materials that still exist remain the main threat to U.S. survival today.
Therefore, protecting the balance is crucial, and, by implication, the
ABM Treaty and associated MAD doctrine are also essential to U.S.
security.[38]

Those who focus on numbers are also more inclined to be critical of
BMD and to reject any move that undermines the rigid and well-
defined requirements of traditional bipolar strategic stability. The emphasis on numbers is apparent in virtually every major critique of BMD
by academics, Russian generals, European officials, editors of major
newspapers, and sceptics in the U.S. Senate and House – all of whom
focus almost exclusively on existing numbers of nuclear weapons and
the parity principle underlying bipolar strategic stability. And all point
to the devastating consequences of an imbalance caused by withdrawal
from the ABM Treaty and MAD via deployment of BMD.[39] If the balance is not protected, the relationship deteriorates.

The response to George W. Bush's speech, following his
administration's 2001 Nuclear Posture Review (NPR), highlights the
emphasis Democratic critics place on numbers when evaluating the
implications of departing from strategic stability by deploying BMD.
Fuerth writes, 'What is missing, however, is an appeal to the concept of
strategic stability ... Depending on how it is done, reducing nuclear

launchers and warheads in and of itself might make this relationship more rather than less dangerous ... If you combine sharply reduced numbers of nuclear weapons and increasingly effective defenses, one way of looking at the result is that it creates an increased temptation for launching a first strike in a crisis.'[40] Contrast this statement with typical reactions to the same speech by Senate Republicans who supported BMD, like senator Jesse Helms: 'The idea of deliberate vulnerability to missile attack is a folly and Russia must come to grips with the fact that the Cold War is over. It is time to scrap the ABM Treaty. But the United States must update its thinking as well, which is why I believe it is appropriate for the President to consider significant nuclear reductions ... [We must] modernize the deterrent that was built during the 20th Century to meet the evolving threats and challenges of this one. In particular, we urgently need new weapons designs to address the problem of biological plagues, and deeply buried targets.'[41] The position put forward by proponents of BMD tends to emphasize the cooperative relationship with Russia and deteriorating relationship with states that are likely to acquire nuclear or biological weapons in the future.

The numbers debate is significantly more complex than what is described above – only the broad parameters of thought concerning stability in a bipolar world are presented here. My purpose is not to provide a primer on deterrence theory or a comprehensive review of the many nuanced debates surrounding the precise numerical prerequisites for strategic stability. While many analysts believed that parity in nuclear weapons was important to stability and deterrence, others were far less concerned. The possession of assured second-strike forces capable of inflicting unacceptable damage in a retaliatory strike, rather than assured destruction, was the real key to a stable nuclear rivalry. In other words, although assured destruction was sufficient for strategic stability, it wasn't necessary – far fewer weapons were required to establish the unacceptable damage benchmark. As McGeorge Bundy argued in 1969, 'In the real world of real political leaders – whether here or in the Soviet Union – a decision that would bring even one hydrogen bomb on one city of one's own country would be recognized in advance as a catastrophic blunder.'[42] Others, such as Colin Gray, stated in 1979 that many more weapons were required for war-fighting dominance, that is, to assure survival of a first strike and to allow retaliation of adequate force to win a nuclear war.[43] The main difference between the two sides of the cold war numbers debate was described by Rajesh Rajagopalan in 1999: 'Assured Destruction suggested that the Soviet

Union would be foolish to attack because nobody could possibly win a nuclear war, while victory theorists proposed that the Soviet Union would not attack only if it was convinced that not only would it lose the war, but that the US would win it.'[44] Those who believe that assured destruction is ample, or that the effects of even a single nuclear bomb are unacceptable, are less likely to be concerned about BMD because no defence system would be robust enough to ensure protection against the thousands of missiles currently held by the U.S. and Russia. In fact, even at significantly lower numbers, the strategic forces on both sides would be more than sufficient to overcome even a very advanced BMD system.

Changes in Relationships

But deterrence stability is not, and perhaps never was, about numbers (regardless of one's views on how much is enough). The issue has always been about relationships.[45] Focusing exclusively on numbers is misleading for several reasons, not the least of which is that existing stockpiles in Russia have more to do with bureaucratic inertia, public indifference to nuclear matters, and the costs of dismantling huge arsenals.[46] Current levels of nuclear weapons are not the determining feature of the U.S.-Russia strategic relationship – they are the legacy of cold war hostility.[47]

The degree to which numbers matter depends entirely on whether the relationship is stable. And stability depends not on the balance of numbers but on the balance of incentives to cooperate – low numbers of nuclear weapons are dangerous under the wrong conditions, high numbers are stabilizing under the right conditions, and high or low numbers are potentially irrelevant under changing conditions.[48] One need look no further than U.S.-U.K.-France relations to appreciate the irrelevance of nuclear numbers under the right conditions. Even though deployment of an American BMD would undermine the deterrent value of British and French nuclear forces, this concern does not enter the consciousness of European officials, for obvious reasons – the relationship renders such calculations meaningless, if not absurd, because the probability of a nuclear exchange is zero.

This is not to suggest that U.S. relations with Russia are as healthy or as stable as those with America's European allies, despite the fact that traditional alliances with Europe in NATO and the UN were seriously strained as a result of the 2003 Iraq war. To the extent that the post–cold war, post-9/11, and post–second gulf war relationship with Russia continues to improve, the relevance of numbers will continue to diminish. Expanding levels of economic cooperation, interdependence, and, in Russia's case, vulnerability have created an environment in which a serious, large-scale conflict with the U.S. is increasingly remote and, for many reasons, obsolete. Economics and the prospects for increased trade are far more useful than military competition in predicting interactions between the United States and Russia, and there is no compelling reason to expect this to change. Indeed, Russian officials are now more inclined to define strategic stability in terms of assured economic viability, not assured destruction. Survival of the Russian state depends less on the balance of nuclear forces and more on the Russian economy and foreign investment from the United States, Europe, and Asia.

Perhaps the best illustration of the transformation in American-Russian nuclear relations was Moscow's quiet acquiescence to Washington's decision to withdraw from the ABM Treaty. Given the incredibly dire warnings from critics that Russia would be forced to proliferate in response to the death of the treaty, the relatively benign reaction from Moscow is perhaps the clearest indication that Russian officials understand the true motivations driving U.S. preferences and priorities – namely, terrorism and the proliferation of WMD to rogue states and regimes. Again, this trend was explicitly recognized in Bush's 2002 National Security Strategy: 'With the collapse of the Soviet Union and the end of the Cold War, our security environment has undergone profound transformation. Having moved from confrontation to cooperation as the hallmark of our relationship with Russia, the dividends are evident: an end to the balance of terror that divided us; an historic reduction in the nuclear arsenals on both sides; and cooperation in areas such as counterterrorism and missile defense that until recently was inconceivable.'[49] In response, the Russian foreign minister conceded the emergence of a 'fundamentally new relationship of strategic partnership,' and that 'Moscow expects that the United States will give priority to implementation of this [BMD] program ... and will involve its friends and partners in it rather than in a destabilizing race of strategic defensive arms.'[50] Moscow's preference for a partnership with the United States in building missile defences is certainly a far cry from the dire warnings of automatic proliferation at the centre of the critics' case.

James Lindsay and Michael O'Hanlon correctly caution that claims about a new strategic climate can be pushed a little too far: 'although

Russia and the United States have better relations, they are not allies. Substantial suspicion still marks the relationship – witness the tensions over the North Atlantic Treaty Organization's 1999 war against Serbia and Russia's ongoing war against Chechen rebels.'[51] I suspect the authors would also include the 2003 Iraq war in the list of recent crises that provoked a higher level of conflict in the relationship. But the tensions associated with contemporary crises (including fundamental disagreements over the intervention in Iraq) cannot compare to the level of conflict and stress experienced by both sides in their many cold war crises. In fact, the lesson from Kosovo (1999) is that despite NATO bombing of a Russian ally for seventy-eight days straight, Russia refused to send even a single ship to the region and, in the end, demanded from Slobodan Milosevic the same concessions requested by NATO. The lesson to be derived from NATO's expansion eastward is that Russia's political and security interests can be accommodated in the new NATO-Russia Council (replacing the NATO-Russia Permanent Joint Council). And the lesson from Operation Iraqi Freedom (2003) is that despite fundamental disagreements and associated threats of United Nations Security Council vetoes, the war in Iraq did very little to undermine the post–cold war U.S.-Russia relationship, as exemplified in Russia's ultimate support for UN Security Council 1483 assigning to the U.S. government the 'Authority' to control economic and political reconstruction in post-war Iraq.

Like Russia, China is fully (and constructively) engaged with the U.S. in an international economic system that is obviously beneficial to both sides. There are powerful incentives to ensure the relationship remains stable and both sides are highly motivated to prevent crises from escalating out of control. That is why, notwithstanding blatant human rights abuses by China in Tibet and Beijing's response to the demonstrations in Tienanmen Square in 1989, the U.S. repeatedly assigned most-favoured-nation status to trade with China. It also explains China's accession to the World Trade Organization in 2001 and why it will host the 2008 Olympics, without even a hint of U.S. opposition. U.S.-China relations take place in a world in which American diplomats struggle to select the perfect phrases for meticulously worded diplomatic communiqués to de-escalate tensions during the U.S./NATO B-2 stealth bombing of the Chinese embassy in Belgrade on 7 May 1999, or over the downing of the U.S. Navy EP-3 spy plane in 2001. With respect to the latter, a decidedly conciliatory approach was deemed by U.S. officials to be essential, even though the Chinese pilot was responsible for risking the lives of twenty-four American crew members who were flying over international waters.

Again, the objective is not to overstate the extent of U.S.-China cooperation or to exclude the possibility of future crises over, for example, Taiwan or North Korea. However, the overwhelming body of evidence appears to support the view that U.S.-China (and U.S.-Russia) relations are driven by great pressure to cooperate to resolve crises peacefully and as quickly as possible. In contrast, U.S. relations with new and aspiring nuclear powers are unstable, unpredictable, and far less manageable because the balance of incentives does not yet favour cooperation.

That is not to say that established nuclear powers will always conduct their regional diplomacy responsibly or maintain nuclear arsenals that are fully safe and secure. On the other hand, if it is so difficult for nuclear powers to maintain strategic arsenals that are perfectly safe, that is just one more reason that BMD deployment makes sense – it creates another layer of defence against failed technology, and provides additional options that avoid exclusive reliance on massive retaliation following an accidental launch.

Globalism and the Diminishing Relevance of Vertical Proliferation

Even if claims of improved U.S. relations with Russia and China are exaggerated, and, for whatever reason, Moscow and Beijing decide to expand their arsenals to re-establish mutual vulnerability in response to U.S. BMD deployment, the issue for U.S. policy makers will always be one of comparative risk.[52] If China doubles or triples the number of inter-continental ballistic missiles (ICBMs) in its nuclear arsenal over the next ten to fifteen years, or deploys MIRV (multiple independently targeted re-entry vehicle) technology to create a more robust retaliatory threat, the security risks to the United States will always be less than the risks of even one nuclear weapon deployed by Iraq, Iran, North Korea, or any other 'state of concern.' When it comes to comparing risks, Russian or Chinese proliferation is easier to deal with and certainly less threatening, which explains why the U.S. will likely withdraw objections to Chinese plans to modernize its nuclear force – a relatively minor concession, considering that Chinese modernization is inevitable regardless of U.S. BMD deployment.

Economic interdependence and mutual vulnerability associated with globalized trade and financial markets will increasingly emerge as the dominant force in relations between major powers, including (and perhaps especially) major nuclear powers.[53] Consistent with expectations derived from an expanding web of interdependencies and vulnerabilities, we can expect ongoing improvements in cooperative relations between East and West; any other outcome makes no rational sense for either side, at least for the foreseeable future. Thus, there is little need for large numbers of nuclear weapons to stabilize the relationships, and to the extent that large numbers of weapons exist they are likely to become increasingly meaningless.[54] Benjamin Rinkle sums up the issue very well: 'In the end, lamentations against the Bush administration's withdrawal from the ABM Treaty rest upon false assumptions. Arms races are merely symptoms of a larger disease, and given that most leaders will come to Mr. Putin's realization that it is better to be an ally of the United States than a strategic competitor, a U.S. missile defense is unlikely to trigger a new arms race.'[55] On the other hand, security against proliferation of WMD to hostile states and regimes, and related threats associated with globalized terrorism, will continue to drive U.S. foreign and security policy.

Two important implications follow. First, the U.S.-Russia and U.S.-China relationship will continue to evolve throughout this transition. Although levels of cooperation will ebb and flow, the threshold for escalation of crises with China or Russia is significantly higher today than it has ever been, and it is likely to become higher in the future. There is no conceivable scenario that would reproduce a cold war crisis today that would come close to reproducing a situation in which nuclear use would be contemplated. Second, policies that would have been provocative and dangerous during the cold war are now entirely conceivable and far less threatening. There is no reason today to expect a unilateral reduction of U.S. offensive weapons, along with simultaneous increases in expenditures on defensive systems, to be destabilizing. The reason is obvious – there is no credible scenario that would realistically take us to the brink of a contemporary (or future) crisis in which American officials would consider (for a second) a pre-emptive first strike against Russia and/or China – even assuming the United States develops and perfects a shield capable of destroying every single missile, decoy, and countermeasure Russia and China would launch in retaliation. The probability of a crisis provoking a pre-emptive first strike is virtually zero. Russia and China are not likely to give the United States reason to contemplate nuclear use, and the U.S. has no rational incentive to pursue a course of action that would provoke a nuclear response from China or Russia.

Sceptics who concede that deterrence stability is indeed more about relationships than numbers might still argue that any action that threatens to remove one state from the condition of mutual vulnerability (whether in the form of assured destruction or unacceptable damage) is qualitatively different from minor disparities in force levels. Actions that impinge on the condition of mutual vulnerability, they might argue, could affect relationships profoundly. Thus, one should caution against the conclusion that numbers are somehow completely irrelevant to the character of deterrent relationships. Several points should be noted in response to these important observations.

First, the most relevant question from the perspective of risks is whether the U.S. BMD will come close to removing states from the condition of mutual vulnerability. Nothing in current or future U.S. deployment plans justifies that concern. Second, even if mutual vulnerability (in the bipolar strategic stability sense) is jeopardized, U.S. policy makers will always compare the risks of that environment with the costs and risks assigned to the status quo. Doing nothing to address proliferation of WMD to new and aspiring nuclear powers will never be perceived (by current or future U.S. administrations) as cost-free. Third, risk and cost assessments are rarely measured in isolation, or in terms of specific threats (for example, the risks of Chinese proliferation in response to BMD deployment). Rather, they are assessed in terms of their capacity to deal with a multiplicity of interdependent threats and enemies. Strategies designed to address one set of threats (for example, BMD to defend against proliferation of WMD by rogue regimes) could conceivably exacerbate other threats (Chinese proliferation in response to BMD deployment), but choices have to be made. In comparative terms, Chinese proliferation is less threatening than proliferation by North Korea because it is thought that leaders in Beijing are more attuned to (and motivated by) the economic, political, and military forces that typically guide relations between rational, mutually dependent, and mutually vulnerable states.

SECTION III: Summary and Policy Implications

As the preceding discussion demonstrates, globalism provides a superb description of contemporary international politics, a brilliant and vivid picture of the complex transformations that are occurring as a result of the death of geography, the death of distance, increasing sensitivities and vulnerabilities, and the death of independence. But these descriptions, in and of themselves, do not provide very useful explanations for international behaviour, and they help us predict even less.

The evidence appears to suggest that globalism simultaneously enhances one dimension of security while detracting from another. It promises to increase and decrease fundamentalism; it makes terrorism more and less likely, the use of encryption more likely but more risky (given advances in decoding technology), and the effects of terrorist attacks more and less severe when they occur; it makes proliferation of WMD technology to terrorists more and less likely, the pace of proliferation more and less rapid, and the success of counter-proliferation and counter-terrorism more and less probable.

Among the by-products of the U.S. response to terrorism after 9/11 is the creation of new alliances with, for example, Pakistan. Arguably, Pakistan's participation in the war on terror can enhance global security by facilitating counter-terrorist operations against the Taliban and Al Qaeda networks inside and outside of Afghanistan. But the U.S.-Pakistan alliance could conceivably undermine global security by contributing indirectly to horizontal proliferation: Pakistan's nuclear program is likely to be legitimized in return for its help in the war on terror. In fact, the United States is currently looking at ways to transfer command, control, and communication technology to both India and Pakistan to help them stabilize their nuclear rivalry. While stabilizing any nuclear rivalry can be viewed as a positive step, the negative implications are obvious for the legitimacy and relevance of multilateral arms control. In addition to accepting Pakistan's status as a nuclear power in exchange for their help in Afghanistan, the strategy also necessitated looking the other way when evidence confirmed that Pakistan helped Pyongyang construct a 'secret centrifuge system of uranium enrichment.'[56] According to Jim Hoagland, this was a 'strategic joint venture' between North Korea and Pakistan in which 'they conspired to ignore all rules and agreements' – including the 1994 Framework Agreement with the U.S. to freeze North Korea's nuclear program. Pakistan's support for U.S. efforts in Afghanistan also created more instability within Pakistan, and between India and Pakistan, as extremists escalated attacks in Kashmir and against American political and military targets in the region.[57] Yet, despite these negative consequences, the more immediate American security priority at the time was the war on the Taliban and Al Qaeda.

This is but one illustration of the multiple interconnections across different dimensions of security. However, the paradoxical effects of globalization are not limited to questions of security – they apply across all issue areas and across all levels of analysis. As Thomas Friedman observed in his insightful research on the subject, 'Globalization can be incredibly empowering and incredibly coercive. It can democratize opportunity and democratize panic. It makes the whales bigger and the minnows stronger. It leaves you behind faster and faster, and it catches up to you faster and faster. While it is homogenizing cultures, it is also enabling people to share their unique individuality farther and wider.'[58] The implications of this complexity are profound; existing research on the subject of globalization does not yet allow us to make reliable, definitive, consistently valid, and empirically verifiable statements about globalism and security or about globalism and terrorism more specifically. With respect to terrorism, the causal sequencing is difficult to pinpoint, the long- and short-term effects produce both positive and negative consequences, and these consequences vary from one level of analysis to another (e.g., international, regional, and domestic terrorism), and from one dimension of terrorism to another (e.g., recruitment, financing, communication, counter-terrorism). In addition, existing literature on terrorism rarely addresses the connection to globalism but typically focuses on specific case studies, root causes, motivations, tactics, state sponsorship, acquisition of WMD, counter-terrorism, and so on.[59] Aside from a few notable exceptions, comparatively less attention has been devoted to the globalism-terrorism-globalism linkage.[60]

These deficiencies, in turn, create hurdles for policy makers searching for informed, efficient, and effective foreign and security strategies. Both Graham Allison and Warren Walker point to some of the obvious challenges confronting policy makers in the United States:

Whether at the Justice Department in the Anti-Trust Division, or at the Federal Trade Commission and Federal Communications Division, or at the Defense Department, or at the National Security Council, not to mention Congress, *the undeniable fact is that people are making policy choices about issues that they do not understand and whose consequences they cannot understand*. The consequences of 'self-accelerating technologies' and of unlimited bandwidth in human interaction are impossible to predict [emphasis added].[61]

The world in which governments must make policy is changing rapidly in unpredictable ways. Changes in information and communication technologies are eroding national borders and creating global markets ... Furthermore, *because of the globalization of issues and the interrelationships among systems, the consequences of making wrong policy decisions are becoming more serious – even catastrophic* [emphasis added].[62]

Current research on the relationship between globalism and security remains dangerously superficial. The field requires a new approach that goes beyond existing work on the subject by building on simplistic and outdated assumptions about the linkages among complex social phenomena. The paradoxical and mutually exclusive implications of globalization directly affect our capacity to accurately assess the policy implications of emerging trends. This is occurring at a time when the demand for high-quality global-security analysis is becoming imperative. A new approach requires identification of other variables, other causal mechanisms, other factors, and intervening forces that combine to direct the effects of globalism one way or the other, in positive or negative directions. Without a more sophisticated framework of analysis derived from a more systematic and rigorous research program, our theories and predictions about the impact of globalization on security (and, more specifically, its impact on terrorism and proliferation) will be incomplete and incorrect.

Perhaps the clearest illustration of the fundamental errors that flow from superficial treatments of the linkages between globalism and security is the emerging consensus that multilateralism has become essential for U.S. security in the aftermath of 9/11, and that unilateralism is not only ineffective and counterproductive, but, in the context of globalism, obsolete. This overarching policy recommendation illustrates the superficial nature of the arguments and evidence used to derive foreign and security policies to deal with proliferation and terrorism in a post-9/11 environment. The next chapter highlights the many deficiencies that plague this conventional wisdom.

Chapter 2

Linking Globalism, Unilateralism, and Multilateralism

Conventional Wisdom

According to several prominent globalization experts, the terrorist attacks on 9/11 destroyed, once and for all, the myth of American independence. According to this view, U.S. officials can no longer remain complacent in the belief that they are somehow isolated from global conflict, or that they have the power to independently protect the United States from external (and internal) attacks. As the world continues to transform, state-centric models of international politics will become increasingly obsolete. These outdated frameworks no longer provide a useful analytical tool for predicting international behaviour and have become almost useless as a guide for foreign and security policy after 9/11.[1] As Benjamin Barber writes, 'The American myth of independence is not the only casualty of September 11. Traditional realist paradigms fail us today also because our adversaries are no longer motivated by "interest" in any relevant sense, and this makes the appeal to interest in the fashion of realpolitik and rational-choice theory seem merely foolish.'[2]

The death of independence, in turn, will have a profound impact on U.S. foreign and security policy. American unilateralism will inevitably be replaced by a strong preference for multilateralism, because only multilateral strategies and institutions can provide the coalitions and international cooperation required to address the security threats created by the forces of globalization. The six quotations in table 2.1 are included here to illustrate the emerging consensus in the literature. These arguments, predictions, and associated policy recommendations, from leading experts in the field, represent the conventional wisdom on

TABLE 2.1
Globalism, Terrorism, Proliferation, and Multilateralism:
The Emerging Consensus

1 It could hardly escape even casual observers that global warming recognizes no
sovereign territory, that AIDS carries no passport, that technology renders national
boundaries increasingly meaningless, that the Internet defies national regulation, that
oil and cocaine addiction circle the planet like twin plagues and that financial capital
and labor resources, like their anarchic cousins crime and terror, move from country
to country with 'wilding' abandon without regard for formal or legal arrangements –
acting informally and illegally whenever traditional institutions stand in their way ...
Terrorism's network exists in anonymous cells we can neither identify nor capture.
Declaring our independence in a world of perverse and malevolent interdependence
foisted on us by people who despise us comes close to what political science rough-
necks once would have called pissing into the wind. – Benjamin Barber, *Washington
Times*, 30 May 2001.

2 Globalization means, among other things, that threats of violence to our homeland
can occur from anywhere. The barrier conception of geographical space, already
anachronistic with respect to thermonuclear war and called into question by earlier
acts of globalized informal violence, was finally shown to be thoroughly obsolete on
September 11 ... The globalization of informal violence means that we are not so
insulated. We are linked with hateful killers by real physical connections, not merely
those of cyberspace. Neither isolationism nor unilateralism is a viable option ... The
terrorist attacks on New York and Washington force us to rethink our theories of world
politics. Indeed, we need to reconceptualize the significance for homeland security of
geographical space. – Robert Keohane, 'The Globalization of Informal Violence,
Theories of World Politics, and the Liberalism of Fear,' 2001.[a]

3 The likely effects of the September 11 terrorist bombings will be to usher in an era
where U.S. foreign policy is more multilateralist than before, an era that indicates
both the essential interconnectedness of world politics and the fact that the U.S. can
neither act as world policeman nor retreat into isolationism. – Steve Smith, essay for
the Social Science Research Council, Washington D.C., 2001.[b]

4 September 11 also brought a realization of how much America's well-being depends
on the international order and on having friends and allies around the world that
share our basic values. Americans took comfort from the supportive words and
actions from our allies. The attacks of September 11 made concrete the certain
knowledge that America cannot effectively combat the terrorist threat alone. America
cannot alone track and stem the financial lifeblood of international terrorist organizations, combat the roots of terrorism, or win a war against a shadowy enemy about
whom much is known but not by us, in a bleak landscape far from our traditional
bases of operation. Perhaps for the first time since the end of the Cold War, Americans may see the international institutions in a new light as directly relevant to our
own well-being. – Lael Brainard, Analysis Paper #12, Brookings Institution, 2001.

5 [M]ilitary power alone cannot produce the outcomes Americans want on many of the
issues that matter to their safety and prosperity ... The problem for Americans in the
21st century is that more and more things fall outside the control of even the most

TABLE 2.1
Globalism, Terrorism, Proliferation, and Multilateralism:
The Emerging Consensus (*concluded*)

---

powerful state. Although the United States does well on the traditional measures, there is increasingly more going on in the world that those measures fail to capture. Under the influence of the information revolution and globalization, world politics is changing in a way that means Americans cannot achieve all their international goals by acting alone ... And in a world where borders are becoming more porous to everything from drugs to infectious diseases to terrorism, America must mobilize international coalitions to address shared threats and challenges. – Joseph Nye, *The Economist*, 23 March 2002.[c]

6 Equally, it became difficult to see how America's vulnerability could be reduced through isolationism, or indeed unilateralism. With international terrorist networks extending globally, security from further terrorist attacks seemed possible – if possible at all – only through a careful courting of America's present circle of allies, and moreover, through an expansion of that circle to transform states that were once part of the problem into parts of the solution. And it was difficult to see how any of this was possible without, first, carefully courting Europe, and second, working with the EU to extend and embed multilateralism in the international order. – John Peterson, *Irish Studies in International Affairs* 13, 2002.

---

[a] Paper delivered at the Annual Meeting of the American Political Science Association, Boston, 2002, available at http://www.iyoco.org/911/911keohane.htm. See also chapter 5 in Keohane's *Understanding September 11* (Washington: New Press Social Science Research Council [forthcoming], available at http://www.ssrc.org/sept11/toc11a.htm).
[b] 'The End of the Unipolar Moment: September 11 and the Future of World Order,' available at http://www.ssrc.org/sept11/essays/smith.htm.
[c] See also Joseph Nye, *The Paradox of American Power: Why the World's Only Superpower Can't Go It Alone* (London: Oxford University Press, 2002). Nye develops perhaps the strongest case for multilateralism in the aftermath of 11 September, but his arguments are balanced with the acknowledgment that unilateralism, like multilateralism, is a necessary but insufficient approach to the security threats produced by globalization.

globalism and the inevitable (and rational) trend towards multilateral solutions to security after 9/11. These are powerfully compelling accounts of the death of American independence and the decline of state power as it relates to an independent capacity to combat proliferation and to fight, let alone win, the war on terror.

David Malone and Yuen Foong Khong provide additional evidence (and illustrations) of this emerging consensus, compiled in a collection of twenty contributions from international experts on American foreign policy.[3] The chapters on security, for example, focus on the mounting costs of U.S. unilateralism in the context of the United Nations (Kishore Mahbubani), peacekeeping (Ramesh Thakur), unilateral and multilateral uses of force (Ekaterina Stepanova), nuclear policy (Qingguo Jia), and arms control and non-proliferation (Kanti Bajpai). Without exception, these international scholars draw similar conclusions to those outlined in table 2.1: the costs and risks of unilateralism are far too great in a globalizing world for it to be considered a rational approach to security. In each case the core recommendation is for the U.S. to seriously consider re-engaging multilateral approaches and institutions, presumably because they hold the most promise for effective and efficient responses to emerging security threats.

Unilateralist Response

But predictions about the inevitable (and rational) preference for multilateralism in a globalizing world simply do not match the U.S. response to 9/11, nor are they consistent with the emerging trend in American security policy as exemplified in the wars in Afghanistan and Iraq. The conventional wisdom, in other words, appears to be wrong. The evidence confirms instead that the more insecure the United States becomes as a result of the globalization of terrorism and the proliferation of WMD technology, the more effort, money, time, and energy the United States will invest in re-establishing independent, self-directed, sovereign, and unilateral control over American security and economic affairs.[4] Despite the reality of *inter*dependence, increasing levels of U.S. vulnerability and sensitivity, the death of geography, the death of distance, and the myth of American independence, American officials continue to implement policies that prioritize re-establishing American independence. Consider some of the unilateral initiatives after 9/11, outlined in table 2.2.

Without exception, all these efforts are designed with one overriding objective in mind: to acquire autonomous control over U.S. security. Washington is committed to becoming less dependent on other states and international organizations for the safety of American citizens; less dependent on the United Nations, European allies, and NATO for multilateral arms control; and less dependent on Russia for bilateral arms control. In contrast, Washington is becoming more dependent on strategic coalitions that fall outside traditional alliance structures, more dependent on homeland security and more willing to pursue international interventions (Afghanistan and Iraq), more open to Mideast di-

TABLE 2.2
American Unilateralism after 9/11

- Billions of dollars invested in the war (and reconstruction) in Afghanistan to change Taliban and Al Qaeda regime (2001)
- Billions of dollars invested in the war (and reconstruction) in Iraq to change Saddam Hussein's Baath Party regime (2003)
- Short-term (unilateral) shifts in U.S. alliances and coalitions:[a]
  - often to combat immediate security threats (terrorism)
  - often at the expense of other security interests (proliferation)
- Proliferation Security Initiative (announced by G.W. Bush in Poland, 31 May 2003):
  - established sweeping mandate to search ships and containers for WMD
- Accelerated deployment of Ballistic Missile Defence:
  - requested increase of $3 billion (to $8.3 billion) for BMD (FY 2002–2003)
- Withdrawal from the ABM Treaty (and multilateral arms control more generally):
  - refusal to ratify Comprehensive Test Ban Treaty
  - refusal to negotiate Biological Weapons Convention Treaty
- Unilateral demand that Palestinian Liberation Authority replace Arafat
- Global deployment of U.S. Special Forces to track Al' Qaeda and Taliban
- Interpretation of Geneva Convention re status of Al' Qaeda and Taliban prisoners
- Substantial increase in U.S. defence budget by $48 billion to $396.1 billion (FY 2002–2003):
  - largest single increase in defence budget since Korean War[b]
- Revising regional command structure – Northern Command (NORCOM):
  - designed to facilitate homeland defence and continental security
- New cabinet-level Department of Homeland Security
- $30 million/day, $1 billion/month for the war on terrorism
- $90 billion economic stimulus bill to deal with economic impact of 9/11
- $39 billion for homeland defence
- $20 billion increase for Intelligence (to approx. $40 billion)
- $23.8 billion on Border and Transportation Security (156,169 employees)
- $15 billion emergency assistance package for airline industry (cash and loans)
- $8.4 billion on Emergency Preparedness and Response (5,300 employees)
- $7.8 billion for Defense Department anti-terrorism efforts
- $5.9 billion to enhance defenses against bioterrorism, including:
  - $1.2 billion to increase capacity for health delivery systems
  - $2.4 billion for research and development on bio-terrorism responses
  - $420 million for the Pentagon to study bioterrorists
- $3.6 billion on WMD Countermeasures (598 employees)
- $3.5 billion to enhance response capabilities of America's first responders:
  - includes firefighters, police officers, and emergency medical workers
- $1.4 billion to secure diplomatic facilities:
  - $755 million for security-driven construction
  - $553 million for upgrades for worldwide security
  - $52 million for a new Center for Anti-terrorism Security Training
  - $60 million for public diplomacy through international broadcasting
- $1.2 billion for the Secret Service (6,111 employees)
- $364 million on Information Analysis and Infrastructure Protection (976 employees)

TABLE 2.2
American Unilateralism after 9/11 (*concluded*)

- Over one hundred new bills, acts, and other legislation passed by U.S. since 9/11:
  - most of which assign new powers to FBI/CIA/NSA
  - designed to enhance surveillance and law enforcement (Appendix 1.1)
- Established more state control over traditional non-security areas:
  - air transportation, trans-national finance, and refugee and immigration laws
- Diplomatic pressure on Canada to:
  - invest $5 billion to improve Canada-U.S. border security
  - rationalize refugee and immigration policies
- Rejection of Land Mines Treaty; International Criminal Court; and Kyoto Treaty

---

[a] For an excellent report on the unintended consequences of short-term unilateral alliance shifts (towards, for example, Pakistan, Iran, and other coalitions of convenience), see Robert Kagan and William Kristol, 'The Coalition Trap,' *Weekly Standard* 7, no. 15 (15 October 2001). They ask readers to think about the message the president is sending: 'Terrorism works. Prior to September 11, Bush had said not a word about a Palestinian state. After September 11, he was declaring it his vision. To the Arabs and Palestinians who danced and cheered as the twin towers fell, Bush's statement told them they were right to celebrate. Kill enough Americans, and the Americans give ground ... In pursuit of the coalition, we have encouraged Palestinians and Arab radicals to believe that terrorism works.' See also Jim Hoagland, 'An Ally's Terrorism,' *Washington Post*, 3 October 2001, A31. According to Hoagland, 'the coalition Bush and Powell have assembled in all necessary haste ... [was] recruited to fight terrorism regimes that practice or tolerate terrorism as a matter of policy.'
[b] Consider the following facts about U.S. military spending compiled from the Council for a Liveable World (http://www.clw.org/milspend/fy03facts.html): 'The U.S. increase of $48 billion is larger than the annual military budget of any other country in the world ... Military spending accounts for 17.8 percent of the entire federal budget ... Military spending comprises over half (53 percent) of total discretionary spending ($755 billion) ... U.S. military budget is greater than the entire economies (Gross Domestic Product) of each of the following countries: Austria, Belgium, Chile, Colombia, Denmark, Egypt, Finland, Norway, Greece, Hong Kong, Ireland, Israel, Saudi Arabia, Peru, Poland, Portugal, Singapore, South Africa, Sweden, Switzerland, Pakistan, Vietnam and Venezuela ... The proposed military budget of $396.1 billion is 15 percent higher than the average Cold War budget.'

plomacy, and more likely to apply coercive diplomatic techniques to opponents and allies alike in order to protect their foreign and security interests. Compare the combined investments listed in table 2.2 to the $870 million in outstanding UN dues owed by the U.S. government (2003) – a relatively straightforward illustration of American strategic priorities and corresponding commitments to multilateralism.[5]

The clearest indication of how seriously the U.S. administration takes the threat of terrorism, and the priority it assigns to unilateral initiatives, can be found in the 2002 *National Security Strategy of the United States of America*. The relationship between the forces of globalization and security is unambiguously central to the document, with repeated references to increasing levels of vulnerability brought on by the complexity and uncertainty of asymmetric threats. These threats have become the core security concerns of the U.S. government.

> The gravest danger our Nation faces lies at the crossroads of *radicalism* and *technology*. Our enemies have openly declared that they are seeking weapons of mass destruction, and evidence indicates that they are doing so with determination ... In a globalized world, events beyond America's borders have a greater impact inside them. Our society must be open to people, ideas, and goods from across the globe. The characteristics we most cherish – our freedom, our cities, our systems of movement, and modern life – are vulnerable to terrorism. This vulnerability will persist long after we bring to justice those responsible for the September 11 attacks. As time passes, individuals may gain access to means of destruction that until now could be wielded only by armies, fleets, and squadrons. This is a new condition of life. We will adjust to it and thrive – in spite of it [emphasis added].[6]

With respect to policy prescriptions and priorities for addressing these threats, there is no disputing the fact that unilateralism, pre-emption, and preventive war are now viewed as absolutely essential to U.S. security. Consider the following excerpts from the *USNSS* 2002 included in table 2.3. Contrary to popular opinion, the principles underlying this grand strategy are not the products of a right-wing conservative administration determined to rid the world of the scourge of multilateralism, as some critics continue to claim; the approach is far more evolutionary than revolutionary. The same underlying themes run through the previous U.S. National Security Strategy produced by the Clinton administration in 1999: 'The United States will do what we must to defend our vital interests including, when necessary and appropriate, using our military *unilaterally* and decisively ... We act in alliance or partnership when others share our interests, but *unilaterally* when compelling national interests so demand ... The decision whether to use force is dictated first and foremost by our national interests. In those specific areas where our vital interests are at stake, our use of force will be decisive and, if necessary, *unilateral* ... We act in concert with the international community whenever possible, but do not hesi-

TABLE 2.3
U.S. National Security Strategy, September 2002
Privileging Unilateralism, Preemption, and Preventive War

1 We will build defenses against ballistic missiles and other means of delivery. We will cooperate with other nations to deny, contain, and curtail our enemies' efforts to acquire dangerous technologies. *And, as a matter of common sense and self-defense, America will act against such emerging threats before they are fully formed ... History will judge harshly those who saw this coming danger but failed to act* [emphasis added]. – George W. Bush, Introductory Letter, 17 September 2002, *The National Security Strategy of the United States of America,* 2002.

2 The conflict was begun on the timing and terms of others. It will end in a way, and at an hour, of our choosing. – George W. Bush, Introductory Letter.

3 We will disrupt and destroy terrorist organizations by: *defending the United States, the American people, and our interests at home and abroad by identifying and destroying the threat before it reaches our borders.* While the United States will constantly strive to enlist the support of the international community, *we will not hesitate to act alone, if necessary, to exercise our right of self-defense by acting preemptively against such terrorists, to prevent them from doing harm against our people and our country* [emphasis added]. – George W. Bush, speech at the National Cathedral, Washington, 14 September 2001.

4 As was demonstrated by the losses on September 11, 2001, mass civilian casualties is the specific objective of terrorists and these losses would be exponentially more severe if terrorists acquired and used weapons of mass destruction. The United States has long maintained the option of *preemptive actions to counter a sufficient threat to our national security. The greater the threat, the greater is the risk of inaction – and the more compelling the case for taking anticipatory action to defend ourselves, even if uncertainty remains as to the time and place of the enemy's attack. To fore-stall or prevent such hostile acts by our adversaries, the United States will, if necessary, act preemptively.* The United States will not use force in all cases to preempt emerging threats, nor should nations use preemption as a pretext for aggression. Yet in an age where the enemies of civilization openly and actively seek the world's most destructive technologies, the United States cannot remain idle while dangers gather [emphasis added]. – George W. Bush, Introductory Letter.

5 In exercising our leadership, we will respect the values, judgment, and interests of our friends and partners. *Still, we will be prepared to act apart when our interests and unique responsibilities require. When we disagree on particulars, we will explain forthrightly the grounds for our concerns and strive to forge viable alternatives. We will not allow such disagreements to obscure our determination to secure together, with our allies and our friends, our shared fundamental interests and values* [emphasis added]. – George W. Bush, Introductory Letter.

tate to act *unilaterally* when necessary.'[7] Washington is unlikely (and apparently unwilling) to heed the concerns expressed by Barber, Keohane, Nye, Smith, and others regarding the futility of contemporary unilateralism (see table 2.1). Nor are American officials likely to accept as fact the obsolescence of geographic boundaries or suddenly acknowledge the death of their own independence. When it comes to protecting Americans after 9/11, there is little evidence to indicate that American officials are in favour of becoming more dependent on the UN or, for that matter, any other state, alliance, multilateral coalition, organization, institution, or regime. Charles Krauthammer offers perhaps the most insightful interpretation of U.S. priorities after 9/11:

It took only a few hours for elite thinking about U.S. foreign policy to totally reorient itself, waking with a jolt from a decade-long slumber. After the apocalypse, there are no believers. The Democrats who yesterday were touting international law as the tool to fight bioterrorism are today dodging anthrax spores in their own offices ... When war breaks out, even treaty advocates take to the foxholes ... This decade-long folly – a foreign policy of norms rather than of national interest – is over ... On September 11, American foreign policy acquired seriousness. It also acquired a new organizing principle: We have an enemy, radical Islam; it is a global opponent of worldwide reach, armed with an idea, and with the tactics, weapons, and ruthlessness necessary to take on the world's hegemon; and its defeat is our supreme national objective, as overriding a necessity as were the defeats of fascism and Soviet communism.[8]

Critics continue to warn that unilateral, state-centric approaches to the war on terror and proliferation will not succeed, because of the diminishing capacity of any country, including the United States, to effectively control homeland and international security in a globalizing world. How, then, can one explain this ever-present and powerful American fixation with maintaining independent control over their own security, notwithstanding the evidence that successful unilateralism is a chimera?[9] The obvious implication is that American leaders simply misunderstand the realities of contemporary international politics and prefer strategies that are not only irrational but may actually make things worse.

On the contrary, what appears on the surface to be an irrational response to the contemporary realities of globalization is in fact a rational strategy derived from an objective assessment of the costs, benefits, and risks of available alternatives. The difficulties of achieving absolute security through unilateralism are irrelevant – whether these strategies are sufficient is beside the point. Unilateral approaches to security are rarely evaluated (and selected) in isolation – they are always compared with the successes, failures, and overall potential of multilateral alternatives. With respect to that comparison, it is becoming increasingly apparent to U.S. officials that multilateral approaches to security have not succeeded, and that unilateral strategies offer a better return for security investment with fewer costs and risks. Indeed, multilateral approaches have not achieved the kind of success that would warrant giving proponents the moral or intellectual authority to dismiss unilateral alternatives. Without evidence of success there is no logical, factual, legal, moral, or policy-relevant foundation for exclusive reliance on multilateral alternatives. Consequently, major powers will forever struggle to re-establish independent control over their security even in the face of difficulties and challenges that prevent ultimate success.

This fact should be the starting point for our theories, explanations, and predictions of international behaviour after 9/11. Far from being obsolete, the logic of geographic boundaries remains a central feature of international politics. A state-centric framework may no longer provide a particularly accurate description of contemporary international politics (perhaps it never did), but it is still the best explanation we have for the foreign and security policies we see playing out today.

Multilateralist Reply and Unilateralist Rejoinder

Citing Washington's return to the UN Security Council immediately after the attacks, Keohane argues that multilateralism offers a better explanation for the U.S. response to 9/11. The Bush administration needed desperately to legitimize its war in Afghanistan and required institutions and international law to accomplish this – 'only the U.N. can provide the breadth of support for an action that can elevate it from the policy of one country or a limited set of countries to a policy endorsed on a global basis.'[10] But Keohane's interpretation of U.S. actions and motivations is misleading, for several reasons.

The U.S. response required very little 'elevation' to be endorsed as legitimate by other leaders. The deaths of just under three thousand innocent Americans provided more than sufficient justification for American retaliation. Expressions of support from almost every other country and international organization on the planet fully endorsed the

American right of self-defence, as entrenched in the UN charter. That support was immediate, unanimous, and virtually guaranteed, for the same reason – the destruction and associated devastation in New York and Washington. European leaders were competing with each other to provide whatever assistance the United States requested, and all reaffirmed their NATO charter commitments to support Washington. In contrast to the British prime minister, Tony Blair, who received praise for his reaction to 9/11, the Canadian prime minister, Jean Chrétien, lost credibility and enraged the Canadian public when his expression of support was slow, equivocal, and ambivalent. European and Canadian leaders, international organizations, and multilateral institutions needed the U.S. to legitimize *their* reaction to 9/11 more than the United States needed them – Keohane got it backwards. While the Bush administration welcomed any and all support it received after 9/11, that support was never perceived as a precondition for responding, for the same reason NATO's response to ethnic cleansing by Milosevic in Kosovo did not require a UN Security Council resolution for legitimacy. Similarly, any unilateral response to the Rwandan genocide in 1994 to save even a fraction of hundreds of thousands of lives lost would have been a legitimate intervention, despite the absence of a UN Security Council resolution and mandate.

The typical argument favouring multilateralism is a simple one, summarized by Ramesh Thakur: 'Because the world is essentially anarchical, it is fundamentally insecure, characterized by strategic uncertainty and complexity because of too many actors with multiple goals and interests and variable capabilities and convictions. Collective action embedded in international institutions that mirror mainly U.S. value preferences and interests enhances predictability, reduces uncertainty, and cuts the transaction costs of international action.'[11] With respect to peacekeeping, for example, Thakur argues that if 'the UN helps to mute the costs and spread the risks of the terms of international engagement to maximise these benefits, the United States will need to instil in others, as well as itself embrace, the principle of multilateralism as a norm in its own right: states must do X because the United Nations has called for X, and good states do what the United Nations asks them to do.'[12] But there are several problems with Thakur's defence of collective action and associated policy recommendations, particularly in relation to multilateral approaches to security in a post-9/11 setting.

First, and foremost, state leaders often refuse to do what the UN asks of them, are often more than prepared to have their publics suffer the consequences of whatever sanctions the UN can mount, and are rarely directly affected by the sanctions that are implemented – assuming the permanent members of the Security Council find it in their collective interest to implement a sanctions regime in the first place. The lessons from UN intervention and sanction efforts over the past decade are not at all encouraging in this regard.

Second, many state and non-state actors fall outside the institutional constraints imposed on the system through global norms and regimes. As the capacity spreads for smaller and smaller groups to inflict increasingly devastating levels of damage on larger states, international institutions will lose the capacity to force or coerce compliance with international law. Consequently, leaders of major powers, such as the United States, will be compelled to respond to security threats through unilateral initiatives. This compulsion will force other powers to push that much harder to control American impulses by demanding that multilateral consensus remain the sole guarantor of legitimacy. These tensions will be exacerbated by the prevailing perception in the United States that these same multilateral institutions are constraining the power and capacity of the U.S. government to protect American citizens from emerging threats of terrorism and proliferation.

Third, the collective-action argument put forward by Thakur typically (and erroneously) assumes that most states are governed by a similar set of political priorities, share common concerns about similar combinations of security threats, are stimulated into action (or inaction) by the same set of economic imperatives, are inspired by a common set of interests and overarching values (such as peace, security, stability), and are encouraged by their respective publics to meet their demands for a common set of public goods. But the differences, tensions, and overall level of competition among states in the system are far greater than proponents of multilateralism acknowledge. Some states are more threatened by terrorism and proliferation than others, have more substantial and direct economic interest in particular regions, are less interested in securing peace, and experience pressure from their respective publics to pursue very distinct foreign and security policies. Consequently, there is no guarantee that a collection of states will have the same motivation to change the status quo, or experience the same imperative to address the same security threats with the same level of resolve, commitment, or resources (relative to their size). In sum, multilateral organizations are less likely today to act with the same level of urgency to address security threats that Washington considers impera- tive. The costs of inaction (derived from exclusive reliance on multilat- eral consensus) are now perceived as being higher than the costs of unilateralism. Although similar threats may have guided collective action through multilateral alliances for much of the cold war, these imperatives were a product of a common Soviet threat. But threats today are many and varied, and few states share the same concerns or face the same obligations to respond. No case more clearly illustrates the growing divisions among former allies than the 2003 Iraq war.

Fourth, decreasing transaction costs may be a valid argument in favour of multilateral cooperation in some cases (e.g., to facilitate post- conflict reconstruction, political reforms, democratization, elections run by the Organization for Security and Cooperation in Europe, food aid, water distribution, and the provision of medical supplies and facilities), but this is not true for all security challenges. In a post-9/11 environ- ment, the transaction costs that are saved through joint efforts will always be compared with the costs of depending exclusively on collec- tive-action mechanisms that ultimately may fail – multilateralism is not free of costs or risks.

For example, one of the many important lessons of the 2003 Iraq war, at least for American officials, is that there are no collective-security guarantees any longer, even from traditional allies. The UN Security Council did not function as a separate entity committed to facilitating and coordinating diplomatic exchanges towards a common good. The UN functions in a highly competitive environment in which traditional power politics plays out. Proponents of multilateralism through the UNSC do not espouse that doctrine in the interest of global security; their efforts are typically designed to use the institution to limit the capacity of the U.S. to act unilaterally to protect American interests. That level of competition, itself driven by competing interpretations of interests, values, and threats, does not lend itself well to the kind of multilateralism its proponents aspire to achieve. Of course, if France shared the same concerns about terrorism, or if leaders in Paris were equally motivated to address the potential for WMD proliferation in and through Iraq, the transaction costs incurred by responding through the UN would be more acceptable. But as threat perceptions continue to diverge, the risks associated with waiting for multilateral consensus are simply too high. The complex nature of contemporary security threats virtually guarantees that similar conflicts will plague multilateral insti- tutions in the future.

Summary and Policy Implications

Proponents of multilateralism rarely offer a balanced accounting of the costs, risks, deficiencies, and unintended consequences of exclusive reliance on multilateral approaches to security in a contemporary setting. This is not meant to suggest that multilateralism carries no benefits. Samuel Berger draws a distinction between 'authority' and 'power' to highlight one of the obvious and most important benefits of multilateralism – legitimacy: 'We must remember there is a difference between power and authority. Power is the ability to compel by force and sanctions; there are times we must do so, but as a final not a first resort. Authority is the ability to lead, and we depend on it for virtually everything we try to achieve. Our authority is built on very different qualities than our power: on the attractiveness of our values, on the force of our example, on the credibility of our commitments and our willingness to work with and stand by others.'[13] The common theme, once again, is that going it alone is impractical because it cannot generate legitimacy. But there is an equally important distinction to be drawn between the authority that people assign to organizations such as the UN simply by virtue of the mistaken assumption that the UN represents multilateral (collective) interests, and the authority that should be earned by the UN on the basis of the same standards Berger demands from American foreign policy and leadership. A little less multilateral apathy in Rwanda in 1994 and a little more unilateral (independent) initiative could have saved hundreds of thousands of lives and was well worth the risks; a little less dependence on UN multilateral authority in Bosnia circa 1990–95, and a little more support for an American proposal to lift the arms embargo against the Muslims and to strike Serb targets earlier (rather than in 1995) would have saved potentially tens of thousands of lives; U.S. unilateral pressure to push for a military solution in Kosovo without the authority from a UN Security Council resolution proved to be the right strategy at the right time against the right person (Milosevic) for the right reasons. The record of UN failures to stop ethnic cleansing and genocide in the Balkans (Bosnia, Croatia, Kosovo), Africa (Sierra Leone, Rwanda, Congo), and the Middle East (Iraq) throughout the nineties is a product of the prevailing (yet mistaken) belief that humanitarian interventions can only receive legitimacy if processed through the authority of multilateral consensus. These catastrophes raise serious questions about whether the UN Security Council deserves the authority to lead by example, or by the values espoused by many of its members, or by the credibility of its commitments, or by the willingness of its members to work with and stand by others. 'The UN is not an association of peaceful democracies,' observes Allan Gotlieb. 'If it were, it could have great moral authority. Rather it is a collection of states that individually may or may not have any moral legitimacy. Some are totalitarian governments that repress their people and cannot be regarded as representing public opinion. Moreover, decision-making in the Security Council is based on an archaic formula that gives inordinate power to some (the unilateral power to paralyze it) and little to others with greater democratic traditions ... In matters of peace and security, there are advantages in obtaining UN endorsement. But the failure of the UN to authorize the use of force could hardly, in itself, delegitimize its use.'[14] Authority and, by extension, legitimacy should never be automatically assigned to the UN by virtue of its status as a multilateral forum, because the consensus required to do the right thing is typically elusive and almost always derived from the competition among states struggling to protect individual economic, political, and military self-interest. The product of this competition rarely produces a common good.

The accession of Libya to the chairmanship of the UN's human rights commission, or Iraq's to the disarmament commission, raises obvious questions about the capacity of multilateral consensus to produce legitimacy and authority.[15] Reports from leading human rights organizations confirm that Libya's human rights record is among the worst on the planet: 'It is a dictatorship where compliance to the will of the state is enforced by the country's secret police and a network of revolutionary committees and people's committees that are an outgrowth of the political writings of the country's strongman, Col. Moammar Gadhafi,' writes Adrian Karatnycky.[16] The U.S. was voted off the commission by the requisite multilateral voting bloc that included several other regimes – China, Cuba, Libya, Saudi Arabia, and Syria. Since 1997, as one would expect under these conditions, the commission 'has taken no notice whatever of extensive rights violations in such countries as Belarus, China, Egypt, Indonesia, North Korea, Laos, Pakistan, Saudi Arabia, Tajikistan, Turkmenistan, Uzbekistan and Vietnam. Several countries, most notably China, have succeeded in defeating censure, in most cases by preventing a resolution from even reaching the floor for discussion.' It is not simply that Libya has emerged as chair of the UN's human rights commission, but that the process included those who indirectly supported the decision through their abstentions – France,

Germany, and Great Britain. As Marian L. Tupy explains, '[I]ncapacitated by colonial guilt, political correctness, and hypersensitivity to criticism emanating from the developing countries, the Europeans have continuously ignored human rights abuses by some of the world's most unsavory regimes.'[17] Perhaps most disconcerting of all was South Africa's decision to nominate Libya and then actively work towards generating the requisite support.

Staunch UN supporters typically dismiss these examples as beside the point or inconsequential, and then move on to what they consider to be far more important issues, such as the deficiencies of unilateralism. But the Libya case is relevant to understanding the fundamental deficiencies that plague multilateral organizations – states do not share the same values, priorities, and security threats. If an international organization of states collectively places Libya at the centre of its authority to make decisions about human rights, and if Libya's success at obtaining that position was a function of the standard policies and selection practices of the organization in question, how likely is it that this same organization will function in a way that would take seriously American security interests or the collective human rights and security interests of the international community?

To the extent that we can identify common goods in the contemporary global system, Michael Mendelbaum is correct:

> The contemporary world is dominated by three major ideas: peace as the preferred basis for relations among countries, democracy as the optimal way to organize political life within them, and the free market as the indispensable vehicle for producing wealth. Peace, democracy, and free markets are the ideas that conquered the world. They are not, of course, universally practiced, and not all sovereign states accept each of them. But for the first time since they were introduced – at the outset of the period that began with the French and Industrial Revolutions and is known as the modern era – they have no serious, fully articulated rivals as principles for organizing the world's military relations, politics, and economics. They have become the world's orthodoxy.'[18]

The author goes on to note that defending, maintaining, and expanding peace, democracy, and free markets has emerged as the 'central purpose of the United States in the twenty-first century and the principal use for American power.' But like many others, Mendelbaum concludes that 'American power, great though it is, is not necessarily sufficient.'[19] Although correct, his argument is incomplete – unilateralism may not be sufficient, but the more relevant question is whether it is necessary and, alternatively, whether multilateralism is sufficient in and of itself to achieve those same goals and objectives. Moreover, security is left out of his list of guiding principles, but in a post-9/11 environment security trumps some of the other goals, at least for the foreseeable future. Alliances with Pakistan in the war on terror, for example, are not likely to be conducive to the spread of democracy in the country; policies to enhance border security will not enhance the prospects of establishing free markets; and the war in Afghanistan and Iraq may improve democracy in these states, and perhaps the region more generally, but it may also undermine peace and security by increasing the probability of terrorist reprisals and escalation of conflict in the Middle East.

The central question addressed in the remainer of the book is whether unilateral or multilateral approaches to security are better suited to addressing emerging threats of proliferation and terrorism. What are the costs, benefits, and risks of each strategy, and how do we determine the best approach in any given situation? Unilateralism should not be evaluated (or dismissed) in terms of whether it undermines the ability to establish and proliferate multilateral institutions and bureaucracies, but it should be evaluated in terms of whether, in comparison with available multilateral alternatives, it can address the challenges of proliferation and terrorism with fewer risks, or unintended consequences.

Unilateral-Multilateral Continuum(s): Measurement Errors and Correctives

Several points should be noted regarding the 'choice' between multilateral and unilateral approaches to security before introducing the case studies. The choice is rarely driven by preferences alone; the decision is often a product of systemic pressures that push leaders in one or another direction – uncontrollable imperatives, not preferences, often explain behaviour. 'People and countries might shape systems,' Tim Hames observes, 'but systems shape countries and people. It is impossible to divorce the exercise of power from the context in which it is set ... A singularly unipolar political structure will produce, absolutely inevitably, a unilateralist outcome ... The sole viable alternative to unilateralism is not multilateralism, but isolationism.'[20]

There are no pure unilateralists or multilateralists. Preferences are likely to vary from issue to issue, region to region, and threat to threat. The application of these strategies is also likely to vary for specific states in different contexts. For example, unilateral strategies were entirely acceptable to France when dealing with ethnic conflict in its former African colonies (Ivory Coast), to Russian officials when combating ethnic violence in Chechnya, and to China when confronting challenges to its unilateral control over Tibet and Taiwan. Yet these same three states were quick to emphasize the importance of multilateralism in the context of U.S. policies in Iraq circa 2003. There is nothing particularly surprising about the selective application of these strategies. Historically, American foreign policy has exhibited elements of both, although Washington tends to receive far more criticism for its unilateral initiatives than praise for its contributions to multilateralism. This bias often creates an exaggerated impression that Washington is decidedly unilateral, even if the record is far more balanced (see chapter 3, under Unilateralism as a Prerequisite for Multilateralism).[21]

Presence and Absence

There are two distinct dimensions to disputes among proponents of multilateralism and unilateralism, both illustrating the difficulties with classifying any major foreign policy initiative in these terms: (1) debates about whether the U.S., its allies, and their opponents are (or are not) acting unilaterally or multilaterally, and (2) debates about whether unilateral or multilateral approaches are more/less effective at enhancing global, regional, or domestic security.

The first set of debates raises several difficult questions about how to actually measure unilateralism or multilateralism, and the second raises both normative and empirical questions about the probability of successfully achieving a set of specific goals. These two debates are taking place simultaneously without any effort to apply a more systematic framework of analysis.

With respect to the first, it is important to begin by noting that there are no purely unilateral or multilateral strategies or policies – virtually every major foreign policy initiative falls somewhere on a continuum ranging from purely unilateral, to bilateral, to multiple bilateral, to coalitional, to à la carte multilateralism,[22] to purely multilateral. John Gerard Ruggie defines multilateralism as 'an institutional form that coordinates relations among three or more states on the basis of generalized principles of conduct.'[23] But even this popular definition raises more questions than it answers. What principles? How general? What kind of institutions? How much coordination? For example, there is nothing in this definition that would exclude the 'coalition of the willing' organized by the U.S. in 2003, yet most have assigned the term *unilateral* to describe the U.S.-U.K. campaign. Unilateralism, in contrast, has been defined as 'a tendency to opt out of a multilateral framework (whether existing or proposed) or to act alone in addressing a particular global or regional challenge rather than choosing to participate in collective action,' according to David M. Malone and Yeun Foong Khong.[24] The authors admit that 'there is no clear dichotomy' and that there are 'many possible gradations between the two orientations' in which 'elements of unilateralism and multilateralism co-exist,' but the writers make no effort to explore the implication of this fact in more detail. In an effort to clarify those gradations and to address the complexity head on, consider the categories described in figure 2.1. Several points about the figure are in order. First, these gradations are by no means exhaustive, nor are the examples meant to be definitive representations of each gradation. The figure is useful insofar as it provides a focus for debates about specific crises, states, actions, and foreign policy initiatives. Others have attempted to apply different terms – Ekaterina Stepanova prefers 'broad' and 'narrow' and describes NATO's 1999 intervention into Kosovo as 'narrow multilateral unilateralism.'[25] The main problem with the literature is that the implications of these nuances are almost never systematically explored.

Second, establishing whether any state is acting unilaterally or multilaterally would require specification of the case in question (e.g., the first gulf war vs. the second gulf war); substantive issue-areas (e.g., terrorism, humanitarian intervention, international trade disputes, arms control); specific alliances (e.g., NATO, UN); regions (Middle East, North Korea); and so on. States could conceivably be classified in different categories for different crises, or at different points in time during the same crisis.

Third, at what point does a multilateral initiative become unilateral (and vice versa)? For example, if Hans Blix declared in his report that Saddam Hussein was in material breach of UNSC 1441, yet France and Germany remained convinced that a breach of 1441 did not amount to justification for war, how would a subsequent decision by the U.S. to intervene in Iraq be classified? Does support for the intervention become multilateral if several states join in, or is it only multilateral if a majority of the permanent members of the Security Council support the

Figure 2.1
Gradations: Unilateralism and Multilateralism (with examples)

Hyper-Unilateral ⟶ Iraq's invasion of Kuwait in 1990

↓

Instrumental Unilateral ⟶ U.S. withdrawal from ABM Treaty and associated BMD deployment; U.S. rejection of Kyoto and International Criminal Court; Russia's intervention into Chechnya; French intervention into Ivory Coast; China's policy on Taiwan

↓

Principled Unilateral ⟶ Any unilateral intervention to prevent genocide in Rwanda, 1994; U.S. intervention in Somalia (sponsoring UNOSOM) 1993

↓

Hyper-Bilateral ⟶ Canadian dependence on U.S. economy and security guarantees

↓

Instrumental Bilateral ⟶ Tied aid and development assistance; U.S. alliances with Pakistan after 9/11

↓

Principled Bilateral ⟶ US$15 billion investment to fight AIDS in Africa

↓

Instrumental Multilateral ⟶ France, Germany, and Russia position in Gulf War II; U.S.-U.K. coalition in Gulf War II

↓

Principled Multilateralism ⟶ First Gulf War; Bosnia 1990-95 (failure); Bosnia 1995 (success); Kosovo 1999 (success)

↓

Hyper-multilateralism ⟶ International apathy, Rwanda 1994; Canada's insistence on UN Security Council resolutions to establish 'legitimate' grounds for Iraq intervention 2003

initiative? How many states, offering what kind of implicit or explicit support, constitute multilateralism?

Fourth, the examples listed in the figure 2.1 are likely to generate debates about how a specific foreign policy initiative (or grand strategy) satisfies the conditions for being 'principled' and/or 'instrumental.' One possible approach would be to focus on the outcome – who benefited from the initiative (*outcome legitimacy*)? Was the policy designed to accommodate collective security and economic interests or merely self-interests? Another approach could focus on the process, regardless of outcome, and whether the initiative was sanctioned by a multilateral organization, such as a United Nations Security Council resolution (*process legitimacy*). To further confound the issue, the process is not necessarily related to the outcome – either in terms of the probability of success or who ultimately benefits from the initiative. Instrumental unilateralism can have a positive outcome that benefits many states, and principled multilateralism can have a negative outcome for all but one state. Consider the first gulf war (1991) as an example. As Robert Kagan observed, '[M]ore than 90 percent of the military forces sent to expel Iraq's army from Kuwait were American. Were 90 percent of the interests threatened American? In almost any imaginable scenario in which the United States might deploy troops abroad, the primary purpose would be the defense of interests of more immediate concern to America's allies – as it has been in Bosnia and Kosovo. This can be said about no other power.'[26]

As a way of highlighting the distinction between process vs outcome legitimacy, consider some of the facts surrounding the two gulf war cases (1991 and 2003). In 1991 the UN Security Council passed a series of resolutions (including UNSC 660 and 678) establishing the multilateral consensus 'to use all necessary means' to invade Iraq. Dozens of countries provided political, financial, and military support to the coalition effort, and it received widespread public backing. The 1991 war was quintessentially multilateral.

In an effort to sustain the legitimacy assigned to the multilateral consensus he worked so hard to put together in 1991, George Bush Sr. decided against moving troops into Baghdad. The failure to change the regime in 1991 arguably produced a decade of pain and suffering for Iraqis, by any measure of social welfare – child mortality rates, disease and malnutrition, diminished political and religious freedoms, crimes against humanity, and genocide of the Kurdish and Shiite minorities. The evidence for these failures can be found in virtually hundreds of documents produced not by conservative think tanks determined to slam multilateralism but by dozens of UN organizations (e.g., UNICEF) tasked with evaluating the social conditions of Iraqis throughout the UN-sponsored-multilateral-sanctions decade that followed 1991 war.

Now, if one was to track the same social, economic, and political conditions in Iraq over the next decade, the 2003 gulf war (characterized by critics as unilateral) will prove to be considerably less costly (in lives) and more successful, constructive, and humane than its 1991 counterpart. These conditions are likely to prevail despite the absence of process-related legitimacy that would have come from a second UN Security Council resolution. If, for whatever reason, the international community fails to improve conditons in Iraq over the next ten years, a second UNSC in March 2003 would have done nothing to alter the success/failure of post-conflict reconstruction.

When evaluating the success or failure of an intervention there is nothing in the just-war doctrine that assigns a higher moral weight to process over outcome criteria. There are obvious risks associated with any just war, and, as Michael Kelly wrote in the *Washington Post* in February 2003, '[A]ny rational expectation has to consider the probable cost to humanity to be low and the probable benefit to be tremendous. To choose perpetuation of tyranny over rescue from tyranny, where rescue may be achieved, is immoral.'[27] But immorality applies equally to unilateral and or multilateral decisions, and it applies both to interventions and non-intervention where states and institutions remain as bystanders to genocide. The implication of exclusive reliance on process-related legitimacy is that fewer states will have the power to quash any multilateral consensus that is inconsistent with their economic and political self-interests. This, in turn, increases the probability of hyper-unilateralism by granting authority to a single state with a crucial veto.

Presumably, many of these concerns would be shared by leading figures in the human security community in Canada. Andrew Mack and Oliver Rohls point out that 'Canada's conception of human security ... is clear and policy focused. In essence, it is about protecting individuals from repression and violence – notably terrorism, civil war and genocide. Human security, in other words, is about the protection of people. National security is about defending the territorial integrity of states.'[28] The U.S. decision in 1991 to prioritize multilateral consensus over regime change should be evaluated in this context. Although that decision was legitimate in terms of process, it was anything but legitimate in its outcome. Arguably the American decision to accept the multilateral consensus that opposed regime change in 1991 in favour of protecting Iraq's sovereignty achieved considerably less human security than the 2003 gulf war. Aside from a few Baath party officials, most Iraqis prefer the Iraq they have now to the one they had in 1991, precisely because the prospects for enhancing human security are much greater.

Most states practice some combination of the strategies listed in figure 2.1 for different policies, across different issues with different partners for a variety of reasons. It is simplistic in the extreme to assign a single title of unilateral to encapsulate, for example, U.S. pre- or post-9/11 foreign policy. Unfortunately, we tend to be very selective (and decidedly unsystematic) when applying these terms, often because our impressions are formed by our views about the foreign policy initiative in question. Proponents of U.S. intervention in Iraq, for instance, are likely to interpret U.S. behaviour in multilateral terms (e.g., unanimous support for UNSC 1441), while critics are likely to characterize U.S. foreign policy in unilateral terms (e.g., American officials went ahead despite the absence of a second resolution).

A more systematic approach would acknowledge several ways to measure the presence and/or absence of multilateral support for any foreign policy initiative – by compiling evidence on, for example: (1) multilateral diplomacy and/or dialogue through international organizations and institutions, (2) explicit (public) offers of political, economic, and/or military support for the foreign policy initiative in question, (3) implicit (non-public, back-channel, secure diplomatic) offers of multilateral political, economic, and/or military support, and (4) tactical (e.g., intelligence sharing), logistical (e.g., communications), operational (e.g., overflight access), and/or strategic military support. Moreover, evidence of assistance across each of these dimensions of multilateralism must be measured on a continuum ranging from no support to extensive support. This approach obviously poses significant challenges for anyone interested in providing a truly systematic (and accurate) assessment of the extent and nature of multilateralism in the international system. In fact, U.S. officials are often required for diplomatic reasons to be careful about acknowledging direct political and military assistance because of the implications for the leadership in states providing that support (e.g., Saudi Arabia). If the U.S. is successful at conveying to the media, for its own political interests, that they have received support from, say, Turkey, there is likely to be a counter media blitz by Turkey to convey to its own public that such support is not forthcoming. Given the diplomatic dance that typically plays out in the press in crisis situations, the challenge for anyone attempting to measure multilateral support is obvious: how much more or less weight should we assign to explicit public support versus implicit (behind-the-scenes) support? How does one distinguish the value of political vs military support for American foreign policy in the period leading to the 2003 war in Iraq? How does one measure the value in multilateral terms of the letter signed by eight European leaders (or the subsequent letter signed by another ten Eastern European leaders a few days later) in support of the American war effort? Where on the unilateral-multilateral continuum does one place the quiet military and operational support offered to the U.S. by several allies in Europe and the Middle East (including opponents of the war)?

Success and Failure

A second but related dimension of the debate between proponents of multilateralism and unilateralism focuses on the capacity of these respective strategies to enhance global, regional, domestic, or human security. There is nothing inherent in unilateralism or multilateralism that makes either approach more efficient or more likely to succeed. With respect to efficiency, multilateral interventions are cumbersome and time consuming – they entail coordination problems that are not easily overcome, especially if there are significant discrepancies in the size, training, and overall quality of the militaries being deployed. As Stepanova notes, '[M]ultilateralism slows down the use of force and, in this sense, is not a force multiplier; it also exacerbates the problem of optimal division of labour between participants,' many of whom rely on 'the exclusive military capabilities of the United States.'[29] The only real benefits to the U.S. are financial (e.g., sharing the $50 billion cost of Desert Storm) and political (i.e., process-related legitimacy – see this chapter, under Presence and Absence).

Some threats will be addressed more effectively through unilateral strategies, while others will require a multilateral response. Consequently, valid assessments (or predictions) of success and failure must be based on the specific threats in question, the policy initiatives put in place to address those threats, and an objective assessment of success and failure. Three brief examples highlight some of the unintended consequences of exclusive reliance on multilateral solutions to border security, human rights abuses and genocide (International Criminal Court), and the global battle against the AIDS pandemic.

Border Security. When it comes to U.S. investments in border security and immigration surveillance, the choice of whether to emphasize multilateral or unilateral approaches will depend, in part, on how American officials compare the costs, benefits, and risks of, on the one hand, increasing the number of highly trained customs inspectors, immigration officials, guards, and X-ray machines at the U.S. border and, on the other, working with (and ultimately depending on) other states to invest their finite resources on these and other security measures, such as improving their capacity to examine containers at their origins and to coordinate visa strategies to track potential terrorists. Obviously, working with other states to develop effective multilateral approaches will help, but in comparing levels of overall confidence, there is a natural tendency to be much less confident in strategies that depend on the expertise, motivation, good will, and priorities of other states, particularly if they are not the ones being targeted by terrorism. Consider the problems noted by Brainard regarding effective coordination of visa strategies – terrorists often 'calibrate their visa strategies to take advantage of different levels of scrutiny across countries.'[30] Unless the approach is comprehensive and inclusive, a steadfast terrorist who is committed to entering the United States, Europe, or Canada can accomplish this task by travelling through countries that are not embedded in the multilateral coalition, are not efficient at coordinating their visa and surveillance strategies, or are simply less committed to spending their finite resources on these particular security challenges (they often have their own problems to worry about). The point is that other states are not likely to share the same priorities as those who are responsible for the security of American citizens. And the resources required to establish a truly multilateral approach to the problem are prohibitive, not only in financial terms but also in the need to be appropriately intrusive in the security affairs of other states to verify compliance with multilateral standards. The current system is not even close to reaching that level of multilateralism, but anything less will never persuade American officials to relinquish unilateral approaches.

These challenges are not exclusive to border security – they apply to virtually every decision regarding the choice between multilateral and unilateral security measures. The assumption that collective actions, by their very nature, contribute to multilateralism and global security is decidedly superficial, because the same alliances that provide benefits for one dimension of security can produce unintended consequences for others. Peterson's observation illustrates the point about simplistic assumptions: 'With international terrorist networks extending globally, security from further terrorist attacks seemed possible – if possible at all – only through a careful courting of America's present circle of allies, and moreover, through an expansion of that circle to transform states that were once part of the problem into part of the solution. And it was difficult to see how any of this was possible without, first, carefully courting Europe, and second, working with the EU to extend and embed multilateralism in the international order.'[31]

The problems with Peterson's arguments (and predictions) are even more apparent in the aftermath of the 2003 Iraq war: despite the potential benefits outlined by Peterson, even traditional alliances will not take precedence over American security priorities, especially when key members of that alliance demand compliance with multilateral institutions and procedures that constrain the capacity of American officials to deal with terrorism and proliferation. Officials in Washington are not likely to prioritize, extend, or embed multilateralism when their threat perceptions diverge from those of their European allies. In fact, contrary to Peterson's predictions, alliances with France and Germany are no longer as crucial to American security efforts as, for example, alliances with Pakistan and Uzbekistan.

With respect to Peterson's claim that the U.S. is expanding the 'circle of allies,' transforming states 'that were once part of the problem into part of the solution,' he is correct. But it is not the case that these new alliances demonstrate a renewed commitment to building multilateral institutions and approaches. An alliance with Pakistan in the war on terror could conceivably produce a net loss for multilateralism by granting to Pakistan key concessions on its nuclear weapons program, or by looking the other way when North Korea sells Pakistan ballistic missile technology. Although these compromises are typical in international relations, and although they were considered by U.S. officials as essential to winning the war (and peace) in Afghanistan, they nevertheless undermine the value and relevance of multilateral regimes tied to nonproliferation and arms control. Alliances do not necessarily produce a net gain in security or multilateralism, because they will always produce compromises and unintended consequences – the more relevant question is this: on balance, how do the costs, risks, and benefits of alternative strategies play out in specific cases, for specific threats at specific points in time. American alliances today are likely to be far more complex than those typically depicted in the 'institution building' / 'liberal internationalist' traditions, as exemplified in the quotation from Peterson.

International Criminal Court (ICC) and the Rwandan Genocide. Similar challenges apply to the ICC and other multilateral efforts to respond to genocide and human rights abuses. Jose E. Alvarez identifies the core problem: 'International lawyers share an appealing evangelistic, even messianic, agenda. We are on a mission to improve the human condition. For many ... this mission requires preferring the international over the national, integration over sovereignty. Multilateralism is our shared secular religion. Despite all of our disappointments with its functioning, we still worship at the shrine of global institutions like the UN (and reject) conservative nationalism, sovereignty and power politics.'[32] The problem with this approach (and its agenda) is that it excludes a more balanced assessment of the costs, risks, and unintended consequences of exclusive reliance on multilateral regimes. Nationally based, domestic alternatives are often rejected, not because they fail to achieve justice, but because they do little to enhance multilateral institutions designed to disseminate multilateral justice. Alvarez uses UN Security Council resolution 955, established after the Rwandan genocide in 1994, to illustrate the point: while the Security Council recognized 'the need for international cooperation to strengthen the courts and judicial system of Rwanda, having regard in particular to the necessity for those courts to deal with large numbers of suspects,' the resolution also established an international tribunal 'that has managed to absorb most of the intellectual and material capital that could have gone to Rwandan institutions.'[33] The annual tab for the tribunal now exceeds $50 million, and the process, Alvarez explains, 'has taken precedence over stemming ongoing violence or bringing the rule of law within reach of the average Rwandan.'[34] Perhaps the most serious problem with this (and other examples of exclusive reliance on building multilateralism) is that the burden of proof is typically placed on unilateralism to prove its worth. But Alvarez is correct to point out that 'the burden of proof should be on those responsible for the proliferation of multilateral treaties, regimes and institutions, including some 55 international or regional entities devoted to dispute settlement alone.'[35] I would argue that the burden of proof should be shared by both sides to demonstrate the utility of their respective alternatives, based on a balanced assessment of the evidence and a comprehensive evaluation of the cost, risks, and consequences (see chapter 5 for more on the burden of proof issue).

AIDS. A similar set of challenges confront those engaged in the global fight against AIDS. Elizabeth Nickson's comparative analysis of data on AIDS in various African states offers a useful set of benchmarks for analysing the relative utility of different approaches. In Uganda, for example, President Yoweri Museveni 'chose not to pull in UN experts [and] invited Christian and Muslim clergymen to preach forthrightly to Ugandans about the need to abstain from premarital sex and stay faithful to their partners.' This approach was augmented by a public relations campaign, lectures to schoolchildren on the benefits of abstention, and condom distribution. The impact of these strategies is noteworthy: 'The AIDS rate has dropped steadily since 1991 ... [by 2002] the number of pregnant Ugandan women testing positive for HIV antibodies had fallen from 21.2% at the height of the epidemic in 1991 to 6.2%. By contrast, in neighbouring Kenya the rate is roughly 15%; in Zimbabwe it stands at 32%; and in Botswana fully 38% of mothers-to-be are HIV-positive – with rates continuing to rise in each country.'[36] The statistics for Uganda, Nickson notes, represent the most dramatic drop in the transmission of HIV in the developing world. Obviously a more comprehensive assessment of various AIDS strategies would be useful, but this doesn't appear to be a priority in many African states or within the UN. The US$15 billion AIDS program introduced by George W. Bush in 2003, for example, was criticized by UN officials because it prioritized using bilateral aid to control the distribution of funds and to maintain compliance with specific guidelines and standards. The special envoy for the UN's AIDS program in Africa, Stephen Lewis, was critical of Washington's refusal to turn the $15 billion over to the UN to sustain its multilateral approach to the problem, but the precise criteria for assessing and comparing relative successes and failures remained unclear. In essence, the ideal of establishing multilateral approaches to global problems is so enticing that the institutions themselves become the benchmarks for success and failure, rather than our capacity to save lives or to reverse the trend lines for AIDS. Critics of Bush's policy will no doubt claim that bilateral assistance serves the ideological and/or political interests of the donor state, but what are those interests? How do they compare with comparable interests and preferences of relevant UN organizations responsible for AIDS? What are the costs and risks of distributing aid through each approach, and what are the results? The fact that the United States or the UN gain more/less from one or the other approach should be secondary to the capacity of the strategy to stop the spread of the HIV virus and to save lives.

The point to the preceding analysis is simple: proponents of multilateralism have an obligation to provide more credible evidence to demonstrate that their preferred policy options are better suited to the task of resolving contemporary military or human security crises. They also have an obligation to admit that successful resolution of some crises can be hindered by demands for multilateral consensus. The expanding list of cases in which dependence on multilateral consensus caused delays that escalated the violence and killing illustrates the point – Bosnia (1990–95), Kosovo (1998–99), Rwanda (1994), and Congo (1999–2003). Far from providing a solution to many intractable security crises, the presence of multilateralism often creates the very impasses and impediments that escalate crises and human suffering. As Mark Steyn points out, the UN appears to have very little interest in 'solving problems, only in establishing bureaucracies to manage them.'[37] The prevailing image among critics is that the United Nations is essentially 'an experiment in self-anointment'; a 'conclave of functionaries ... unelected governments or unelected non-governmental organizations'; and 'an endeavor by ailing, peevish powers ... Third World tyrannies [and] ... mandate-free bureaucrats.'[38] For proponents of multilateralism these criticisms, and the long list of deficiencies, simply represent the costs they are willing to accept (or overlook), perhaps because they have invested so much time and intellectual capital in defending the ideal and are simply not prepared to contemplate alternatives that threaten to undermine that commitment. But defending a process and associated institutions simply for the sake of protecting those institutions is no longer a morally sustainable position. Exclusive reliance on multilateralism in every case, for every crisis, and for every challenge and threat is simply wrong.

While some challenges will be more effectively resolved through multilateral cooperation (assuming states are willing and equally motivated to provide it), other threats will demand unilateral initiatives in light of fundamental disagreements among states about what constitutes a compelling threat. The more relevant question is this: On balance, how do the costs and risks of alternative strategies play out in specific cases, for specific threats at specific points in time? Multilateralists should spend less time pointing to flaws with unilateralism and a little more time demonstrating (as opposed to simply asserting) the utility of multilateralism in comparative terms; that has always been a more challenging project, but one that the remaining chapters of this book will address head on.

Crucial Case Studies: Gulf War II and Ballistic Missile Defence

To accomplish this task, the remainder of the book will compare unilateral and multilateral approaches to security, with specific emphasis on evaluating appropriate responses to proliferation and terrorism. While one could conceivably write similar books dealing with other security threats (e.g., small-arms proliferation, international crime, drug trafficking, ethnic conflict, peacekeeping), or other substantive issue/policy areas (e.g., environment, human rights, international trade and development), this volume will not address these other questions.

An important prerequisite for comparing unilateral and multilateral strategies requires that we unpack (or disaggregate) the many different security threats that leaders face today. A bottom-up case study approach is the only way to explore the utility of policy alternatives in the context of globalization. Traditional methods of evaluating these alternatives often approach the subject from the top down – for example, the systemic forces tied to globalization are assumed to be an appropriate basis for recommending multilateralism as a solution to contemporary security (see the first four sections of this chapter). The assumption that these global forces have a meaningful and easily identifiable impact on the utility of policy options is mistaken, as explained in chapter 1, because the effects are multiple, interdependent, and often mutually exclusive. It is the threat itself (WMD proliferation and terrorism) that determines the utility of policy options – a point often overlooked in the literature.

The next three chapters will focus, therefore, on two crucial case studies from the field of international security that continue to fuel contemporary debates among proponents of unilateralism and multilateralism – the U.S.-U.K.–led war in Iraq in 2003, and the U.S. BMD program. The objective is to use the evidence derived from both cases to highlight fundamental weaknesses and errors in the literature on multilateralism, and to provide a more compelling and rational explanation for why unipolarity, and associated preferences for unilateral approaches to security, will remain the dominant feature of the international system and U.S. foreign policy for the foreseeable future.

The two cases are crucial for several reasons. The 2003 Iraq war represents an important departure from post–cold war military interventions as support from traditional allies was bypassed in favour of strategic coalitions with states that, until now, were not considered essential to major U.S. military operations. Indeed, the effort by European powers to balance American hegemony was perhaps as unexpected as the collapse of the Soviet Union; the impact on international politics and traditional alliances (including NATO) is likely to be equally significant. The Iraq case is ideal because the two dominant positions defended by the French-German-Russian and American-British camps encapsulate the arguments put forward by proponents and critics of multilateralism.

With respect to the second case, proliferation of ballistic missiles and WMD, and associated threats tied to terrorism and rogue regimes (which amplify the threats of WMD proliferation) constitute the two most serious security threats of our time. The overall potential for unimaginable levels of destruction renders this particular combination of threats far more important than, for example, small-arms proliferation or drug trafficking. Debates among proponents of BMD (BMD-unilateralism) and critics who favour reliance on the non-proliferation arms control, and disarmament regime (NACD-multilateralism) represent mutually exclusive solutions to proliferation. Few issues, especially in the security field, lend themselves so well to this kind of dichotomy. While BMD may not encompass the only unilateral strategy in the U.S. national security arsenal, it is certainly the clearest and most explicit form of unilateralism as it applies to proliferation.

The U.S. administration is criticized today for excessive unilateralism based, in part, on the prevailing belief that multilateralism holds the only promise for security in an age of globalization. The patterns that emerge from the case studies will challenge this conventional wisdom.

Chapter 3

Gulf War II: Unilateralism and Multilateralism in Practice

Perhaps the most popular myth about the United Nations is that it is a separate entity that competes with nation-states to constrain their behaviour, to prevent defection from international norms, rules, and regimes, and to serve as a multilateral alternative to unilaterally motivated power politics. The UN represents the most prominent multilateral check on relentless efforts by states to protect their own national economic, political, and security self-interests. But in reality the UN is merely another venue through which power politics plays out, and the Iraq war of 2003 is but the latest (and perhaps clearest) in a long list of crises that confirms this view. Far from being distinct and separate, the UN plays an important role in an international process through which the powerful compete and the most powerful prevail. The reason for discrepancies in perceptions of the UN, as John Humphrys points out, 'is that there is a great gulf between the UN of our dreams and the UN of the real world.' As he explains, 'The ideal UN is a powerful and strictly impartial body, dispensing truth and justice (or at least our notions of truth and justice) without fear or favour. It would command universal respect for its collective wisdom and judgments ... The great illusion is that the UN is a world power in any real sense. It is not. It is a forum, a talking shop. And the most powerful of the talkers has the loudest voice. Nor is this talking shop a place of high-minded debate, preoccupied with the best interests of humanity as a whole. It is a place of horse trading between narrow national interests, of mutual back-scratching and Machiavellian diplomacy.'[1] This characterization may not apply to all aspects of the UN, but it is an accurate portrayal of multilateral approaches to global security. In essence, the UN serves to perpetuate the relevance and centrality of power politics and competition – unilateralism prevails in virtually every interaction, and the Iraq war of 2003 is no exception.

French and Russian Unilateralism (a.k.a. Dishonest Multilateralism)

With respect to French and Russian interests in Iraq, consider the following data compiled in a report by James A. Paul in 2002 on estimates of Iraq's 250 billion barrels in oil reserves.[2] Assuming a 50 per cent recovery rate (a conservative estimate, according to Paul), 'recoverable Iraqi oil would be worth altogether about US$3.125 trillion.' Paul goes on to explain the following implications:

> Assuming production costs of $1.50 a barrel (a high-end figure), total costs would be $188 billion, leaving a balance of $2.937 trillion as the difference between costs and sales revenues. Assuming a 50/50 split with the government and further assuming a production period of 50 years, the company profits per year would run to $29 billion. That huge sum is two-thirds of the $44 billion total profits earned by the world's five major oil companies combined in 2001. If higher assumptions are used, annual profits might soar to as much as $50 billion per year ... Two Russian companies, Zarubeshneft and Rosneft, were preparing in 2002 to develop Iraq's Nahr Umr field that they estimated was worth about $570 billion. Reliable estimates for the value of the fabulous Majnoun field go up to $400 billion and beyond ... In 1997 TotalFinaElf, China National Oil Company, and Lukoil of Russia signed agreements with the Iraqis for deals worth hundreds of billions of dollars. Lukoil's deal concerned development of the West Qurna field ($20 billion), while TotalFinaElf obtained rights to Majnoun and China Nations to North Rumailah (the latter is the huge field that lies astride the border with Kuwait) ... Chinese economists estimate that China may have to import as much as 5.5 million barrels a day from the Gulf by 2020.[3]

Diane Francis notes that the production estimates for the Majnoon oil field range from 450,000 to 1 million barrels a day.[4] The obvious potential for significant economic gain explains why, by 2001, France, Germany, and Russia signed deals with Baghdad to spend up to US$38 billion for exploration, improvements, and expansions to increase Iraqi oil production to five million barrels a day. In addition to these very substantial oil interests, France also profited from the UN's oil-for-food program – controlling 22.5 per cent of Iraq's imports, 13 per cent of its arms imports, and $1.5 billion in trade with Iraq through sixty French companies.[5] Like France, Russia also had substantial economic interests above and beyond oil – including $7 billion in Soviet-era debt largely accumulated by Iraq during its war with Iran.[6] According to Helle Dale, Russia's trade with Iraq was between $500 and $1 billion; Moscow controlled 5.8 per cent of Iraq's imports, and from 1981 to 2001 Russia supplied close to 50 per cent of Iraq's arms imports.[7] And in 2002, Russia signed a $40 billion economic agreement with Iraq to explore for oil in western Iraq.

These significant economic interests explain the decidedly unilateralist pressure France and Russia imposed on other European allies to reject the call for a U.S.-U.K.–led intervention in Iraq. Alan Freeman's excellent report on the 2003 EU summit (organized to discuss Iraq) included several references to a fascinating series of exchanges between the French president, Jacques Chirac, and thirteen future European Union member states.[8] At the time, Poland, Hungary, and the Czech Republic (with 2004 membership dates) signed a letter sponsored by Spain, the U.K., and five other EU members to support U.S. policy in Iraq. This was followed by a similar letter of support signed by ten Central and Eastern European countries a few days later. In response, President Chirac stated that this was 'not really responsible,' it was 'dangerous' and 'not well brought-up behaviour. They missed a good opportunity to keep quiet.'[9] Bulgaria and Romania were also singled out by Chirac, whose warnings included repeated references to their 'very delicate' status in the EU – 'if they had wanted to reduce their chances of entering Europe, they couldn't have found a better method.' The responses by several European leaders to whom the outburst was directed highlight the nature and perception of French unilateralism. The Romanian president, Ion Iliescu, stated that Chirac's comments were 'irrational' and 'totally unjustified, unwise and undemocratic ... [W]hat we have is the arrogance of the aristocracy, and we no longer live in the context of the 19th century.'[10] In his response, the Bulgarian deputy foreign minister, Lubomir Ivanov, added that 'this approach will not help create unity in the [United Nations] Security Council.' And the Polish deputy foreign minister, Adam Rotfeld, noted that Poland, like France, 'also has a right to decide what is in its own good, and France should in its turn consider it with respect.' This is precisely what the French were demanding of the Americans, and for many of the same reasons.

It is unlikely that these (and similar exchanges) would ever be serious enough to permanently damage European relations, but consider the concerns expressed in an editorial by one of Europe's most prominent newspapers: 'If France and Germany were prepared to ditch Turkey (and, by extension, the United States), would they also be prepared to veto NATO countermeasures to help the fledgling democracies of eastern Europe in the event of a crisis?' [11]The editorial went on to point out that the Franco-German position was 'especially threatening to the new member states because they are being asked to give up much of their 'all-round' defensive capabilities at NATO's behest in order to contribute specialist skills. They are, therefore, particularly dependent on its collective security guarantees.' Shortly after these exchanges, the German foreign minister, Joschka Fischer, tried to address the growing divisions in Europe with the following acknowledgment: 'I fully understand the reaction of Central European countries that survived 50 years of Soviet occupation and have a different attitude towards the United States.'[12] But these fundamental differences in European attitudes were rarely acknowledged by critics of the U.S.-U.K.–led intervention who characterized the policy in unilateral terms without acknowledging the decidedly unilateralist position of France and Germany, within Europe. Only three out of nineteen members of the NATO alliance (France, Germany and Belgium) were critical of U.S. policy and, at least initially, vetoed contingency plans to mobilize NATO troops in support of Turkey. The fact that NATO cannot act unless support is unanimous is not a policy conducive to multilateralism; it is a policy in support of unanimity and, in essence, gives more power to fewer states.

The unilateralist predispositions of France and Russia are not limited to the Iraqi crisis – both states have a long record of unilaterally intervening in military conflicts without first obtaining the automatic 'legitimacy' that comes from a UN Security Council resolution. Consider, for example, the deployment of twenty-five hundred troops to the Ivory Coast in 2003, France's most significant military presence in Africa.[13] In addition to protecting French nationals from escalating fighting in the region, France also has an interest in protecting a very lucrative trade in cocoa, rubber, and timber. Russia's ongoing ten-year war in Chechnya is another case in point.

Several other reports offer details of the many economic benefits to France and Russia generated by the UN's oil-for-food program, benefits that explain their preferences for the status quo in Iraq and their initial opposition to earlier drafts of UNSC resolution 1483 on post-war Iraq reconstruction.[14] The largest share of goods and products were from France and Russia. For the latter, trade with Iraq provided perhaps the best hope for getting at least some of Iraq's debt to Russia repaid. The point here is not to criticize France and Russia for protecting their investments – all states inevitably prioritize their own self-interests.

Horse Trading: The Unilateral Self-Interests of Smaller Powers

Unilateral economic and political self-interests also explain the decisions by many smaller states who supported (and opposed) the U.S.-U.K.–led war in Iraq. Consider a portion of the evidence of horse trading by France and the U.S. as they worked on attracting coalition partners to multilateralize and legitimize their respective campaigns – outlined in table 3.1.[15]

Perhaps the most obvious and open horse trading took place between the U.S. and Turkey. These negotiations were particularly important, given American military plans to move ground troops through Turkey into Kurdish-controlled northern Iraq. The U.S. offered Turkey $15 billion in loans and grants, but the more interesting aspect of the negotiations is how the bargaining played out, described in detail in a report by Glenn Kessler and Philip P. Pan: 'The week before the parliamentary vote that U.S. officials expected on Feb. 18, a delegation led by Yasir Yakis (former Turkish foreign minister who played a key role in the talks with the United States) arrived in Washington to discuss Turkey's financial package for agreeing to the troop request. The administration had offered $4 billion – $2 billion in grants and $2 billion in military credits.'[16] According to Kessler and Pan, negotiations began with an initial meeting with House Speaker J. Dennis Hastert (R-Ill.), followed by a meeting with Secretary of State Colin Powell. Yakis rejected the $4 billion offer and asked for $92 billion over five years (or at least $22 billion in year one). After Powell informed Yakis that the entire U.S. foreign aid budget was $18.5 billion, the secretary of state went on to increase the U.S. offer by $6 billion – with $1 billion immediately and $8 to $10 billion in loans. The negotiations failed, not because Turkish leaders were particularly concerned about American unilateralism, or because they were worried about the risks and consequences for the Iraqi people of a U.S.-U.K.–led intervention. Negotiations failed, according to a senior U.S. official, because 'the Turks came to think we would pay anything for their cooperation [and] got to believe they were indispensable, and it colored their capacity to decide when

TABLE 3.1
Unilateralism, Bilateralism, and Horse Trading Over Iraq

Afghanistan: Received US$178 million under the Taliban but has already received US$450 million since the fall of the Taliban and liberation of Afghanistan.[a]

Angola: France promised 'home in exile' for Angola's president; in an effort to counterbalance U.S. offers, France offered to help rebuild infrastructure after civil war.[b]

Bulgaria (non-permanent member of the UN Security Council during Iraq campaign): Commerce Secretary Donald Evans announced that U.S. 'officially' considers Bulgaria a 'market economy' (opening up the country to foreign investment); also discussed with U.S. officials ways to facilitate membership in NATO.[c]

Cameroon and Guinea: (non-permanent member of the UN Security Council during Iraq campaign): Eligible for 'preferential access' to U.S. markets through the Africa Growth and Opportunity Act; access granted only if countries *refrain from engaging in activities that undermine US national security or foreign policy interests* [emphasis added].[d]

Georgia: US$64 million for training special forces to fight rebels from Muslim Chechnya.[e]

Hungary and Poland: Stationing of American military bases (to be relocated from Germany).[f]

Indonesia: Arms restrictions lifted.[g]

Jordan: US$250 million in economic aid; expected to receive US$198 million in military aid.[h]

Oman: Sold US$1.2 billion in fighter jets.[i]

Pakistan: Restrictions on arms exports (implemented after the 1998 nuclear tests) lifted; military-related aid increased from US$3.5 million before 11 September to US$1.3 billion afterwards; U.S. pressure on IMF and World Bank to forgive US$1 billion in loans to Pakistan (Cienski 2003); following Turkey's decision to deny access to American ground troops, the U.S. House of Representatives granted Pakistani rug makers U.S. aid and denied aid to Turkish rug makers.[j]

Philippines: US$100 million in 'planes, helicopters, patrol boats and other arms, including 30,000 M-16 rifles, to hunt down members of the Muslim Abu Sayyaf gang, linked to al-Qaeda.'[k]

Poland: Signed deal in December 2002 'to buy 48 F-16 jet fighters, paid for by a US$3.8 billion loan below market rates, one of the largest such loans ever granted, as well as an additional US$10 billion in "offset" investments promised by Lockheed Martin, which builds the F-16.'[l]

Tajikistan, Yugoslavia, and Azerbaijan: All had arms embargoes lifted.[m]

United Arab Emirates: Sold US$40 million in Harpoon Block II anti-ship missiles.[n]

Uzbekistan (instrumental in successful U.S. war in Afghanistan): Received US$25 million in loans for U.S. weapons and equipment, US$40.5 million 'for economic and law enforcement assistance,' US$18 million 'for anti-terrorism, de-mining and non-proliferation programs' (along with an additional US$11 million in 2002), and US$8.7 million in 2003 in military loans.[o]

Yemen: Received increased aid and assistance.[p]

[a] Ian Cienski, 'Washington Trades Arms for Friends' Military Aid: War on Terrorism Has Redefined America's Interests,' *National Post* (Toronto), 27 January 2003.

[b] Marguerite Michaels and Karen Tumulty, 'Horse Trading on Iraq,' *Time*, 10 March 2003.

they had negotiated enough.'[17] Yakis himself acknowledged that '[we] thought the United States needed the northern front. We made bargaining plans based on this. We did not consider the possibility that they would apply Plan B.'

The claim that the U.S. is the only unilateralist power simply misses the point that international relations, including multilateral activity and international cooperation, are driven by unilateral interests and competition. Without doubt, the U.S. has worked hard to achieve unilateral objectives by strategically applying its wealth and power to enhance what it considers to be its primary directive. In fact, this is an explicit objective: 'While our focus is protecting America, we know that to defeat terrorism in today's globalized world we need support from our allies and friends. Wherever possible, the United States will rely on regional organizations and state powers to meet their obligations to fight terrorism. Where governments find the fight against terrorism beyond their capacities, we will match their willpower and their resources with whatever help we and our allies can provide.'[18] This explains the 'bazaar economics which lavished up to $500,000 (£312,000) of aid on Pakistan, but denied Turkey a cent,' prompting Watson and Beeston to dub Bush's coalition 'the coalition of the billing.'[19] The key point is that unilateral self-interests define the behaviour of all states *and* international organizations.

Even the UN is not immune to acting in its own unilateral self-interests. Consider some of the evidence from a comprehensive report

---

[c] Ibid.

[d] Roland Watson and Richard Beeston, 'America Pulls the Rug from under Turkey,' the Times Online, 7 March 2003.

[e] Cienski, 'Washington Trades Arms for Friends' Military Aid.'

[f] Vernon Loeb, 'New Bases Reflect Shift in Military: Smaller Facilities Sought for Quick Strikes,' *Washington Post*, 9 June 2003, A01.

[g] Cienski, 'Washington Trades Arms for Friends' Military Aid.'

[h] Ibid.

[i] Ibid.

[j] Watson and Breston, 'America Pulls the Rug from under Turkey.'

[k] Cienski, 'Washington Trades Arms for Friends' Military Aid.'

[l] Ibid.

[m] Ibid.

[n] Ibid.

[o] Ibid.

[p] http://www.usaid.gov/locations/asia_near_east/countries/yemen/yemen.html and Maher Chmaytelli (2002) http://www.metimes.com/2k2/issue2002-42/reg/yemen_gets_aid.htm.

on the UN oil-for-food program, by Claudia Rosett: the UN collects 2.2 per cent commission on every barrel sold, ostensibly to cover 'administrative costs.'[20] According to Rosett, 'The program's bank accounts over the past year (2002) have held balances upward of $12 billion ... With all that money pouring straight from Iraq's oil taps ... the oil-for-food program has evolved into a bonanza of jobs and commercial clout. Before the war it employed some 1,000 international workers and 3,000 Iraqis.' But the UN rarely makes the financial records of these transactions public and there is very little accountability. The report goes on to document 'rampant' abuse in which deals were made between the Baath party and the UN to allow non-essential products (not sanctioned by the UN) to go directly to party officials.

American Multilateralism in Iraq

Although unilateralism is universal, France managed to co-opt the term during the 2003 gulf war when describing American foreign policy – a description that was immediately accepted by critics. As Christopher Hitchens points out, '[A]n accusation of "unilateral" behavior can be made to stick, almost by axiom, by any power that withholds consent. When that consent is eventually given, the prize of "multilateralism" has been attained, again by definition.'[21] Hitchens goes on to make two other very important observations. First, Saddam violated seventeen multilaterally derived (many unanimous) resolutions passed over twelve years, but rarely was the title *unilateral* assigned to his behaviour; and rarely was the term *multilateral* assigned to the coalition the U.S. put together to combat that unilateral record. Second, 'a majority can in theory and practice act one-sidedly, just as a single state may have more respect for pluralism than a dozen rival states put together.'[22] A single state can be well disposed to protecting the very multilateral rules and laws established through UNSC resolutions, despite the decision by one or two permanent members to veto that decision. Support for multilateralism, then, appears to be in the eye of the beholder – as such, critics can claim to be supporting multilateralism, because the UN hasn't authorized the action, at the same time the United States can claim to be supporting the very multilateral resolutions that were passed (including 1441) promising serious consequences for Saddam's decade of unilateralism. 'Despite what many believe,' Robert Kagan argues, 'there really isn't a debate between multilateralists and unilateralists in the United States today. Just as there are few principled multilateralists, there are few genuine unilateralists. Few inside or outside the Bush administration truly consider it preferable for the United States to go it alone in the world. Most would rather have allies. They just don't want the United States prevented from acting alone if the allies refuse to come along.'[23] In reality, multilateralism does not exist in opposition to unilateralism. Both processes act concurrently in almost every situation, in every crisis, and with every resolution passed by the UN Security Council. There is no multilateralism distinct from the product of competing and aggregated unilateralisms. The two processes are interconnected and mutually reinforcing – a fact often missed (or intentionally overlooked) by critics of U.S. foreign policy.

The U.S.-U.K. intervention in Iraq, for example, had significantly more multilateral support than critics were prepared to acknowledge. Contributions included military, logistical, and intelligence support, specialized chemical and biological response teams, overflight rights, and humanitarian aid. The coalition included every major race, religion, and ethnic group in the world and represented populations totalling 1.18 billion people from every continent on the globe and a combined gross domestic product of $21.7 trillion.[24] A more detailed account of the multilateral support on the U.S. side of the zero-sum equation is compiled in appendix A.[25]

Although Germany and France were opposed to the U.S. intervention they nevertheless provided indirect tactical, logistical, and operational support to the U.S. campaign through NATO – including AWACS support and Patriot missiles to Turkey. Comments by the German ambassador to Canada, Christina Pauls, illustrate some of the problems confronting anyone interested in measuring multilateral support in this case; she claimed that Germany would not participate in any military operation against Iraq, but when asked about German airmen flying in AWACS and the logistical support Germany would provide U.S. troops deployed from German bases in coalition operations, her response was less decisive: 'you will not see "German" *ground forces* or *naval forces* or *air force* participating in armed conflict against Iraq' (emphasis added).[26] How exactly should one measure Germany's indirect military support when calculating the extent and nature of U.S. unilateralism or multilateralism? While Germany did not provide troops, those troops were neither requested nor required for a successful U.S. campaign. On the other hand, Germany did provide the U.S. with important, although indirect, support.

Saudi Arabia is perhaps the clearest example of a state officially 'opposed' to the war effort (and excluded from the list of coalition members) but nevertheless one that provided substantial assistance to the coalition. Karen DeYoung reports that the Saudi government agreed to allow 'expanded US air operations' and 'full use of Prince Sultan Air Base as an air operations center.'[27] The agreement also allowed for full use of the 'air command and control center at Prince Sultan,' permitted U.S. pilots to 'fly refuelling aircraft, AWACS surveillance planes and JSTARS battlefield radar aircraft from Saudi airfields ... [and] use Saudi airfields to base fighter jets that undertake interception missions against Iraqi aircraft and that enforce the "no-fly" zone over southern Iraq.' The agreement also obtained a commitment from the Saudis to increase oil production by 1.5 million barrels a day to compensate for any shortfall in production resulting from the war. This obviously amounts to a very impressive contribution to a war the Saudi government did 'not' support. Several other countries in the Middle East (listed in appendix A) expressed very little outright support for the U.S. but offered an equally impressive level of assistance to the U.S.-U.K. campaign.

Conversely, Saddam Hussein received no explicit support from any state (as distinct from indirect political support through opposition to U.S. policy in the UN) and obtained no direct military assistance from any government. This is not an insignificant point – the lack of direct support for Baghdad is perhaps the clearest indication of what opponents of U.S. policy were not willing to risk to defend their position. Obviously, a French deployment of troops to Iraq in support of the Baath regime would constitute far more serious opposition to U.S. unilateralism than philosophical debates about multilateralism and global peace around the UN Security Council table. Although France, Germany, and Russia were all opposed to the war, the values and interests they were trying to protect did not constitute sufficient reason to even threaten to 'fight' for their cause. What is even more telling about the level of opposition to the U.S.-U.K. intervention is the fact that France and Russia ultimately supported UNSC resolution 1483 after the war, which assigned to the U.S. and U.K. the 'Authority' to run the reconstruction efforts after their victory.

Some critics will no doubt argue that U.S. multilateralism in Iraq was qualitatively distinct from the kind of multilateralism they espouse, especially considering some of the countries included in the U.S. coalition – Honduras? Macedonia? Marshall Islands? Panama? Rwanda? Yet these same critics are quick to emphasize the importance of Cameroon,

Angola, and Guinea when it comes to establishing the multilateral consensus in the UN Security Council.

Unilateralism as a Prerequisite for Multilateralism

Several important points follow from the preceding analysis and evidence. American unilateralism is not simply a product of what Washington officials decide to do on their own – unilateralism is a two-way street. It is as much a product of what others decide not to do as it is about what the U.S. decides to do. Understanding the self-interests, motivations, and incentives that compel critics to reject U.S. initiatives puts American unilateralism (and multilateralism) in its appropriate moral and strategic context.

Multilateralists are rarely (if ever) driven by some higher moral calling to create a global order grounded in international justice and multilateral regimes. They are motivated by the same fundamental imperatives that drive American foreign policy (and that compel other states to join American multilateral coalitions) – some combination of power, status, influence, domestic political support, economic and financial gain, or enhanced security. Each aspiring multilateralist brings to the diplomatic game a set of very specific interests and objectives. These unilateral interests determine a nation's position regarding which multilateral coalitions to join, which ones to ignore, and which ones to challenge by forming competing coalitions.

Following on from the previous two points, multilateralism is not something that miraculously appears when major powers acknowledge a global injustice or perceive a major security threat – there are far too many fundamental disagreements among states regarding 'justice' and 'threats' to make this a likely model for international behaviour. Few states share the same concerns at the same time for the same reasons, which explains why collective action through the UNSC is so anomalous – it is the exception, not the rule. Paris can afford to wait years for Iraq to disarm, but officials in Washington (who really are worried about a repeat of 9/11) are slightly more motivated, for obvious reasons. 'Europeans have concluded,' Kagan writes, 'that the threat posed by Saddam Hussein is more tolerable for them than the risk of removing him. But Americans, being stronger, have reasonably enough developed a lower threshold of tolerance for Saddam and his weapons of mass destruction, especially after September 11 ... The incapacity to respond to threats leads not only to tolerance but sometimes to denial.

It's normal to try to put out of one's mind that which one can do nothing about.'[28]

Consequently, multilateralism almost always requires a major push, and that push often relies on the application of coercive threats of unilateral action by major powers acting alone, at least initially. Indeed, threats of unilateralism have emerged as an essential prerequisite, a necessary condition, for multilateral support. This is not a new phenomenon – the foreign policy strategy pursued by Bush, Powell, Rumsfeld, and Rice in Iraq (2003) was virtually identical to that pursued by Clinton, Christopher, Cohen, Berger, and Albright in Bosnia (1995), Iraq (1998 bombing campaign), and in Kosovo (1999).

Consider how multilateralism unfolded in the period leading to the latest Iraq war. The essential prerequisite for a return of UNMOVIC inspectors to Iraq was the deployment by the U.S. and U.K. of close to 100,000 troops to Kuwait, Qatar, and Turkey. The impetus for such a substantial and sustained deployment of American troops to the region began with the multilateral support George W. Bush managed to obtain in the historic midterm elections in 2002 – the first time since 1934 that a sitting president's party picked up seats in the House during their first term, and the first time ever to pick up enough seats in the Senate to become the majority party. This level of public support for Bush's foreign and security policies succeeded in motivating/coercing Democrats in the U.S. Senate to attach their support to Bush's multilateral coalition by voting, on 11 October 2002, 77–23 authorizing the president 'to use all means that he determines to be appropriate, including force, in order to enforce the United Nations Security Council Resolutions [regarding Iraq], defend the national security interests of the United States against the threat posed by Iraq, and restore international peace and security in the region.' The House joined Bush's evolving multilateral coalition by approving an identical resolution (296-133).

The passage of these resolutions was driven by a very strong conviction held by members of the U.S. Congress that the only way to get Europe, Russia, China, and other members of the UN Security Council to accept UNSC 1441 was to covey to these leaders that the United States was resolute in its determination to respond (alone if necessary) to what Americans perceived as a growing threat to their security after 9/11. The subsequent deployment by the U.S. and U.K. of troops to Kuwait and Qatar established the credibility of U.S. resolve and moved the bargaining space towards the multilateral consensus that produced a stronger and more robust inspections regime through UNSC 1441.

Without those threats, and the fears held by officials in Paris and Moscow that their substantial economic interests in a post-Saddam Iraq were in jeopardy, the UN would not have come close to achieving a unanimous resolution.

And, of course, the only reason Saddam Hussein allowed UNMOVIC inspectors back into Iraq (with a significantly more emphatic mandate than UNSCOM) was the successful application by the U.S. administration of coercive diplomacy throughout this entire period. Hard power prevailed. Take away the application of coercive diplomacy by the U.S and U.K. at each stage and none of this multilateral activity would have unfolded. There was no soft-power, multilaterally initiated alternative that would have come close to achieving what coercive and credible threats of U.S. unilateralism achieved in Iraq. These lessons applied equally well to earlier crises, as Kagan points out: 'The lesson President Bill Clinton was supposed to have learned in the case of Bosnia is that to be effective, multilateralism must be preceded by unilateralism. In the toughest situations, the most effective multilateral response comes when the strongest power decides to act, with or without the others, and then asks its partners whether they will join. Giving equal say over international decisions to nations with vastly unequal power often means that the full measure of power that can be deployed in defense of the international community's interests will, in fact, not be deployed.'[29] Despite the fact that the evidence is overwhelming, proponents of multilateralism are not likely to acknowledge the role of coercive diplomacy in kick-starting the process. All multilateralism is local; it evolves over time and has to be built from the bottom up piece by piece. Consider the multilateral consensus that evolved into the unanimous support for UNSC 1441: it began with building consensus within Bush's cabinet, progressed to building support among the Republican Party leadership in the House and Senate, moved to building support among rank and file Republicans, then on to members of the Democratic opposition and the American public, and ended with support from allies (all of whom face similar challenges and political pressures at home), states who initially opposed intervention but were persuaded to sign on nonetheless, and, ultimately, to a unanimous UN Security Council Resolution. Unless compelled into action by coercive military threats, or until the interests and priorities of all these actors are sufficiently accommodated by offers of political rewards, financial assistance (grants and loans), or economic investments in foreign and military aid, there is very little chance the United States would have succeeded in producing any multilateral coalitions in 2003 (or 1999, 1998, 1995, 1991, and so on).

Multilateralism as an Impediment to Successful Coercive Diplomacy

The essential nature and overall utility of coercive military threats is not a new phenomenon, as Walter B. Slocombe notes:

> Not the sincere efforts of leaders and citizens to substitute international institutions and international diplomacy for military power, not the terrible costs of two massive Europe-based world wars, not those of countless smaller internal and international wars throughout the world since 1945, not even the potential consequences of war fought with nuclear, chemical, and biological weapons, as well as the massively increased potential lethality of conventional technology, have fundamentally changed the fact that the threat and use of force are the ultimate instruments of international relations ... By contrast, the credibility and costs associated with the consequences that can be imposed by other means of pressure are puny compared to those of military force.[30]

To work well, coercive diplomacy requires effective communication (e.g., an explicit threat of serious consequences and a timeline for compliance), a clear commitment to the issue(s) at stake (e.g., counter-terrorism; counter-proliferation), the capability to impose painful military costs on the opponent (e.g., the military power to impose regime change), and the resolve to carry through on threats (e.g., the deployment of hundreds of thousands of troops to the region; a record of fighting similar wars, for similar reasons, in the past).[31] The theory stipulates that if these conditions are met, the coercive threat should be perceived by the opponents as credible, and it should work – the threat should either successfully deter the specified act or compel the opponent to act in ways consistent with the demands in the threat.

But consider some of the impediments to credibility in a multilateral context in which the U.S.-U.K. coalition was attempting to send a multiplicity of simultaneous signals about their commitments, resolve, and credibility to different target audiences: Saddam Hussein, the Iraqi military and other members of his regime, the Iraqi people, other terrorist groups and rogue states, other opponents in the Middle East, friends and foes in the European alliance, and other members of the UNSC.

Heavy reliance on multilateral consensus as the benchmark for legitimacy undermines the potential effectiveness of coercive threats because key opponents may assume that legitimacy and credibility are
linked, when in reality they are not. Legitimacy relates to whether the
threatened action is sanctioned by, for example, a UN Security Council
resolution. Credibility is solely a function of whether the four conditions have been met. The mistake that ultimately led to the fall of both
Saddam Hussein and Slobodan Milosevic was their assumption that
the U.S.-U.K. threat lacked credibility because it did not satisfy the
requirements for legitimacy. When multilateral consensus is priori-
tized, potential opponents are more likely to underestimate the probability of an attack that does not obtain UN endorsement and are more
likely to underestimate the risks of non-compliance. In the absence of
process-related legitimacy, 'coalitions of the willing' are expected by
opponents to collapse more quickly. The demand for multilateral consensus means that leaders who might otherwise fold are given reason
to reject the demands specified in very credible military threats. This
often leads to serious miscalculations because decisions about whether
to follow through on a threat are almost exclusively a function of
resolve and the credibility that would be lost if the U.S.-U.K. backed
down. Ironically, the need (indeed, obligation) to demonstrate U.S.-
U.K. resolve to carry through on their threats became even more pronounced, if not essential, in light of increased efforts by France and
Russia to deny legitimacy to their actions; U.S. credibility and
Washington's unwavering resolve to protect its vital interests were
significantly enhanced when the constraints imposed by the lack of
legitimacy were ignored or rejected. In essence, this crisis became the
precedent-setting case for establishing the credibility required to successfully manage and prevent future crises. Optimists believe that U.S.
actions in Iraq will make it much easier in the future to achieve multilateral consensus, because, ironically, Washington conveyed a strong
willingness to act unilaterally. The relatively quick passage after the
war of UNSC 1483 and 1511, assigning power to the U.S. and U.K. for
Iraq's post-war reconstruction, is a case in point.

Multilateral impediments to coercive diplomacy also explain the paradox that played out between the two sides in the Iraq war. British and
American officials were convinced that a war became more likely as the
demands for multilateral consensus increased. Critics believed, on the
other hand, that the surest way to provoke a war was to mobilize such a
large contingent of troops to the region, because once those troops are

deployed credibility dictates the imperative to use the troops to enforce compliance. On the other hand, there would have been absolutely no compliance to any UN resolutions on disarmament and inspections without the deployment of a credible force to the region. Saddam Hussein's mistake was to assume that the two positions held equal weight, but, as he found out, credibility trumps multilaterally mandated legitimacy.

European and American Differences: Illusions and Realities

Illusions: Culture, Values, and Ideology

Despite the ubiquitous nature of unilateralism in international politics and the role of coercive diplomacy in generating the multilateral coalitions that emerged over Iraq, the conventional wisdom remains solid: there are important cultural and ideological differences between European and American approaches to global security. Steven Everts describes the difference this way: 'When Europeans debate foreign policy, they tend to focus on "challenges," whereas Americans look at "threats." European concerns are challenges such as ethnic conflict, migration, organized crime, poverty and environmental degradation. Americans, particularly conservatives such as Bush and his team, tend to debate foreign threats such as the proliferation of weapons of mass destruction, terrorism and "rogue states."'[32]

Evert goes on to note that 'Europe's support for multilateral regimes is actually the consequence of a deeper conviction that most of the world's problems – ranging from economic instability to environmental degradation to security threats – can almost always be solved only through robust multilateral efforts.'[33] Kagan describes European preferences in terms of 'principled multilateralism.' As he explains, 'U.N. Security Council approval is not a means to an end but an end in itself, the sine qua non for establishing an international legal order. Even if the United States were absolutely right about Iraq, even if the dangers were exactly as the Bush administration presents them, Europeans believe the United States would be wrong to invade without formal approval. If the Security Council says no, the answer is no.'[34] In contrast, most Americans are 'instrumental multilateralists' – the motivations for collective action are derived from rational and pragmatic calculations of the costs and benefits of creating strategic coalitions to share the financial burden and to decrease political costs and risks.

But these distinctions are incomplete and misleading, for several reasons. First, as demonstrated here throughout the Iraq case study, the unilateral self-interests of France and Russia were clear and unmistakable. Every state is a unilateralist before joining one or another multilateral coalition. Second, France, Germany, and Russia do not share the security threats of Americans. When security threats diverge, policies and priorities are expected to diverge as well. A true measure of cultural or ideological distinction would require evidence that the threats are identical, yet, despite these shared security concerns, the approaches diverge. Conversely, if the threats, interests and priorities are different, we should expect the strategies to differ; when threats converge (e.g., the Soviet threat during the cold war; escalating ethnic violence and NATO credibility problems in the Balkans from 1990 to 1995) similarities in strategies are more likely. This has nothing to do with culture. Third, as explained in chapter 2, there are no purely unilateral or multilateral strategies or policies – virtually every major foreign policy initiative falls somewhere on a continuum (and often includes components of many strategies co-existing simultaneously). Moreover, as described in figure 2.1, there are 'principled' and 'instrumental' forms of unilateralism, bilateralism, and multilateralism, and these foreign policy options are often applied concurrently and strategically. For example, multilateral apathy in Rwanda was unprincipled, and any effort by any state to act unilaterally in a way that would have prevented the loss of hundreds of thousands of lives would have been principled, despite the absence of multilateral consensus derived from UNSC Chapter VII mandate. What distinguishes principled or instrumental forms of the strategy is not the process, but its outcome; both processes can produce positive and negative outcomes (and often produce both). There is nothing about the process itself that precludes or privileges one outcome over another.

Realities: Competing Multilateralisms (Weakness vs Power)

In contrast to the view that culture and ideology explain American and European differences, Brendan O'Neill draws a more useful distinction between 'strong' and 'weak' multilateralism in which the 'old-style clash of the great powers has been replaced by a "clash of multilateralisms."' The American approach is characterized as 'strong, militaristic multilateralism' while the Europeans prefer 'determined diplomatic multilateralism.' But in both cases, 'the big powers appear to be pursuing their own interests under the guise of an increasingly fictitious "international community.'"[35] Multilateralism, then, is used strategically to buttress the support and legitimacy assigned to efforts by major powers to enhance the very same unilateral economic and political self-interests for which most states compete. The differences in approach, says Kagan, are not derived from culture or values but from relative power:

> American military strength has produced a propensity to use that strength. Europe's military weakness has produced a perfectly understandable aversion to the exercise of military power. Indeed, it has produced a powerful European interest in inhabiting a world where strength doesn't matter, where international law and international institutions predominate, where unilateral action by powerful nations is forbidden, where all nations regardless of their strength have equal rights and are equally protected by commonly agreed-upon international rules of behavior. Europeans have a deep interest in devaluing and eventually eradicating the brutal laws of an anarchic, Hobbesian world where power is the ultimate determinant of national security and success.[36]

What is so important about Kagan's research is that he demonstrates the validity of his thesis throughout history: there is an almost automatic tendency for large and small powers to prefer different forms of multilateralism, with the latter preferring more rules that constrain powerful states, the former preferring fewer institutional limitations, yet both remain convinced that their security is enhanced through these competing (almost mutually exclusive) imperatives.[37] American and European priorities were reversed in the period between the two world wars, as exemplified in Woodrow Wilson's push to establish a multilateral security regime through collective adherence to his fourteen points.

Chapter 4

WMD Proliferation: The Case for Unilateral Ballistic Missile Defence

The debate between supporters of ballistic missile defence (BMD-unilateralism) and their critics who favour reliance on multilateral arms control and disarmament encapsulates the policy divide in the security community and, like the previous case study, helps to explain the post-9/11 preference in Washington for unilateral solutions to the proliferation of weapons of mass destruction (WMD).[1]

There are essentially three prerequisites for establishing a strong case in favour of U.S. BMD. First, the evidence must highlight the fundamental logical, historical, and factual errors critics make when developing their case against BMD. Disconfirming claims about BMD's technological limitations, economic costs, and proliferation risks indirectly strengthen the case in favour by diminishing concerns about the political, economic, and military costs of deployment. Second, the evidence should point to the contributions of interceptor technology to deterring ballistic missile attacks, minimizing the damage associated with such attacks if they occur, and preventing WMD proliferation and associated security threats to the United States, Canada, and their allies. In other words, deploying BMD makes sense if one can demonstrate that WMD and ballistic missile technology is spreading. The third approach, to be developed more fully in chapter 5, is to demonstrate the fundamental deficiencies with multilateral alternatives embedded within the non-proliferation, arms control, and disarmament (NACD) regime.

The strongest case for BMD requires that all three standards be met. To that end, the remainder of this chapter will focus on the first two components by responding in detail to eight core arguments put forward by critics of BMD deployment:

- Demise of Arms Control and Disarmament
- Automatic Proliferation by Russia
- Automatic Proliferation by China
- The Myth of an American First-Strike Advantage
- Technological Limitations of BMD Interceptors
- Financial and Political Costs
- Powerful Influence of the U.S. Military Industrial Complex (MIC)
- Exaggerated Rogue Threats from Rational Rogue Leaders

Demise of Arms Control and Disarmament

According to critics, a U.S. BMD system would fundamentally undermine international stability and destroy decades of disarmament negotiations. Consider the following example of the argument, by Representative Richard A. Gephardt:

> By announcing his intent to move forward with as yet unproven, costly and expansive national missile defense systems, the President is jeopardizing an arms control framework that has served this nation and the world well for decades. [This] approach is likely to increase threats to the US and decrease global stability, as exhibited by the likely consequences: Russia's preservation and China's construction of large stocks of nuclear weapons to counter US missile defenses; an end to transparency and verification of other nations' nuclear arsenals; and the continued proliferation of weapons of mass destruction as other nations follow America's lead in taking unilateral steps that may serve their own immediate interests.[2]

The most common criticisms of BMD are associated with warnings like Gephardt's about the demise of the Non-Proliferation Treaty (NPT) and Anti-Ballistic Missile (ABM) Treaty. Signed in 1968 and 1972, respectively, these treaties are the two pillars of nuclear disarmament and arms control that, according to proponents, were responsible for slowing the pace of proliferation during the cold war and stabilizing the longest nuclear rivalry in history. Included among the most popular myths associated with the ABM Treaty, for example, is that the agreement kept the nuclear peace, prevented a large-scale conventional war between the U.S. and Russia, and helped both sides avoid an expensive and dangerous arms race. According to a *New York Times* editorial in 2001, the ABM Treaty not only restrained arms race competitions between the United States and the Soviet Union but, since the end of the cold war, 'it has allowed Russia and America to dismantle significant portions of their nuclear arsenals without fear that they would be unable to respond effectively to a surprise attack.'[3] Apparently, the ABM Treaty accomplished everything short of curing cancer and solving Mideast conflict. Because these treaties continue to be essential for controlling proliferation and maintaining a stable nuclear environment, it is imperative that their underlying principles and fundamental logic not be undermined.

What critics fail to acknowledge is that the worst abuses of horizontal and vertical proliferation of nuclear weapons and technology occurred after these treaties were signed. Regardless of the indicator used to track nuclear proliferation – overall nuclear stockpiles, numbers of strategic warheads in submarine-launched ballistic missiles, intercontinental ballistic missiles, strategic bombers, production and stockpiles of weapons-grade plutonium, thefts of fissile material, trade in dual-use technology tied to the atomic energy industry, trade in ballistic missile technology, and so on – the evidence of an increase in the pace of proliferation is clear.[4] The numbers indicate that the Soviets had approximately two thousand warheads in 1972 (the year the ABM Treaty was signed) and about twelve thousand by 1989. The only thing the ABM Treaty really accomplished was to guarantee a substantial increase in the number of Soviet and American warheads, permit the development of a missile defence system around Moscow and Washington, and prevent both sides from deploying technology neither could afford (or wanted to deploy) in the first place. The missile interceptor, sensor, computer, and satellite technologies were not ready and were exceedingly expensive.

With respect to related claims about BMD's impact on the NPT, this treaty is also on the verge of collapse, not because of missile defence or any other deployment decision that the Americans, Europeans, Russians, or Chinese have made, but because political and military leaders in Israel, Iraq (until 2003), Iran, India, and Pakistan (and many other states) are convinced that nuclear and other WMD weapons serve their security interests. These regional concerns will continue to exist regardless of whether the United States has a defensive shield. Recent nuclear tests by India and Pakistan, and declarations by North Korean officials that they intend to restart Pyongyang's nuclear program, are but the latest illustration of this patterned consistency. For critics of BMD to claim that the NPT and now defunct ABM Treaties represent essential components of an international arms control structure does not say much for these treaties or the prospects for serious arms control in the future. The failure of these treaties to control proliferation may also explain why both the Canadian and American disarmament communities are split on the implications of BMD for arms control, many of them concluding that an effective system could serve the global desire for significant cuts in stockpiles of nuclear weapons.

Proponents of multilateral arms control will argue that the more relevant measure of the NPT's effectiveness is the number of nuclear weapons states in the world today, and will likely reject my criteria for judging the NPT's success. These critics argue that there were five nuclear weapons states when the NPT was signed in 1968, and there were still five at the beginning of 1998. There are now seven with India and Pakistan. If the undeclared states (Israel) and the wannabes (Iran and North Korea) are included, the number rises to ten. The expectation when the NPT was negotiated was that there would be far more nuclear weapons states, but that prediction was obviously wrong. The NPT would, therefore, seem to be an overwhelming success. Even as late as 1974 U.S. intelligence experts expected close to two dozen nuclear powers would emerge in the next twenty years.

Applying the same logic, perhaps the success of the New York Police Department should be judged on the basis of how many millions of New Yorkers do not commit murder or rob liquor stores, regardless of whether they ever intended to commit those (or other) crimes. (I suspect the police commissioner and/or mayor of New York would be very happy if the public evaluated their performance this way.) But this approach – and the corresponding percentage of law-abiding citizens – obviously biases the case in their favour; it does not hold them accountable for the crime rate (the number of crimes that are actually committed in a typical year). Similarly, an overwhelming majority of non-nuclear states are not interested in acquiring nuclear weapons, not because their leaders find the weapons repulsive or morally reprehensible (although some probably do) but primarily because nuclear weapons provide no added security (at this time), or because security guarantees from allies are more than adequate (as is the case in Canada). Most states, in other words, have the luxury of being able to say no to nuclear weapons. The success of the NPT must be measured instead by (a) how many aspiring nuclear states (signatory and non-signatory) are prevented from acquiring or developing the technology to deploy nuclear weapons and/or their delivery vehicles, and (b) how many states (signatory and non-signatory) continue to provide the requisite technology to aspiring nuclear powers. It takes only one nuclear weapon to produce the catastrophe the NPT was designed to prevent, and the capability to deliver and inflict that kind of damage continues to spread on a daily basis.

The fate of the NPT (and perhaps its ultimate demise) will continue to depend on political, military, and security environments in places such as India, Iran, Iraq, Israel, Libya, North Korea, Pakistan, and Syria. Leaders in these states are sure that ballistic missiles and nuclear weapons technology are essential to their security and national interests and are more than willing to accept related technology transfers from China, Russia, France, the United States, and Canada to accommodate their concerns. Critics in the arms control community refuse to acknowledge that these states will continue to perceive security threats for reasons that have nothing whatsoever to do with U.S. BMD, or with U.S. and Russian adherence to arms control limits stipulated in Strategic Arms Reduction Treaties (START) II, III, IV, or V. The limits (assuming they are ever achieved) will never be enough to convince officials in aspiring or new nuclear states to roll back their programs for one simple reason – decreasing nuclear stockpiles from thirty thousand to roughly twenty-five hundred is far easier than going from one to none.

Leaders of new and aspiring nuclear weapons states are typically perceived by many critics as passive observers who learn from established nuclear powers and, in true Pavlovian fashion, respond accordingly. The best way to control proliferation and maintain a stable nuclear environment, critics assume, is for established nuclear powers to teach other states to adhere to the principles, logic, and moral (read civilized) standards entrenched in multilateral arms control treaties. That assumption is not only naïve but also dangerously superficial when it comes to understanding the proliferation puzzle and the myriad factors that explain most (if not all) proliferation decisions. It is particularly insulting to officials in, for example, India and Pakistan to be told that their capacity to make informed decisions about their own global and regional security interests is limited and in need of American, Russian, or Canadian advice. The notion that established nuclear powers or the international arms control community command the intellectual high ground on these issues and, by extension, have a moral obligation to guide other cultures and peoples towards the enlightened path of nuclear sanity is reminiscent of the worst extremes of Manifest Destiny and its associated policies. While this obviously is not the intention of the arms control community, it is precisely how its pronouncements are perceived in many states.

Finally, it is particularly ironic for critics in the arms control and disarmament community to embrace the ABM Treaty (prior to its demise) as the cornerstone of non-proliferation and nuclear sanity when the same critics vehemently criticized the treaty throughout the 1970s and 1980s for entrenching the logic of mutually assured destruction (MAD) in United States–Soviet rivalry. 'Positions have certainly reversed over the years,' according to Andrew Coyne. 'It used to be the hawks that stood for the grim logic of deterrence, and doves who insisted that mutual assured destruction was immoral. Now those same liberals and peace activists cling to MAD as the only basis of a stable international regime.'[5] That logic was used by both sides to justify huge defence expenditures, ever-increasing nuclear stockpiles, and an exponential increase in the number of nuclear targets – all with reference to the benefits of maintaining a robust, stable, and credible second-strike capability.

Perhaps the most important point missed by critics of BMD in the arms control community is that throughout this entire period, the absence of BMD did not prevent the most significant arms race in history. Contrary to the critics' claims, therefore, excluding a BMD system from the picture is not enough for arms control and disarmament, and the presence of BMD is neither necessary nor sufficient for an arms race.

Automatic Proliferation by Russia

Building on claims about the impending demise of decades of disarmament negotiations, critics predict that Russia will respond to BMD with a major build-up and deployment of more sophisticated weapons as it attempts to regain its security by re-establishing its second-strike capability. This automatic action-reaction sequence will, in turn, create security threats for China, India, and Pakistan, who will inevitably retaliate by setting up greater numbers of nuclear weapons for their own security. A huge spiral in defence spending and arms production will result and propel the international community back to square one of the arms control and disarmament agenda – all because of a relatively minor investment of about 2 per cent of the United States defence budget (or 0.3 per cent of its federal budget).

There are three interrelated assumptions (assertions) that underlie these predictions:

Assumption 1:   Proliferation by Russia is directly related to (and
will be a primary consequence of) American BMD
deployment.

Assumption 2:   Russian warnings convey a real fear that the United
States is attempting to gain a first-strike advantage.

Assumption 3:   Russian proliferation, even if it occurs, represents for
American officials a greater threat to U.S. security than
WMD proliferation by aspiring nuclear powers or
terrorists.

Each assumption reveals logical inconsistencies and factual errors. With
respect to assumption 1, for example, the causal chain cited by critics to
predict almost automatic proliferation by Russia ignores evidence compiled over years of research on why states proliferate. Without exception, the research points to a complex set of political, economic, and
military-strategic prerequisites.[6] If the findings and patterns identified
in this body of work are correct, Russia can be expected to acquire and
set up more advanced systems if and when the technology becomes
available and affordable, and as long as Russian officials perceive as
unfair the imbalance between their nuclear capabilities and those of the
United States, Europe, and NATO.

Russia has already created advanced weapons designed to circumvent missile defence systems, so critics' fears of this kind of proliferation following BMD deployment are moot. The new SS-X-27 ballistic
missile, for example, has an accelerated boost phase of 100 seconds
(down from 180 seconds, making it harder to detect on launch), can
carry three warheads, and is highly manoeuvrable.[7] It is particularly
interesting that this combination of technologies serves as an effective
countermeasure not only to BMD but to the alternative missile defence
system that Vladimir Putin, the Russian president, offered the Europeans in June 2000. Self-interest dominates the Russia agenda – Russian
leaders, in other words, have not suddenly become more enlightened
than the Americans over how to achieve global security, and no serious
observers doubt that Russia intends to do precisely the same thing in a
few years as soon as the technology becomes available to them, regardless of what the Americans do today.

Critics who predict that Russia will accelerate the pace of proliferation in retaliation for BMD should at least explain what Russia would
gain and why its leaders are likely to ignore the political, military, and,
especially, economic costs of returning to a cold war footing. Most if not

all concerns about the inevitably destabilizing response by Russia should now be laid to rest; Moscow understands that U.S. security priorities will, for the foreseeable future, continue to focus almost exclusively on emerging threats of horizontal proliferation by rogue states and terrorist groups.

With respect to assumption 2, we should expect statements about BMD from Moscow to be less than conciliatory and occasionally quite threatening. But warnings about the negative impact of BMD, and Russian opposition more generally, should be understood in their proper context.[8] There is nothing particularly earth-shattering about Russia's preferences here – they will always oppose (explicitly and implicitly) any and all decisions by the United States to deploy an advanced defensive and/or offensive system, under any circumstance. This is to be expected, even if the system in question does nothing to undermine the logic of mutually assured destruction or the foundation of bilateral strategic stability.

Russian officials can be expected to play on those fears, regardless of the impact of BMD on the stability of their deterrent capability. As Frank J. Gaffney Jr. explains, 'This suits the Kremlin, of course, which is anxious to retain the last vestiges of superpower status and which, under Vladimir Putin, rarely misses an opportunity in American elite circles and allied nations to threaten increased tensions, or worse, if the United States abandons its present posture of absolute vulnerability to missile attack.'[9] But political rhetoric is distinct from political reality. With respect to rhetoric, Russian officials are playing to a domestic, international (European), and American audience and have an interest in perpetuating the myth of Russian superpower status. The status quo provides Moscow 'with a sense of lingering prestige, and equality with the United States ... a cherished vestige of the days when Moscow was America's peer,' say Kenneth B. Payne, Yuri Chkanikov, and Andrei Shoumikhin.[10] In reality, Russia's preferences have less to do with serious concerns about strategic stability and more to do with gaining bargaining leverage over Washington on several issues, including BMD architecture.

In the aftermath of 11 September, the assertion that U.S. BMD constitutes a diabolical scheme to solidify American strategic superiority over Russia as a precursor to a pre-emptive first strike should be dismissed as absurd, and most Russian officials view such scenarios in this light. Some critics respond to this line of argument with the following refrain: U.S. leaders may not be trying to acquire a first-strike advantage, but several high-ranking military officials in Moscow remain convinced that American military leaders are trying to accomplish this goal. But if we accept the critics' case that perceptions of intentions are far more important than actual operational plans, we are faced with a set of pressures that render progress on any and all arms control and disarmament strategies virtually impossible to achieve. After all, if Russian officials still believe that the United States is secretly hell-bent on acquiring (and possibly launching) a nuclear first strike on Russia, there is absolutely no hope for constructive dialogue, let alone successful implementation of any aspect of arms control, since American threats are completely and entirely disconnected from the signals and intentions American officials think they are sending. Consequently, the trust and rationality required to succeed on arms control and disarmament are missing and doom the regime to failure. If Americans are perceived to be contemplating the launch of a ballistic missile that would kill millions, there is no hope for arms control. Indeed, the stronger the case critics make for automatic proliferation by Russia because of perceived U.S. first-strike threats, the weaker the case for expecting success of the multilateral NACD regime or even bilateral disarmament.

A more realistic take on the relationship suggests that leaders in Moscow know that BMD will go forward regardless of their preferences. Given the inevitability of BMD, Russian officials are left with few real options. They could reject any effort by the United States to deploy defence systems and continue to criticize the unilateral decision by American leaders to withdraw from the ABM Treaty. But aside from modest levels of diplomatic support, which will likely diminish as more European leaders accept the reality of BMD as well, this option is not particularly appealing. A U.S. deployment of BMD in the face of Moscow's protests would provide further proof of Russia's declining influence and signal the end of its status as a superpower worthy of serious consideration – a humbling experience for any leader. President Boris Yeltsin suffered the humiliation of having almost all his demands ignored during NATO's Kosovo campaign, and President Putin had a very similar experience in the 2003 Iraq war. With respect to BMD, Putin or any future Russian leader is not likely to welcome the same plight.

The more appealing alternative for Russian leaders is to engage the U.S. (as apparent equals) by conceding the point that Washington is within its rights to withdraw from the ABM and by pushing for their own preferences in regard to arms control and strategic stability. The key for Russian diplomats is to strike the optimum circumstance that maintains the core of a bipolar strategic balance but allows for significantly more robust BMD testing, development, and ultimately deployment. Negotiations over verification, monitoring, and intelligence sharing and other confidence-building measures will follow, but not in the context of codified rules and regulations common to traditional arms control negotiations. This is precisely what is happening – the United States and Russia have begun to address the danger of misreading signals, including a new Russian early warning system, cooperative arrangements to share information, additional measures to protect against unauthorized launches, installation of cameras or acoustic or seismic sensors in missile fields, and others.[11]

Finally, with respect to the third assumption about competing risks to U.S. security, even if optimists are wrong about improved U.S.-Russia or U.S.-China relations, and even if Moscow and Beijing decide to expand their respective arsenals in response to the U.S. withdrawal from the ABM Treaty, the issue for American policy makers will always be one of comparative risks. The security risks associated with vertical proliferation by these states will always be less significant to those responsible for U.S. security than the risks of even one nuclear weapon being deployed by North Korea, Iran, or any other state of concern that may emerge in the future.

Automatic Proliferation by China

Although the United States and Russia have the luxury as major nuclear powers to disarm seventeen hundred to two thousand warheads, China will continue to build up to the levels befitting an ascending major power. But there are at least six reasons why U.S. BMD deployment will have virtually no measurable effect on China's nuclear weapons program. First, the real incentive for China to upgrade its nuclear capabilities is Beijing's perception that its nuclear deterrent is unstable, for the same reason that American and Soviet officials worried in the 1960s and 1970s – China's retaliatory capability is not sufficiently potent (or credible) in second-strike terms.[12] Whether the United States deploys BMD is irrelevant; Chinese officials will continue to see a capability/credibility gap, at least until their stockpiles approach parity with those in Europe. Even if the United States were to decide to scrap BMD tomorrow, China would continue to upgrade its systems on the basis of worst-case scenarios and assumptions about surreptitious BMD development.

Second, over the past decade China has continued to modernize its nuclear program without BMD to worry about. Its military is currently committed to add by 2015 dozens of additional survivable land- and sea-based mobile missiles to its inventory, all in an effort to build up to a second-strike capability. This capability is the primary and overriding incentive for China's nuclear program, an incentive driven by the same mutually assured destruction logic that was so compelling to American and Soviet officials throughout the cold war. Ironically, it is the same MAD logic that is defended today by critics of BMD as a stabilizing feature of contemporary U.S.-Russia nuclear rivalry. And it is the same MAD logic that will continue to drive China's program. As one Chinese official explains, 'To hold the US credibly deterred is just to reciprocate, to a much lower extent, what the US has long done against China during the nuclear age. In fact, it was US nuclear threats to PRC on a number of occasions that prompted Beijing to start its nuclear weapons programme. Though the US has the most formidable nuclear arsenal and most powerful and sophisticated conventional arsenal, it retains the option of a first-strike with nuclear weapons as its deterrence policy. Now the US would even revise or abolish the ABM which assures nuclear weapons states of their mutual security.'[13]

Indeed, the stronger the arguments in favour of protecting the MAD doctrine embedded in bipolar strategic stability (see chapter 1), the weaker the arguments that BMD is a necessary condition for Chinese proliferation. If the acquisition and protection of second-strike capabilities was (and remains) essential to stabilizing the U.S.-Russia nuclear rivalry (for all the reasons put forward by critics of BMD), the same strategic imperative will drive China to continue to modernize and expand its nuclear program to achieve that end, with or without BMD. Why should we expect Chinese officials to act any differently today, or to develop nuclear programs and strategies that are any less logical? Again, excluding a BMD system from the picture is neither necessary nor sufficient for arms control and disarmament.

Third, the only barriers to China's ability to deploy a second-strike capability today are access to newer technologies and that nation's ability to pay for them. Consider for a moment what Chinese president Jiang Zemin (or future presidents) would choose if given the option of instantly deploying twenty-five hundred land- and sea-based nuclear warheads (i.e., approximate START III levels). Do critics of BMD honestly believe China would pass on the opportunity to establish their second-strike capability, even without a U.S. BMD system to worry about? Wouldn't this enhance the credibility of their retaliatory threats to prevent the U.S. from becoming involved in a China-Taiwan conflict? In sum, while the U.S. and Russia are trying to work their way down to START III levels, China is attempting to establish parity by working up to those levels. The important point to keep in mind is that deployment of U.S. BMD will not create the necessary technology for China or provide its leaders with the funds to develop or deploy more advanced systems.

Fourth, China's modernization plans are related, in large part, to fears associated with theatre missile defence (TMD) and the prospects of this technology spreading to Taiwan. As Kagan observes, 'What China wants to "deter" is American support for Taiwan. Beijing officials now routinely warn that American military intervention to defend Taiwan against Chinese attack could lead to a missile strike on, say, Los Angeles. Whether or not that threat is credible, the Chinese understand that a worried, vulnerable America is more likely to be pliant in negotiations over Taiwan's future than a confident America. A national missile defense system would negate China's entire strategy.'[14] But TMD systems are fully compliant with ABM Treaty limitations – they were permitted under ABM demarcation agreements and memorandums of understanding signed by the U.S. and Russia in 1997. Regardless of TMD's compliance with the ABM Treaty, however, China continues to protest any and all transfers of defence technology to Taiwan. Why? Perhaps Chinese priorities have less to do with protecting the sanctity of the Non-Proliferation Treaty (NPT) or the ABM Treaty and more to do with protecting Beijing's capacity to deter the United States (or NATO) in future crises. And some critics will no doubt welcome the thought of China (or Russia) deterring the United States from intervening in global crises, given their concerns about American foreign interventions in Iraq 1991, Bosnia 1995, Kosovo 1999, Afghanistan 2001, and Iraq 2003. But evaluations of the moral and legal foundations of American foreign policy should be kept separate from assessment of BMD and its implications for vertical and horizontal proliferation.

Fifth, and perhaps most important, several distinguished critics of BMD in the U.S. have argued all along that research and development (R&D) of BMD should continue; the real problem, they argue, is with deployment. As John Steinbruner points out, 'there will be continued research and development efforts, and I would not object to them. *I don't think anybody would object to them.* Clearly one attempts to work on the problem and figure out what the technical difficulty is. The issue does not have to do with technical development efforts so much as with commitment to deployment, and under what terms [emphasis added].'[15] And more recently, Senate Minority Leader Thomas Daschle argued that 'the president may be buying a lemon here ... Now, we're for additional research. We're for finding ways with which to improve the technology, but to deploy and to violate or abrogate the ABM Treaty before we've ensured that it works is absolute silliness, and really has to be addressed.'[16]

On the surface, the positions put forward by these distinguished critics of BMD make perfect sense, particularly from the point of view of public policy. The U.S. president, they admit, has a moral obligation to develop technology to stop incoming ballistic missiles, and it would be irresponsible for any administration to stop these programs. But consider this widely supported R&D policy from the perspective of Beijing officials. The fact that the U.S. government and the American military industrial complex are fully engaged in R&D of BMD and TMD technologies, and that this policy is widely supported by both parties in the U.S. government, is reason enough for China to expand its nuclear weapons program and to deploy advanced missile technology when it becomes available. The logic is simple: U.S. R&D of missile defence systems increases the probability that at some point in the future Washington will be able to rapidly deploy more effective BMD and/or TMD technology – from China's point of view U.S. R&D is justification enough to continue with their defense programs. Are critics of BMD, such as Daschle and Steinbruner, willing to compromise on R&D, given China's preferences, perceptions, and likely response? If not, then what are the implications for their position on BMD more generally, especially as they relate to threats of automatic proliferation by China?

Once again, even if we accept the argument that U.S. BMD is a necessary condition for China's modernization program, the issue for U.S. policy makers is one of comparative risks. The relationship the United States has with new and aspiring nuclear powers is not particularly stable, is far less manageable, and shows no prospects of improving. Chinese proliferation on the other hand is easier to deal with and less threatening.

If China expanded its program over the past decade without BMD to worry about, continues to develop plans to double the number of deployed medium- to long-range ballistic missiles in response to what at the time were ABM-compliant technologies, is prevented only by access to technology and money from moving even faster, and will continue to modernize under conditions of BMD research and development alone, how exactly will ending BMD change any of this? Officials in Beijing will continue to blame BMD for their modernization program, and critics will continue to repeat China's pronouncement as definitive proof of impending doom. But the public deserves a more sophisticated discussion of the proliferation puzzle, nuclear diplomacy, and international politics. Screaming 'I told you so' whenever China blames BMD for its modernization program is not particularly helpful in this regard. If the same dire warnings emanate from Moscow and Beijing regardless of BMD's impact on their capacity to retaliate (or the credibility of their retaliatory threats), why are critics so quick to accept (and repeatedly cite) these pronouncements as if they were proof positive that proliferation is inevitable? Why are critics willing to accept current threats by China and Russia to proliferate in retaliation for BMD when they rejected as logically absurd the identical strategy when it was used by the United States and the Soviet Union to increase nuclear stockpiles throughout the cold war. Their arguments once included reference to the fact that most of the thirty thousand missiles on each side were redundant and entirely useless; a few missiles were enough to inflict unacceptable damage, and this 'minimum' deterrence was more than sufficient to provide effective security from a pre-emptive strike. Who, they argued, would be foolish enough to risk provoking even one retaliatory missile? Yet critics in the arms control community today refuse to apply the same logic and criticism to Russia and China because it undermines the rationale tied to warnings about automatic proliferation. As an unintended consequence of their arguments, nuclear proliferation becomes a reasonable and logical response to BMD.

Sixth, and finally, the fact that BMD could conceivably persuade U.S. officials to intervene in a China-Taiwan dispute, and to do so in ways that undermine China's political, military, or economic interests, is not, in and of itself, a criticism of BMD. The more relevant question for American policy officials in this regard is whether Taiwan is worth the fight if and when it takes place. Debates over the merits of U.S.-Taiwan-China policies are prior to and far more important than the question of whether Chinese officials are worried about BMD. That Chinese officials worry about U.S. intervention is not necessarily a bad thing. The

concerns expressed by Joseph A. Bosco are important: 'Chinese en-
hancement of Iraq's air defense system in violation of U.N. Security
Council resolutions confronts American policymakers with a disturbing question: Has China launched a new Cold War against the United
States? Beijing's global "anti-hegemonism" policy can no longer be
dismissed as mere rhetoric. It has supplied the technology for delivering Baghdad's chemical and biological weapons. It was instrumental in
the emergence of North Korea's and Pakistan's nuclear weapons programs and has transferred missile technology to Iran and Syria.'[17]
Unfortunately, constructive engagement does not appear to have worked
either – 'integrating the People's Republic into the international system, sharing advanced technology, making it a trading partner, even a
strategic partner, has not moderated Beijing's core hostility to the United
States, its values and its role in the world.'[18]

The Myth of an American First-Strike Advantage

To the extent that Russian and Chinese fears of losing second-strike
capabilities are real (as opposed to fabricated for domestic and international audiences) they are based, in part, on a basic and fundamental
cold war concern that effective missile defences will give the United
States a first-strike advantage. BMD deployment, then, will force Russia and China to acquire additional missiles to augment their retaliatory
capability and, by extension, the credibility of their nuclear deterrent
threat. But critics who express these concerns should be prepared to
explain why their fears are legitimate, and, more precisely, why the
United States would sometime in the future threaten or launch a massive first-strike attack against Russia and/or China. Again, what credible scenario would take us to the brink of a contemporary crisis in
which American officials would contemplate (for a second) a pre-emptive
strike – with or without a perfected BMD shield capable of defending
against every single missile, decoy, and countermeasure Russia and
China would launch in retaliation? What conceivable set of circumstances would explain why Moscow and Beijing would follow a course
of action that would give the United States reason to contemplate a
nuclear attack! If critics are not prepared to offer realistic scenarios,
perhaps they should refocus some of their energy towards convincing
China and Russia that their concerns about a pre-emptive, first-strike
attack by the United States are absurd – far more absurd, and significantly less probable, than threats tied to ballistic missile proliferation by

rogue states in five, eight, or ten years. Of course, acknowledging the absurdity of Russian and Chinese fears undermines a central component of the case against BMD because it refutes the logical imperative justifying Russian and Chinese proliferation.

In response to my challenge to identify a single, credible scenario that would take us to the brink of nuclear war, David Mutimer offered the following argument: 'I am not certain that I would be willing to trust the American government's word were I a decision-maker in Moscow with a memory of forty years of cold war, or in Beijing with a memory of a United States so implacably hostile that it refused to recognize my regime for more than twenty years ... What is more, given the state of the political culture in the United States and the possibility of a change in government at least every four years, how could one expect that kind of trust from the Russians and Chinese? Even now, the United States is a country in which Jesse Helms has considerable influence on foreign policy!'[19]

These 'reasons' do not provide a realistic scenario leading to a nuclear exchange; nor do they amount to justification for Chinese or Russian officials to worry about a pre-emptive first-strike attack by the United States. These 'excuses' (and hundreds of others I am sure critics of BMD can come up with) fall far short of the serious thought critics should give to these questions. A credible critique of BMD demands much more than a few quips about Jesse Helms, George W. Bush, or the U.S. electoral process.

In one of the most interesting contributions to the debate over BMD, Fareed Zakaria interviewed Thomas Schelling, the economist who applied game theory to politics and international relations (as Zakaria explains, 'work that should have won him the Nobel Prize, if economists weren't such snobs about political science'), developed the 'basic conceptual structure of deterrence theory,' and whose ideas are 'at the heart of the counterintuitive logic of mutual assured destruction that has underpinned American nuclear and arms-control strategy for four decades.' Zakaria introduces his interview by quoting a line from Woody Allen's *Annie Hall*: 'Woody Allen and Diane Keaton are standing in line at an arty Manhattan movie house while a pompous academic pontificates about Marshall McLuhan. Exasperated, Woody finally goes to the lobby and wheels out McLuhan himself, who turns to the professor and announces: "I heard what you were saying. You know nothing of my work ... How you ever got to teach a course in anything is totally amazing."'

Zakaria's point is simple – critics of BMD who predict automatic proliferation by Russia and China, or who pronounce the end of strategic stability after BMD deployment, or who defend the sanctity of MAD as a cornerstone of arms control, should take the time to cite Thomas Schelling's work on deterrence theory. When asked whether he thought President Bush's proposals undermined strategic stability, Schelling's response was no, 'because missile defense is not likely to be as revolutionary as either its proponents or opponents believe. Both sides are vastly exaggerating the scope of this program.' Schelling went on to explain that 'the current proposals, to the extent we have any details, are really oriented toward defending the United States against small attacks from rogue states. That's why I don't like the way the President is selling his program as a shield to protect the whole nation. It isn't, and I think we have incurred diplomatic costs around the world because of this rhetorical posturing.' And when asked about automatic proliferation by Russia and China: 'I don't see how. Stability between the United States and the Russians depends on the fact that both sides can inflict unacceptable harm on the other even if one were hit by nukes first. That "second-strike capability" will be intact, since no defense system we could develop would protect us against Russia's massive arsenal. I think the Russians understand this, which is why they have stopped being so belligerently opposed to missile defense.' Finally, when asked about the ABM Treaty and arms control: 'The ABM treaty was wonderful for its time. Arms control doesn't depend on negotiated treaties. It depends on both sides restraining themselves out of self-interest. If you can get good mutual understanding, you can actually move faster without treaties. We now have a pretty good understanding with the Russians about arms reductions. And if we both keep reducing nuclear weapons – which we should – how can one say that this is a new arms race?'[20] Many critics of BMD continue to cite Schelling's work without really understanding his central observations about the nature of nuclear rivalry or the underlying logic of deterrence theory.

Technological Limitations of BMD Interceptors

In terms of BMD architecture, the scope of the proposed project is expected to evolve through four stages, beginning with twenty interceptors in 2005 and growing to a larger system by 2011. William Broad's report on the system's initial architecture describes a land-based, midcourse shield requiring '2 launching sites, 3 command centers, 5 communication relay stations, 15 radar, 29 satellites, 250 underground silos and a total of 250 missile interceptors (by 2011).'[21] Key locations for the system include Hawaii, Alaska, California, Colorado, North Dakota, Massachusetts, Greenland, Maine, and Britain. Ten interceptors will be deployed in Alaska and California by 2004, another ten by 2005, and approximately twenty interceptors on Aegis ships. With respect to other details of the BMD program under George W. Bush, the 'how' of BMD will be resolved in favour of a $80–$100 billion, multiple-platform (land, sea, and air), layered (boost, mid-course, and entry level) missile defence system. After 11 September and following another successful fully integrated test in June 2002 (five successful intercepts out of seven attempts), the Bush administration moved quickly to construct radar sites (in Alaska), speed up the BMD testing schedule, and withdrew from the ABM Treaty. Peter Pae's overview of BMD technology is perhaps the best summary to date of the requirements for hitting a bullet with a bullet; excerpts are included in table 4.1.[22]

With this overview in mind, two main assertions underpin the 'technological limitations' critique: current defence technologies don't work, and simple decoys and countermeasures can overcome any shield. According to the critics, an investment of $80–$100 billion simply cannot be justified on the basis of recent tests or the prospects for success in the future. The technology is not capable of satisfying even the most basic requirements for success and, according to criticisms among scientists and engineers, it is unlikely ever to be powerful enough to deal with decoys and simple countermeasures. There are several problems with these arguments.

Critical Contradictions

Consider the contradictions in the critics' case so far. If BMD is indeed plagued by these incredible technological hurdles, and if simple countermeasures can easily offset any investment in defence, then, aside from wasting money, what exactly is the problem for China and Russia? After all, only an effective BMD shield would justify an arms race to re-establish a stable second-strike capability. Conversely, if BMD does not (or will not) work, any security concerns expressed by Russia or China about maintaining the credibility of their nuclear deterrent are misplaced and mistaken, as are related decisions by these states to proliferate in retaliation for a useless defensive system (why waste their time

TABLE 4.1
BMD Technology – How to Hit a Bullet with a Bullet[a]

The kill vehicle has the job of intercepting a warhead at a closing speed of 15,000 mph – more than four miles a second or five times the speed of a bullet. It must measure where it is going, calculate a collision path and operate rocket thrusters to score a hit in millionths of a second at the very end of its flight. The kill vehicle not only has to find its target in the void of space, but distinguish it from decoys and other countermeasures.

The device carries a range of sophisticated items, including optics to navigate by the stars, antennas to receive data from ground radar, a small computer and a refrigeration unit to form krypton ice cubes for cooling sensors. The whole thing can fit on a kitchen table. Because it carries no explosives or other munitions, the kill vehicle destroys the target by the sheer force of the collision. But hitting the cone-shaped warhead, typically about 6 feet tall, isn't enough. The device must hit the warhead at its so-called sweet spot – measured in centimetres – where the nuclear munition is housed to prevent an errant nuclear explosion. As the booster makes its ascent, it uses a telescope with three mirrors and two infrared sensor systems to look out hundreds of miles for its targets, while getting help from ground radar. The telescope is made with beryllium to protect against electromagnetic pulse from potential nuclear blasts that could damage its electronic components.

Not long after launch, the kill vehicle starts its cryogenic refrigeration system, which cools the infrared sensors to hundreds of degrees below zero. Krypton gas is funneled to the area around the sensors. Krypton ice cubes form and stay with the device until it collides with the warhead. Cooling the sensors is vital to detecting and discriminating the target from decoys and countermeasures. In Raytheon's kill vehicle, the infrared sensors take precise measurements of heat radiation, distinguishing the relatively warm warheads against the ice-cold background of space. The readings are digitized, converted to a numerical value and analyzed by computers. The infrared sensors, considered the most vital elements of the kill vehicle, were developed at Raytheon's El Segundo facility. The sensors feed information into infrared detectors that analyze the signals. Those were developed by Raytheon engineers in Goleta, California.

Exactly how it distinguishes a decoy from a real warhead is highly classified, but it uses computer algorithms to make the decision. The algorithm includes preprogramm-ed profiles of enemy warheads compiled by U.S. intelligence agencies. About three minutes into its flight and 1,400 miles from the target, the device separates from the booster in a show of pyrotechnics as more than a dozen cables and connectors are blown off. The kill vehicle itself is gently pushed out by four springs. Immediately after separating from the booster, the kill vehicle uses its thrusters to sharply move to its right or left to avoid getting hit from behind by the booster. It is carried forward to the target by only its own momentum. The four divert thrusters mounted on its sides were tested recently at Edwards Air Force Base, where the device was able to hover about 20 feet above the tarmac for about 30 seconds.

About 100 seconds before impact, the sensors are turned on and begin to detect and discriminate the target, sending signals to the computer that steers the device. Until this point, it navigates by using an inertial navigation system, aided by star sightings through a separate stellar navigation unit. The computer uses Intel processors that are more than 10 years old – though Raytheon says it may upgrade to a newer chip. Most of the number crunching is completed by an electronics unit, Raytheon officials said.

TABLE 4.1
BMD Technology – How to Hit a Bullet with a Bullet[a] (concluded)

Although small compared with other sophisticated satellites and weapons, the kill vehicle takes about 24 months to produce with about half of that time spent undergoing tests and sensor calibrations. There are no moving parts except for the valves on the thrusters. To ensure quality, Raytheon built a high-tech, dust-free manufacturing plant in Tucson.

---

[a]The detailed description of BMD technology is taken from Peter Pae, 'Kill Vehicle Scores a Hit with Proponents of Missile Defense Weapons: The Pentagon Says the Successful Tests May Restore Credibility to the Program,' Los Angeles Times, 26 March 2002.

and limited budgets?). Juxtaposed against the previous warnings about automatic proliferation, the technological critique is, in many ways, mutually exclusive – the stronger the arguments and evidence about the deficiencies of BMD, the weaker the claims about automatic and justifiable proliferation by Russia and China because they have no reason to feel compelled to compensate for losses in security.

For those who remain unconvinced that these obvious contradictions exist and are often repeated, I offer the following illustrations in tables 4.2 and 4.3, all of which should be readily accepted as representative of the views of prominent critics in Canada and the United States, respectively. There are many others, but space constraints preclude a more detailed treatment here.

If it is so clear to these critics that BMD technology is guaranteed to fail, why would it automatically trigger a global arms race? Apparently only critics are bright enough to appreciate the mind-boggling technological hurdles facing BMD. Apparently the Russians and Chinese simply can't grasp the same obvious 'facts' about the insurmountable deficiencies that plague interceptor technology. Critics who don't see the contradiction are either arrogant or naïve, or both. U.S. Senator John Kerry is one of the few critics who understands the logic of BMD and deterrence and thus avoids the error: 'If you can't shoot down 100% of them [incoming missiles], you haven't gotten rid of mutually assured destruction. And if you can, you set off an arms race to develop a capacity that can't be touched by a missile defence system.'[23] Russian and Chinese military and political officials understand that time is on technology's side, and U.S. BMD will be made to work. But they also understand that the American system currently being deployed is incapable of mounting an effective defence against their arsenals (and

TABLE 4.2
Canadian Critics and Contradictions

1 It has long been recognized that constructing such national defence shields (leaving aside the improbability of their working – and two recent BMD tests have failed) would spur opposing states to develop new offensive weapons to circumvent defence systems. Thus arms races would keep accelerating. – Senator Douglas Roche, O.C.[a]

2 The expert consensus appeared to be that the anti-ICBM systems currently envisaged would not work for at least a decade – certainly, they offer no guarantee that the U.S. would be capable of shooting down all incoming missiles in a major attack by a well-armed, determined adversary ... Both Russia and China would be compelled to assume that the American NMD deployment would be operated against them, increasing the U.S. capacity for political intimidation. – Lloyd Axworthy: Preliminary Report on Liu Centre Consultation on NMD[b]

3 The hard reality is that no such BMD system is, or is likely to be, available. Current proposals fail on all three counts ... First, present technology is utterly unreliable ... But, second and more important, BMD would scuttle, rather than promote, arms control, creating enormous incentives to increase efforts to acquire weapons of mass destruction and the missiles to deliver them. – Ernie Regehr[c]

4 The technical problems, however, are mind-boggling. To work, missile defence requires orbiting satellites capable of detecting the launch of attacking missiles, radar systems to track the incoming warheads, and interception rockets that must be launched quickly, find the warhead among decoys, then kill the warhead in space ... Mr. Putin has adamantly refused to reopen the treaty, and he and Chinese President Jiang Zemin issued a communiqué this week denouncing the proposed U.S. missile shield, saying it would upset decades of arms-control agreements that have deterrence, not defence, at their core. – Thomas Axworthy[d]

5 Canadians should be horrified at the prospect that our government might soon enlist as a junior partner in the U.S. Ballistic missile defense (BMD) project. Not only does BMD threaten to trigger a new global arms race, but it is unlikely to deliver on its promise to protect North Americans from a limited nuclear attack. – Peter Sarcino[e]

6 It's dangerous because it could start a whole new missile race. It's illegal because it would violate the Antiballistic Missile Treaty prohibiting comprehensive antimissile defences. It's fiscally irresponsible because it would costs tens of billions of dollars. It's technologically unsound because not even the best radar and most accurate interceptors can reliably shoot down an incoming ballistic missile. It's a threat to all people everywhere because it would upset the established system of nuclear deterrence and raise the chances of atomic armageddon. Canada should resist this pressure. Missile defence poses a direct threat to Canadian security. If the Americans break the taboo on missile defences by building their shield, Russia and China are likely to react by building better offences and aiming them at North America, undoing years of progress on arms control and possibly igniting a new arms race.[f]

[a] *Globe and Mail* (Toronto), 3 April 2000, A13.
[b] *The Missile Defence Debate: Guiding Canada's Role: A Preliminary Report on an International Consultation on U.S. Missile Defence* (Vancouver): The Liu Centre for the

strategic deterrent forces), and never will have that capability. The real question, then, is whether a functioning BMD system (with some probability of success) is worth the financial costs and security risks associated with deployment and whether those costs and risks are acceptable when compared with evidence of costs, risks, and benefits of alternative strategies (which critics rarely offer) for dealing with the problem.

There is one other contradiction in the critics' case as it relates to technological deficiencies. The current BMD system is criticized for being too expensive *and* unable to work effectively. But for any defence system to work more effectively the government would have to invest larger amounts in the program – obviously, more effective defences are more expensive defences, and vice versa (even allowing for Pentagon waste). In comparison with fifty interceptors, for example, ten thousand interceptors will clearly provide a more effective defence against fifty incoming missiles (even allowing for a low hit-to-kill ratio). Efforts to address the problem of overall expense (e.g., by lowering the costs) are likely to make the other problem (BMD effectiveness) worse. The question is whether the system is necessary – the costs are secondary. As David Warren asks, '[W]hat are taxes for? Surely being spared from nuclear incineration is on the short list of public desiderata. And why this sudden concern on the part of the world's pacifists for the beleaguered American taxpayer?'[24]

Perhaps the $80–$100 billion price tag is simply too much for BMD. But the more relevant question is how much is a well-functioning BMD system worth? Or, alternatively, how much is New York or Washington

Study of Global Issues, University of British Columbia, 15–16 February 2001. The list of participants (mostly BMD critics) involved in discussions on which the report was based include Dr. Rajesh Basrur, Professor Michael Byers, Dr. Ann Denholm Crosby, Mr. Anatoli Diakoy, Dr. Gloria Duffy, Mr. Francis Furtado, Mr. Li Genxin, Professor Mary Goldie, Mr. Joshua M. Handler, Dr. Brian Job, Dr. David Krieger, Dr. Paul Meyer, Dr. Don Munton, Mr. John Newhouse, Mr. Daniel Plesch, Professor John Polanyi, Mr. Tariq Rauf, Mr. Ernie Regehr, Professor Douglas Ross, Dr. Jennifer Allen Simons, Dr. Gordon S. Smith, Dr. John Steinbruner, Marianne Lykke Thomsen, Dr. Michael Wallace, Dr. Mitchel Wallerstein, Mr. Jon Wolfsthal, Mr. Stephen Young, Ms. Wang Xiaolin, Mr. Gu Ziping, Ambassador Sha Zukang.
[c]'Canada and Ballistic Missile Defence: Dilemmas and Options,' http://www.ploughshares.ca/content/MONITOR/monm99e.html.
[d]*Globe and Mail* (Toronto), 'Flying Madly Off in All Directions,' 21 July 2000.
[e]Peter Saracino, 'Canada Should Shun Missile System,' *Toronto Star*, 17 April 2000, A13.
[f]Editorial, *Globe and Mail* (Toronto), 'George W. Bush and the Real Missile Threat,' 3 February 2001.

TABLE 4.3
American Critics and Contradictions

1 [T]here is virtually no prospect, virtually no prospect, that the current system as being projected could hunt in the wild, if you will – could actually cope with a realistic scenario. It is easily penetrated … It is, quite frankly, technically and strategically irresponsible to say to our countries that we can protect them by waiting until the missile is fired and then hitting it before it lands. You can't do that. Anybody basing national defence on this basis should be indicted for malfeasance … Any plausible projection of the defence program being advanced by the advocates would put both the Russian and the Chinese deterrent forces in grave jeopardy over a period of time … It therefore virtually compels some kind of reaction both from Russia and from China if it is not contained by agreement. – Dr. John Steinbruner[a]

2 There are compelling reasons for this conclusion (re. serious problems with BMD). First, as currently designed, the U.S. BMD system will be vulnerable to relatively unsophisticated countermeasures. Second, the primary justification for BMD – the potential long-range ballistic missile threat – has been exaggerated. Third, the current U.S. BMD program, if fully realized, will gut the ABM treaty, jeopardize strategic stability and weaken the non-proliferation Regime. Fourth, U.S. BMD deployments will impede further significant nuclear weapons reductions by Russia and force Moscow to rely more heavily on dangerous rapid reaction command and control procedures. Fifth, U.S. BMD deployments will drive China away from arms control negotiations, provoke an increase in the scale of its on-going strategic force modernization, and bring Beijing to question the legitimacy of all space-based activities. Finally, BMD deployment will undercut alternative approaches to combating proliferation, such as arms control (including transparency measures), economic incentives, cooperative programs and export controls, all of which enjoy international support and legitimacy.[b]

3 Spurred on by exaggerated fears of North Korean missiles and Chinese nuclear espionage, the idea of deploying a Ballistic Missile Defense (BMD) is enjoying a political revival on Capitol Hill. The $60 billion spent on missile defense projects since 1983 has produced precious little beyond a string of cost overruns and technical failures. Deploying a missile defense system would violate the Anti-Ballistic Missile (ABM) treaty and risk sparking a new nuclear arms race. – William Hartung[c]

4 Clinton should not, under any circumstances, decide to deploy the proposed missile defense system. This system is a lemon and will never work effectively enough to make it worth the monetary and security costs it would entail … [T]he current national missile defense proposal is architected for an unproven system that may never work and is designed against a threat that does not – and may never – exist. At the same time, deploying that defense may well undermine the entire nonproliferation regime, severely damaging relations with Russia and China – the only two potential U.S. adversaries that already have the ability to attack the United States with long-range missiles … Russia has 6,000 long-range nuclear warheads, a force so large that no defense against it is practical. – Noam Chomsky et al.[d]

[a] *Standing Committee on National Defence and Veterans Affairs / Comité permanent de la défense nationale et des anciens combattants* (recorded by electronic apparatus),

worth to critics of BMD? For 20 poorly functioning interceptors $80 billion may be a waste of money, but $500 billion may be an acceptable investment in 350 interceptors that offer close to perfect protection. Obviously 10,000 interceptors will work even better than 350, and so on – critics have to assess how much we should value the target that 10,000 missiles are protecting. Criticisms about the costs of BMD should be kept separate from questions about success and failure, since success is directly tied to expenditures (if you don't want a lemon, buy a Mercedes). If critics are not willing to increase the BMD budget they cannot simultaneously criticize the system for being less than 100 per cent successful. Spending the entire $390 billion defence budget on BMD, for example, would obviously increase success rates.

Notwithstanding these obvious logical inconsistencies, which critics rarely acknowledge and almost never attempt to address, the limits of interceptor technology are offered as reason to scrap the entire program.

BMD Will Never Be 100 Per Cent Effective – But What Is?

The argument put forward by Burton Richter (winner of the 1976 Nobel Prize for physics) exemplifies the perfectionist critique:

> Assume for the sake of argument that an attack is composed of five missiles (the massive attack we used to worry about from the Soviet Union would have involved hundreds or thousands) and suppose that the chance of one interceptor finding and destroying the real warhead from one of the attacking missiles is four out of five, or 80 percent. Then, the chance of killing all five incoming warheads with five interceptors would be calculated this way: 0.8 for the first interceptor on the first warhead, multiplied by 0.8 for the second on the second, and so on for all five. Work it out, and

---

Tuesday, 29 February 2000. Government of Canada, available at http://www.parl.gc.ca/InfoComDoc/36/2/NDVA/Meetings/Evidence/ndvaev19-e.htm.

[b] Http://www.gsinstitute.org/laws.pdf.

[c] William D. Hartung, 'Star Wars Revisited: Still Dangerous, Costly, and Unworkable,' *Foreign Policy in Focus Brief* 4, no. 24 (September 1999) Washington, DC, and Silver City, NM: Interhemispheric Resource Center and the Institute for Policy Studies, ed. Martha Honey (IPS) and Tom Barry (IRC).

[d] Noam Chomsky, Baker Spring, Lisbeth Gronlund, Stephen Young, Jack Spencer, David Nyhan, 'National Missile Defense System,' The American Prospect Online, 18 July 2000 at www.prospect.org.

the probability of getting all five is about 33 percent, or a two-out-of-three chance that at least one of the incoming warheads will get through. Since one warhead can kill hundreds of thousands of people, that is not good enough.[25]

The assertion that we should scrap a system that approaches even 95 per cent effectiveness because at least 5 per cent of the missiles would get through is not an argument against deployment – for the same reason that similar statistics on lives saved and lost from heart surgery would not be grounds to scrap the procedure. That the U.S. should wait for the system to be fully functional before deploying makes no practical or scientific sense, since deployment is required for testing and improvement of the system. The twenty interceptors planned for Alaska and California in 2004 and 2005 are being deployed as an operational test facility for this very reason. As James Hackett explains, 'In order to get new weapons through the lengthy development process and into the hands of the warfighter, the Pentagon is adopting a new way of acquiring weapons called "evolutionary spiral development." This involves moving a new weapon into the field early and using it in operations for realistic and rigorous testing, then developing its capabilities further on the basis of that experience.'[26] The assertion that perfection is necessary before any deployment decision is made confuses the goals of BMD with the requirements for achieving the ends to which critics aspire. Obviously perfection is a desirable (although unattainable) objective, but the probability of reaching that level of success is enhanced by early deployment and testing.

Critics are fond of citing every technological impediment with BMD to make the point that the system is far from perfect. But the more important fact critics seem to overlook is that the complex technological requirements to sustain and perfect their preferred approach to strategic stability (i.e., MAD) have never been fully tested and continue to be plagued by questions of effectiveness. The report by Lt. Gen. Ronald T. Kadish on the Atlas ICBM program serves well to remind BMD critics that 'Atlas experienced 12 failures in its 2½ year flight-testing history. And the Minuteman 1 program suffered 10 failures in a 3½ year testing program.'[27] Both the Atlas and Minuteman programs were central to the evolution of the American nuclear force posture throughout the cold war and an important feature of MAD. In other words, the alternative MAD system that critics prefer also experienced several technological hurdles in its development phase.[28] Similarly, the managers of the once-secret Corona satellite program 'had to survive 12 failures and mishaps before they orbited [the] first operational reconnaissance satellite (Discoverer 14).' Kadish goes on to point out that some of the parallels between the Corona and BMD programs are striking. 'Among other things, booster development was in its infancy, and today, although we have come a long way, building reliable boosters for our missile programs continues to be a challenge.'[29] Yet, despite these failures, the notion that the Americans should kill or dismantle the U.S. satellite launching program is absurd – the scientific priority will always be to fix the problem, given the crucial importance of maintaining a satellite program.

Failures are not only commonplace and predictable but essential to any development program. To cite these failures as a basis for premature closure of BMD experimentation and development is not only unreasonable, excessively short-sighted, and potentially dangerous, but it is also entirely inconsistent with widely accepted patterns of how technology should develop. In the end, the technological-deficiency argument should always be subordinate to the issue of security – if the system is needed because it promises to enhance security in ways that current strategies cannot, the costs and barriers to technological progress are inconsequential. The same set of principles and strategic imperatives should guide debates over BMD. If the contributions to U.S. and global security outweigh the risk, and if those risks are more acceptable than the risks associated with the status quo (i.e., doing nothing), costs and technological hurdles should remain secondary to development and deployment. Technological failures, at this stage in the program, are irrelevant.

Even if we acknowledge the enormous technological hurdles that remain, and even if we accept the fact that the current system is far from perfect, the more relevant questions for policy makers are, How close to perfect does the technology have to be to be useful? How prudent is deployment today if interceptor technology, speed, and precision will continue to evolve and improve? Critics have never clarified how close to perfect a defence system has to be to be useful or worthy of funding. Even the staunchest critics of BMD assign at least some probability of success. The question is whether that probability is worth an investment of 2 per cent of the U.S. defence budget, and the answer from the current U.S. administration is yes! As Defense Secretary Donald Rumsfeld points out, 'Anyone who's ever been involved with research and development activities knows that it is highly unlikely that in the first try someone will develop something that is perfect ... Most systems are imperfect; that is to say, for every offense, there's a defense, and vice versa. But what we're talking about here is a new set of capabilities to ... dissuade or deter ... as well as to defend against a growing threat in the world. And ... they need not be 100% perfect, in my opinion, and they are certainly unlikely to be in their early stages of evolution.'[30]

Measuring Probabilities of Success and Failure

With respect to probabilities of success, most critics in the scientific community intentionally sidestep the link between the numbers of interceptors launched at a single target and the probability of a successful hit. Current plans are to use three or four interceptors for each target, applying a 'shoot-look-shoot' strategy. If interceptors approach 80 per cent accuracy, two or three attempts would increase the probability of a successful hit to 96 per cent and 99 per cent, respectively. For the sake of argument, accept the most pessimistic assessment of the technology in 2005–2008 – say, a 71 per cent success rate (consistent with the current record of five successful fully integrated BMD tests out of seven attempts from 1999–2002).[31] Using the same logic and mathematical formula used by Nobel laureate Burton Richter (above), and assuming a 71 per cent success rate for a single interceptor, the second, third, and fourth attempts increase the probability of a successful hit to 91.6 per cent, 97.56 per cent, and 99.29 per cent, respectively.[32] Even if we experience a series of cataclysmic testing failures in the next round of tests that drop the percentage to, say, 30 per cent, the second, third, and fourth attempts would still increase the probability of a successful hit to 51 per cent, 66 per cent, and 76 per cent, respectively.

The question American policy makers are facing (and one critics refuse to address) is whether the Pentagon should deploy a system that protects the population from, say, seven out of ten incoming missiles (the equivalent of almost ten out of ten missiles using four interceptors for each attacking missile), or scrap the entire program and, in so doing, face a 100 per cent probability that the population will suffer the effects of all ten missiles if attacked. The technological case in favour of deployment is even stronger when one considers that the current success rate of 71 per cent is far lower than the estimates offered by both critics (Burton Richter suggests 80 per cent) and proponents in the scientific community, and significantly lower than estimates cited by scientists involved in all aspects of the BMD testing program.

Many sceptics who dismiss BMD suffer from static impressions of progress and overlook constant improvements in interceptor and decoy-identification technology. That BMD has not yet reached 100 per cent accuracy is irrelevant. What is relevant is whether the pace of innovation is such that BMD will, at some future point, produce a high-enough probability of success to warrant deployment – 71 per cent for a single interceptor is arguably high enough (in fact, so is 30 per cent). The current program provides adequate promise of security to offset the costs and potential risks, and alternatives (such as constructive engagement, transparency, economic sanctions, and so on) are neither cheaper nor more likely to accomplish the same (or other) security imperatives.

Current and future decisions regarding research, development, and deployment should depend on these calculations; they should not (and most probably will never) depend on the status and limitations of present-day technology, especially if the technology is evolving at such a rapid pace. The current exo-atmospheric kill vehicle, for example, weighs about 130 pounds and is approximately fifty-four inches long. It is designed to move towards the target, register, and analyse movement with miniaturized computer sensors, relay that information back, and fire thrusters to change direction to kill a target on impact. As Broad notes, designers praise the current kill vehicle as 'the apex of miniaturization and accuracy.'[33] By contrast, the first successful hit-to-kill interceptor (1984) required 'a 15-foot-wide steel umbrella to raise the odds of collision.' According to Donald R. Baucom, an antimissile historian at the Pentagon, '[T]he new kill vehicle's deadly agility is rooted in its miniaturized parts and light weight – pounds versus earlier tons. So firings of its four small thrusters produce fast manoeuvres ... and greater accuracy.'[34] In addition, the newer generations of sensors will have chips of up to 1,048,576 pixels, 'far beyond the current level of 65,536 pixels and the 72 pixels of the 1984 success.' Sensor resolution is also expected to increase in the near future, helping the next generation of interceptors to distinguish between warheads and decoys.

Perhaps the most important point about the constantly improving nature of the technology is that failures are essential to the process, and some failures are more relevant than others.[35] For example, in December 2002, the U.S. Missile Defense Agency conducted IFT-10 (the tenth integrated flight test).[36] The tenth attempt failed because the kill vehicle designed to hit the target did not separate from its booster rocket. Technically, the BMD technology was never really tested – what was tested (and failed) was conventional missile and booster technology that has been around for decades. In fact, as James Hackett reminds us, the other three failures 'stemmed from low-tech, quality-control problems that are easily fixed. One was a clogged cooling pipe, while another was a failure to transmit data, although the same equipment had worked fine for years on Minuteman missiles. In the most recent test, early indications are that a loose wire may have prevented the signal needed to fire the mechanism that separates the kill vehicle from the booster rocket.'[37] Everything else worked pretty well.

Countermeasures and Decoys

Perhaps the greatest technological challenge for BMD is the system's capacity to disregard countermeasures and decoys. According to some critics, decoys are extremely hard to detect and cheaper to produce than interceptors. Thus, BMD will always be a waste of time and money. There are at least three problems with this argument. The first is logical: if countermeasures are so easy to develop and deploy (especially for major powers with advanced weapons programs like Russia and China), then, once again, there is no reason to expect automatic proliferation by these states. Relatively inexpensive decoys are more than sufficient.

Second, critics tend to downplay the problems developing countries face when incorporating countermeasures into a well-functioning ballistic missile program. They are also quick to acknowledge the costs and limitations the world's richest and most scientifically advanced country faces when developing interceptor technology. The only states with the capability to develop and deploy decoys easily are the very states that don't need them; they already have a stable and credible retaliatory threat. Not true, claim the critics. Developing countries also have the technology to build and deploy decoys to circumvent BMD interception. If that is true, they most certainly have the capacity to produce ballistic missiles as well.

Third, if ballistic missile threats are real, it makes perfect sense for the United States to increase the costs incurred by states of concern by forcing them to deploy decoys and countermeasures when developing their ballistic missile programs. Just because countermeasures can be deployed does not mean the United States should accept defeat, scrap BMD, and give potential adversaries the option of disregarding those added costs. Scrapping BMD simply means that the ballistic missile threat will continue to proliferate in ways that are more affordable. The harder it is for potential opponents to threaten (implicitly or explicitly) the use of these weapons, the better. On the other hand, if ballistic missile threats are indeed fabrications to justify military expenditure on BMD, and if there is no reason to be worried about the pace of ballistic missile proliferation, critics should be prepared to provide a more persuasive defence of that position by confronting the evidence to be discussed in more detail in chapter 5.

A related set of criticisms focuses on controversies surrounding the BMD testing program and the standards used by the Pentagon to evaluate success and failure. Some critics have accused the Pentagon of fixing the tests in order to facilitate a successful hit and, therefore, congressional funding. I am not a ballistics scientist and do not presume to understand the physics involved in the program, but it seems perfectly reasonable to develop and improve technology through stages by first identifying BMD's limitations; the system's maximum potential is irrelevant at this time. If the initial tests were designed to assess an interceptor's capacity to hit hundreds of targets and twenty to thirty different decoys, and it fails, the results provide virtually no useful information about what went wrong, why certain decoys may have been missed, or what needs to be improved to move the technology forward. On the other hand, if a single interceptor misses/hits a single target, or misses/hits a simple decoy, the information obtained from that test is far more useful and relevant.

Consider the following analogy. Two world-class baseball pitchers are competing for a $1 million prize. The challenge is to hit a two-foot-by-two-foot target, approximately a hundred feet away, with a single marble. Each pitcher is given one month to practise, along with thousands of marbles with which to prepare. One pitcher practises for the entire month by grabbing handfuls of marbles and launching them at the target, hitting the target with one or more marbles each and every time. The second pitcher practises by launching marbles at the target one at a time, judging distance, velocity, angles of decent, and so on. Obviously both pitchers acquire some information about these important components of a successful launch, but one is more likely to acquire more useful information to launch the single marble at the crucial time. Assuming all other skills are equal, which pitcher would you put your money on? Pushing the analogy a little further, imagine a second challenge to hit a moving target with either one of two marbles, both launched at the same time. If the respective training regimens remain the same (with one pitcher launching handfuls of marbles and the other only two at a time), the second pitcher is more likely to obtain the crucial information needed to win again. If the Pentagon's current plans are to use up to four interceptors for each target (not fifty or sixty), it makes perfect sense to learn from the bottom up, not from the top down.

Accusations about simplifying BMD tests to increase the probability of success miss this key point about how scientific knowledge progresses. The same criticism also tends to underplay the incredible scientific achievements associated with even a very simple test. The system cannot achieve its maximum potential today, but we should not evaluate the performance of current technology based on the requirements for a system envisioned in five, eight, or ten years – interceptor technology, like any technology, will continue to improve.[38] Again, the question for policy makers is whether deployment today makes sense considering the probability of success, the probability of improving future systems through trial and error, and the costs and benefits to overall security when compared with alternative strategies.

Leaving aside the obvious problems with the technological-deficiency arguments, and despite the fact that critics rarely evaluate the components of the technology they are so willing to dismiss, BMD is often compared to the Maginot Line. But, as Caleb Carr correctly points out, '[T]he Maginot Line was not a technological failure, but a strategic one. This marvelous system of mammoth bunkers and forts, named for the French minister of war André Maginot, did not fail to stop the German army: Germany's generals were so impressed by the line that they never seriously considered a frontal assault against it. Instead they went around it ... The actual flaws in the Maginot Line were that it was not extensive enough ... These are also the true weaknesses of the missile defense shield.'[39] The prospects for BMD failures must be assessed in similar contexts. For example, the failure in July 2000 had no significant effect on the project, because scientists acquired no new (or crucial) information about the components central to the functioning of BMD. The kill vehicle failed to separate from the booster rocket, but that technology has been functioning successfully for decades in practically every major space mission, satellite launch, and shuttle project in the history of the National Aeronautical Space Administration. Ironically, the only clear failure, aside from booster separation, was the balloon designed to simulate a basic countermeasure. This is likely to be far more upsetting to critics who offer the simplicity of countermeasure technology as a reason to scrap the program – if the United States is having a hard time deploying a decoy, what does that suggest about the capacity of developing countries to accomplish the same task?

Financial and Political Costs

The most straightforward way to evaluate the quality of any significant defence expenditure is to estimate the money involved in the program (financial costs); the contribution the program makes to state security (benefits); the probability those benefits will be realized (technology); the potential dangers associated with deployment (risks); and the probability those costs will be incurred (security costs). The results should then be compared to other security programs that promise to accomplish the same objectives, for the same (or lower) costs, with the same (or a higher) probability of success. Finally, the estimates should be compared to those for alternative programs designed to address a different set of equally or more important security objectives. All things considered, programs with higher utility for a state's security should be funded.

With respect to the financial costs, the current proposed U.S. investment in BMD ranges anywhere from US$80 to US$100 billion over ten years, or about $8 billion a year between 2002 and 2011. That is about 2 per cent per annum of a defence budget of approximately US$397 billion (FY 2002–03), or approximately 0.3 per cent of the total federal budget of $2 trillion. The numbers are likely to be lower as the defence budget increases to FY 2004: $405 billion (1.9 per cent in FY 2004), $426.2 billion (1.8 per cent in FY 2005), $447.5 billion (1.7 per cent in FY 2006), and $469.6 billion (1.7 per cent FY 2007). Nevertheless, the average yearly expenditure of 2 per cent serves as a useful benchmark to make a few important points.[40]

For critics, 2 per cent of the defence budget is a waste of money (high financial costs) for a system aimed at non-existent threats (no benefits) that will not work (technological limitations) and will probably create a host of other problems (security costs) as Russia and China automatically retaliate by proliferating nuclear weapons (high risks). I have already responded to these assertions by outlining above in this chapter important flaws and inconsistencies in the critics' case. The remainder of this section will focus on the four items I have not yet addressed: financial costs; alternative programs that critics claim are more suited to addressing WMD threats; other programs designed to address different (more important) security threats; and more important investments to resolve other non-security problems.

Financial Investment in BMD

How cost-effective is the American investment on BMD? With the preceding analysis in mind, and for the sake of argument, accept an optimistic estimate of evolving threats tied to the proliferation of ballistic missile technology – say, only a 10 per cent probability that by 2005 to 2008 one or more states of concern will have developed and deployed one or more medium- to long-range ballistic missiles capable of reaching the United States or one of its allies (note that even critics of BMD would find these numbers very low). The current annual investment is quite a bargain, especially when one looks at where the remaining 98 per cent of the defence budget ends up and the correspondingly limited contribution much of that investment makes to United States security. Consider, too, that lawmakers, according to Jack Spencer, 'have packed a record amount of "pork barrel" spending into the proposed budget bills, including $18.9 million for the Puget Sound commuter rail project in Senator Patty Murray's state of Washington, $20 million for "railroad rehabilitation" in Senator Ted Stevens' Alaska, and $1 million for the Tuscaloosa City Riverwalk and Parkway development in Senator Richard Shelby's Alabama.' Spencer's review of the defence budget tracked about nineteen thousand in-district projects worth about $280 billion – by comparison the relatively minor investment in BMD represents a small fraction of that amount a year.[41]

These expenditures aside, even a 5 to 10 per cent probability of a ballistic missile threat to the U.S. in the next five years represents a far more significant and relevant concern when viewed in terms of the overall defence profile: that is, all American security, defence, and threat scenarios the government considers relevant and the consequences in damage and lives should those threats become real. If only those threats the United States can tackle with at least some likelihood of success are included, the 10 per cent probability of a ballistic missile threat becomes even more relevant. Put differently, a 10 per cent probability of another terrorist attack in the next three years resulting in 5,000 deaths should be no less relevant to policy makers than a 1 per cent threat with the potential to produce 50,000 casualties. And the latter should be less relevant to budget allocations than a potential attack that could produce 500,000 casualties. The crucial question policy makers will continue to confront is whether the strategy designed to address the 1 per cent threat has a higher probability of success than corresponding funds and programs to stop, for example, terrorism.

Would another $80 billion invested in counter-terrorism provide an equal (or better) return in security (with a higher probability of success) than stopping even a single missile with a well-functioning BMD system at 71 per cent effectiveness – or close to 99 per cent effectiveness using four interceptors? These calculations should be the reference point when assessing the investment in BMD – it should never be viewed in isolation.

Finally, critics rarely if ever include in their calculations the savings generated by BMD deployment, for obvious reasons. Using existing and pending bilateral agreements as a benchmark, the U.S. Defense (and Energy) departments would see significant cost savings in maintaining lower levels of U.S. nuclear and strategic forces. For example, START II levels of thirty-five hundred warheads by the year 2007 would produce a $700 million/yr savings through 2008, and approximately $800 million/yr thereafter; START III levels of twenty-five thousand warheads would save $1.5 billion/yr; and START IV levels of a thousand warheads would save close to $2 billion/yr (1998 dollars). The upshot is that BMD would actually cost only 1 per cent of the U.S. defence budget, not 2 per cent.

Alternative Strategies to Combat WMD Proliferation

Most critics concur that Washington tends to exaggerate the rogue-state threat, but few are prepared to argue that there is nothing to worry about. The real debate, then, is about time and the pace and evolution of the threat. Take, for example, a recent report produced by nine ballistic missiles experts from Germany, Norway, Russia, and the United States. Among other observations the report concludes that new threats from North Korea will be from very few missiles, which will be capable of carrying smaller, less accurate payloads – none of this is likely to reassure proponents of BMD. In any case, these nine experts share the responsibility of estimating the appropriate amount of funding for their preferred solution to the problem, even if their estimate of the probability of a ballistic missile threat in 2005 to 2008 is, say, only 1 per cent or less. How much of the United States defence budget would critics be willing to allocate to address a 1 per cent threat, using whatever program they prefer? If they conclude that 1 per cent (or $4 billion) of the defence budget would be enough to implement their preferred solution, they have an obligation to explain what the money buys, how the proposed investment enhances United States security, how likely it is that the program will succeed, and present at least some evidence that similar programs and investments have worked in the past. Ultimately, critics should be prepared to evaluate their alternatives using the same standards they apply to BMD, with reference to the same criteria for assessing costs, risks, and probabilities of success and failure.

The alternatives proposed by critics usually include some combination of improved transparency, weapons verification, monitoring, import/export controls, and a host of other diplomatic (for example, constructive engagement) and coercive diplomatic (for example, economic sanctions) strategies – a more detailed evaluation of these alternatives will follow in chapter 5. Presumably these strategies are more constructive because they address the proliferation problem from the demand and supply side and, in the process, produce none of the costs associated with BMD. But these popular policy pronouncements rarely include the details policy makers need to compare them to BMD.

If one were to look at the money already invested, not only by the United States but also globally, on programs tied to transparency, monitoring, verification, and economic sanctions, I suspect the total far exceeds the proposed investment in BMD. Spending on the verification of WMD, such as outlined under the START treaties, costs the U.S. approximately $8 billion annually (the current annual investment for BMD). The cost of one satellite and its launch to monitor such agreements is $1.25 billion.[42] What exactly do we have to show for all these efforts, strategies, and investments? Even the most optimistic take on the proliferation record is not particularly encouraging – witness the recent failures of inspection, monitoring, verification, and sanctions regimes in Iraq, North Korea, India, and Pakistan. Where exactly is the proof that another $80 billion invested in these strategies is more cost-effective, or more likely to address (prevent) ballistic missile threats circa 2005 to 2008? How exactly will diplomatic efforts to engage Iran or North Korea prevent them from developing their nuclear programs, and what is the probability of success – is it 71 per cent, 99 per cent? Wouldn't improved economic relations provide the same capital required for these states to augment their respective nuclear and ballistic weapons facilities? How probable is that outcome?

Other More Serious Threats

Critics claim that there are other ways to detonate a nuclear device that cannot be stopped by BMD. There are at least two problems with this argument. First, proponents of BMD would agree that asymmetric threats (e.g., terrorist attacks) represent a serious problem, and that other methods of delivering WMD could be used.[43] The fact that there are other ways to detonate a nuclear device, or other threats to worry about, says absolutely nothing about the utility of BMD – it simply implies that investing solely in BMD and ignoring these other asymmetric threats is likely to be dangerous.[44] It would certainly not be unreasonable to assume that the U.S. government is investing a good portion of the remaining 98 per cent of the defence budget on these other threats, especially after 9/11.

Second, the argument that money would be better spent on these other threats overlooks an important point about the nature of security threats facing the U.S. Terrorist groups who are committed to their respective causes (and who are unified in their collective hatred of the U.S.) will continue to exploit the weakest component of any defensive system, whether it is border security or BMD. Consequently, the more money spent on counter-terrorism by investing in airport and border security, the more likely it is that other avenues will be used by terrorists. Similarly, the less money spent on BMD, the more money will be invested by rogue states and terrorists to develop and deploy ballistic missiles. It's perfectly rational to do so. There is a self-fulfilling prophecy associated with military investments – the more one invests in defending against one type of threat, the more likely it is that opponents will focus on other approaches. If Washington decided to move away from BMD, ballistic missile proliferation would logically become more, not less, likely – rogue leaders would be more inclined to invest in a ballistic missile program when it is less expensive (i.e., when countermeasures are not needed) and when it is more likely to work (i.e., when BMD doesn't exist).

In sum, there certainly are other pressing threats out there, and other methods of delivering strong messages to the United States and its allies, but it would be irresponsible, to say the least, for an American president to focus exclusively on these threats to the exclusion of those associated with ballistic missiles. Even if nuclear threats are less likely the effects would be catastrophic if they were not stopped or deterred.

Other More Useful Investments: Human Security

Given other more pressing security priorities, some have argued that BMD investments should be redirected. Take the recommendation by

Michael Wallace as an example: '[I]t was estimated by a Greenpeace activist from the Cook Islands that the $100 million wasted on the failed July 7, 2000 [BMD] test could have built and run a hospital and provided free university education for the entire population of the Cook Islands for many decades. Surely, American security would be better served by spending money on such worthy projects than by a futile attempt to create an unattainable fortress America.'[45] Every billion spent on BMD, Wallace argues, 'is money that cannot be devoted to improving education, strengthening health care, reducing the debt, shoring up Social Security and Medicare, and developing a sensible non proliferation policy.'[46] Statements like these are assumed to establish the point about a better use of the money. But consider the following.

For fiscal years of 1999–2000 and 2000–1, the U.S. contributed $2.4 billion and $2.56 billion, respectively, to the United Nations budget. Given the pace of current peacekeeping operations, the U.S. expects the UN budget will reach nearly $3 billion by the end of 2003 and increase even more in subsequent years, given new deployment commitments in Afghanistan and Iraq. American peacekeeping contributions alone amounted to approximately $216 million for 1999.[47] Now, where exactly does this money go? What does it pay for? Does this investment make an unambiguous contribution to security? And how do we know? Critics like Michael Wallace are fond of claiming that there are so many other, more relevant, threats that need immediate attention and that should receive the bulk of existing security-related expenditures – health care, education, drug trafficking, environmental degradation, world hunger and disease, AIDS, intrastate ethnic conflict, refugees, terrorism, small-arms trafficking, intelligence, chemical and biological weapons proliferation. But the existence of a threat says nothing about how the problems can be addressed by redirecting these funds, whether transferring BMD's $80 billion from the defence budget is a better and more cost-effective investment in security, or whether the transfer will do anything to address evolving ballistic missile threats (however remote they may be), since these traditional threats will not go away by spending more on AIDS.

These tough questions and associated decisions are almost never addressed by critics who claim to have a much clearer understanding of where $80 billion should be spent. While we're at it, why not invest the entire $396 billion defence budget in development assistance and other human security spending priorities? But before we do, consider the total combined investment in human security by the largest American and European aid agencies over the past ten years (the same time frame for BMD's $80 billion/ten-year procurement). Now, track the trends in drug trafficking, environmental degradation, AIDS, ethnic conflict, and so on, during the same period. Despite the billions of dollars invested globally by several countries, hundreds of international aid agencies, and non-governmental organizations, all these problems have arguably become worse. Exactly how would another $80 billion help resolve these and similar human security problems? Are the benefits (and attendant probabilities of realizing them) likely to make an equal (or greater) contribution to United States security than $80 billion on BMD? And, most important, how does one measure success and failure? Like BMD, decisions about which program to fund, or which security threats to address, must be based on a balanced evaluation of the costs, risks, and the probability of accomplishing core security objectives.

Unfortunately, those who offer these superficial 'solutions' to global problems rarely accept for themselves the obligation to meet the same standards required of proponents of BMD. They also have an obligation to explore in more detail the long-term implications of addressing the security threats they see as more important than prolif-eration. Many of these other security threats (underdevelopment, popu-lation pressures, AIDS, resource depletion), for example, exacerbate levels of international conflict and competition for pieces of the global resource pie. This, in turn, produces additional incentives for some developing nations (e.g., North Korea) to acquire the tools to improve their bargaining leverage. Indeed, one of the most important lessons learned by nuclear-threshold states when assessing the costs and ben-efits of crossing the line is that it pays. Consider for a moment the attention paid today to India, Pakistan, and North Korea. Initially, international reactions took the form of diplomatic pressure, economic sanctions, and condemnation, but the approach has now shifted to one of engagement, dialogue, and economic cooperation. What are the lessons for Iran, Libya, Syria, and others?

This is not to suggest that we should stop funding programs to combat other serious problems. But critics should acknowledge that many of these other problems and related solutions are incredibly complex, perhaps even more complex than interceptor technology. Suc-cess often depends on the support, sacrifice, and sponsorship of ruling political and military elites in these countries who may have other priorities – environmental security and controlling child labour, for example, are not likely to be priorities in a country whose leaders are trying to encourage foreign investment through cheap labour and low production costs. In addition, many human security threats are highly interconnected and interdependent, which further complicates the search for cost-effective solutions. Critics will continue to make reference to other more important security concerns when criticizing expenditures on BMD, but they should at least be honest about the costs and success rates associated with resolving the problems they choose to highlight.

In sum, if the relative importance of these other threats is debatable, and if these problems are very complex, interdependent, and difficult to resolve because of existing international and domestic impediments, and if investing in them instead of in BMD does absolutely nothing to resolve ballistic missile threats, how exactly does transferring $80 billion in defence funds to human security make any practical sense?

Powerful Influence of the U.S. Military Industrial Complex (MIC)

The rogue-state portrait of North Korea, most often put forward by defence and military officials in Washington, is formed by deeply held convictions that this and other states are acquiring ballistic missile technologies and, at some point in the future, will threaten to use them. Among the more popular explanations for Washington's refusal to budge on the rogue-state or 'axis of evil' portraits is, in a word, money: the Pentagon and its powerful military industrial complex (MIC) of defence-related industries need rogue threats to justify expenditures on the latest military technologies.

Put differently, Washington is compelled to exaggerate (even fabricate) non-existent threats for economic reasons. What better way to justify the $80 billion expenditure for BMD than by convincing as many · people as possible that Iraq, Iran, and North Korea are not particularly friendly. Notwithstanding the fact that Washington, Moscow, most European powers, and the UN weapons inspectors for years acknowledged as real the threat from rogue states such as Iraq and North Korea, many critics continue to defend the MIC thesis. There are at least four problems with this argument.

First, major defence contractors tied to the MIC will make money from the $80 billion defence procurement regardless of whether BMD goes forward. The money is allocated from the U.S. defence budget; if BMD is scrapped tomorrow, none of that money will be spent on roads, health care, education, or development assistance. The simplicity of the MIC thesis also precludes identifying any useful answers to why BMD is preferred to the more ABM-compliant sea-based or boost-phase missile defence. Presumably these alternative systems are just as likely to generate impressive profits for U.S. defence contractors.

Second, if rogues are indeed creations of the U.S. MIC in search of profits, it is not at all clear why every major UN arms control report produced by UNSCOM and UNMOVIC between 1991 and 2003 (including Hans Blix's many reports under UN Security Council Resolution 1441 and dozens of similar reports about nuclear, biological, and chemical weapons proliferation worldwide published by the International Atomic Energy Agency) outline very clear and explicit concerns about major violations by Saddam Hussein and other rogue leaders. If the MIC thesis is valid, Russia, the UN, and every European power that signed on to UNSC 1441 and IAEA reports have all been co-opted by powerful American defence contractors. A German intelligence report produced in February 2001 concluded that 'despite the destruction of much of the country's arms industry by United Nations control teams, Iraq was again close to achieving its aim of producing a missile with a 3,000-km (1,864-mile) range, (which) would be capable of hitting the German capital or indeed the German Security Service headquarters in Pullach outside Munich. By 2005 Iraq should be able to launch a missile containing at least 1 kilogram (just over 2 lb) of anthrax bacteria.' The report estimated that 'if such a payload would be dropped on a German town it would kill between 70 and 80 percent of the inhabitants within a few days.'[48] Obviously European intelligence organizations concluded that rogue threats are not entirely America's problem.

Third, if military industries make an $80 billion profit in the process of developing an effective and successful defence system (i.e., one that contributes to U.S. security with a high-enough probability of success), any questions about who profits are entirely beside the point. Would critics be more favourably disposed towards BMD if it was produced by a non-governmental organization in an environmentally friendly plant where profits fund culturally sensitive programs to feed people in less developed countries? Probably not, because, according to critics, 2 per cent of the U.S. defence budget is a waste of money for an untested system aimed at non-existent threats that will probably create a host of other proliferation dilemmas for Russia and China. But none of these issues (i.e., technological limitations, fabricated threats, and security risks) has anything to do with who actually gets paid for BMD. Questions about the utility of BMD, therefore, are prior to and far more important than questions about who makes a profit.

Fourth, the fact that U.S. defence contractors make a profit from selling military equipment to the Pentagon is irrelevant if the Pentagon, CIA, State Department, Congress, and the White House truly believe the project enhances U.S. security. Most officials in Washington (and a large segment of the American public) believe that BMD is an effective way of dealing with ballistic missile proliferation and, after 11 September, the increasing probability that these weapons will be used against American citizens someday. The only way to verify the 'fabricated threats' thesis, then, is to demonstrate that U.S. officials are not at all concerned about the proliferation records of Iraq, Iran, Libya, Syria, and North Korea, and are not at all persuaded by the evidence of evolving nuclear and ballistic missile threats. On the other hand, if Washington's fears in this regard are real, the 2 per cent expenditure (and associated profits compiled by the MIC) is entirely tangential. And with respect to these fears, a five-minute Web search using keywords *nuclear proliferation* or *ballistic missile proliferation* should provide optimists with evidence to reconsider their position on rogue states and the nature of the threat (please refer to chapter 5 for details).[49]

In sum, a balanced assessment of U.S. national security policy demands a sophisticated evaluation of the issues. The analysis must be kept separate from unsubstantiated claims and assumptions about abusive and corrupt defence industries or the hegemonic and imperialist predilections of an evil American empire. Many of the arguments put forward by critics often appear to have less to do with the realities of proliferation as outlined in UN (UNSCOM, UNMOVIC and IAEA) documents and reports, and more to do with the latest efforts to slam American defence spending – not because BMD is wrong or unnecessary, but because it requires funds for the military and, by definition, is a waste of money. I suspect proponents of the MIC thesis have never met a defence expenditure they liked or a weapons system they were prepared to fund. If they have, the system they put forward would almost certainly share many of the same qualities, costs, and security risks associated with BMD. If they have never met a defence expenditure they are prepared to support, I suspect their rejection of BMD has virtually nothing to do with missile defence or WMD proliferation.

Exaggerated Rogue Threats from Rational Rogue Leaders

Exaggerated Threats?

Many critics acknowledge WMD proliferation but choose to interpret this behaviour as relatively benign or assume it can be addressed with the right mix of diplomatic pressure. David Mutimer, for example, acknowledges the 'facts' of the rogue state threat but claims that U.S. officials tend to exaggerate its implications:

> The rogue state, and its newer incarnation, the state of concern, needs, therefore, to be seen for what it is: the creation of the United States military to justify its claim on resources ... The rogue state, however, is a myth. It is not mythical in the sense that it is not real, but rather in the sense that it has been vested with a totemic importance by the United States ... Rogues ... are not a lie told by knowing capitalists in an instrumental fashion to hoodwink Congress into passing over-inflated contracts ... I am not arguing that the United States fabricated evidence, but rather that it produced a particular frame within which to interpret that evidence ... The issue, therefore, is not the evidence but rather how the 'facts' are 'evidence' of a particular form of threat labelled 'proliferation' by actors labelled 'rogue' ... At issue is not 'the facts' but the ways in which those facts are assembled and the interpretation that is given to them.[50]

Mutimer develops an interesting argument about how U.S. leaders make conscious choices about who and what constitutes a 'threat' – it's entirely up to them to decide who is or isn't a threat. He concludes that some (if not most) of the blame for creating the 'rogue' problem rests with the United States. The threat, in other words, is a perfect example of a self-fulfilling prophecy tied to efforts by the American defence community to fill the 'threat blank' after the cold war. Understanding the conditions under which certain threats (and enemies) are 'in' and which are 'out' – and the complex combination of political, military, and economic factors that influence these decisions – is an important area of research. But a debate about the 'origins' of the current or future security crisis is a separate issue – the key facts regarding actual proliferation of nuclear, chemical, biological, and ballistic missile technology are not in dispute. Once you make that fundamental concession, much of the BMD debate is essentially resolved, because the need for defence is a given.

Ballistic missile proliferation is difficult to deny. Immediately after the 2003 Iraq war, sceptics were quick to deny that Saddam Hussein and his sons were committed to developing WMD, and prematurely concluded that the U.S. government fabricated the entire WMD threat. But consider the evidence of material breach of UNSC 1441 outlined in the following excerpts from the interim report submitted to Congress by inspector David Kay (head of the Iraq Survey Group):

- A clandestine network of laboratories and safehouses within the Iraqi Intelligence Service that contained equipment subject to U.N. monitoring and suitable for continuing CBW (chemical biological weapons) research.
- A prison laboratory complex ... that Iraqi officials working to prepare for U.N. inspections were explicitly ordered not to declare to the U.N.
- Iraq concealed equipment and materials from U.N. inspectors when they returned in 2002 (including) a collection of reference strains that ought to have been declared to the U.N. Among them was a vial of live C. Botulinum Okra B. from which a biological agent can be produced. This discovery (was) hidden in the home of a BW scientist.
- The scientist who concealed the vials containing this agent has identified a large cache of agents that he was asked, but refused, to conceal. ISG is actively searching for this second cache.
- New research on BW-applicable agents, Brucella and Congo Crimean hemorrhagic fever, and continuing work on Ricin and Aflatoxin were not declared to the U.N.
- Iraq after 1996 further compartmentalized its program and focused on maintaining smaller, covert capabilities that could be activated quickly to surge the production of BW agents.
- Documents and equipment, hidden in scientists' homes, that would have been useful in resuming uranium enrichment by centrifuge and electromagnetic isotope separation.
- A line of UAVs (unmanned aerial vehicles) not fully declared at an undeclared production facility and an admission that they had tested one of their declared UAVs out to a range of 500 km, 350 km beyond the permissible limit.
- Continuing covert capability to manufacture fuel propellant useful only for prohibited SCUD-variant missiles, a capability that was maintained at least until the end of 2001 and that cooperating Iraqi scientists have said they were told to conceal from the U.N.

- Plans and advanced design work for new long-range missiles with ranges up to at least 1,000 km – well beyond the 150-km range limit imposed by the U.N. Missiles of a 1000-km range would have allowed Iraq to threaten targets throughout the Middle East, including Ankara, Cairo, and Abu Dhabi.
- Clandestine attempts between late 1999 and 2002 to obtain from North Korea technology related to 1,300-km range ballistic missiles – probably the No Dong – 300-km range anti-ship cruise missiles and other prohibited military equipment.
- Systematic sanitization of documentary and computer evidence in a wide range of offices, laboratories and companies suspected of WMD work. The pattern of these efforts to erase evidence – hard drives destroyed, specific files burned, equipment cleaned of all traces of use – are ones of deliberate, rather than random, acts.

Among the many obvious points to emerge from David Kay's interim report is that information about materials proscribed and explicitly prohibited under UNSC 1441 (and every other unanimously endorsed resolution passed by the UN Security Council on Iraq since 1991) were intentionally hidden from UNMOVIC inspectors in much the same way Saddam Hussein's massive WMD program was hidden from UNSCOM inspectors throughout most of the '90s. The patterns and intentions cannot be denied, even by the staunchest critics of American foreign policy.

The significance of this post-war evidence must be measured in terms of the material breach conditions stipulated in 1441. There was overwhelming consensus among intelligence agencies worldwide (including UNMOVIC's chief weapons inspector, Hans Blix, as outlined in every one of his reports) that the Iraqi regime was hiding something (see chapter 7, table 7.1, for details). While the global intelligence community is now confronting the discrepancies between pre- and post-war expectations of stockpiles of proscribed munitions, finding huge stockpiles of WMD *was never the benchmark for establishing material breach or the requisite justification for intervention to enforce compliance with international law.* Hussein's regime was never found by Hans Blix to be in compliance with resolution 1441, and the evidence compiled after the war confirmed that Blix was right to withhold that judgment.

Unfortunately, evidence for material breach is typically accepted or rejected depending on one's position on the war, whereas one's position on the war should, logically (and legally), depend on the facts for or against material breach. It bears repeating that the dispute was not about whether Iraq was attempting to hide proscribed weapons but whether the WMD that most intelligence agencies expected to find were enough to justify a war. U.S. officials looked at the intelligence and respectfully disagreed with France, Germany, and Russia over its relevance to American security, a disagreement that was compounded by the fact that these other states did not suffer the consequences of 9/11 or incur any of the mounting costs of maintaining the only military force in the region that, everyone agreed, was responsible for UNMOVIC's accomplishments. Unlike other members of the UN Security Council, for American officials the central issue was Iraq's relentless attempt to acquire and hide these weapons and Saddam's clear 'intent' to produce more advanced WMD capabilities in the future, despite the presence of inspectors and twelve years of UN resolutions demanding compliance with clear and unambiguous disarmament mandates.

These are 'real' security threats driven by technological progress, the spread of scientific knowledge related to these weapons systems, diminishing costs, ongoing regional tensions in the Middle East and Asia, and, most important, time. Critics who are unconvinced by the evidence in U.S. intelligence reports should track the evidence compiled by the intelligence communities in Russia, France, Germany, the United Nations (IAEA), or any number of publicly and privately funded NGOs or think-tanks with a mandate to track and estimate evolving proliferation threats. Although enemies will change from time to time (for whatever reason) this is not a reason to forgo missile defence. In a world of WMD proliferation and terrorism, enemies and security threats will remain ubiquitous. Sweeping generalizations about how all this evidence constitutes one big lie fabricated by the U.S. military for corporate profits are not particularly persuasive. Among other requirements for testing this particular economic theory of war one would have to demonstrate the following: that there is no serious security threat, that George W. Bush (along with his cabinet, advisers, and members of his re-election team) understood the limited nature of the threat, and that spending $87 billion of taxpayers' money for the war in Iraq and post-war reconstruction was viewed by the White House as a brilliant economic strategy to risk an election in favour of enhancing the profits of America's military industry. Aside from the fact that this makes so little political (or logical) sense, there is obviously much more to international politics than the MIC thesis can accommodate. Those who offer it as an explanation for the war in Iraq or BMD have an obligation to provide the same detailed defence of their position, and to do so by going well beyond implicit assumptions about irrational abuses by the U.S. government in politically suicidal attempts to appease American military corporations. The security threats are real, and, in that context, the foreign policy responses are entirely understandable.

Rational Rogue Leaders?

The U.S. doesn't need a missile defence system, others argue, because rogue leaders will never be foolish enough to launch a missile at the U.S., given the entirely predictable devastation that would be unleashed by a U.S. nuclear (or conventional) retaliation. According to Richard Gwyn in early 2001, 'It's absurd. It's laughable. It's surreal. Why would the leader of any of these backward, near-bankrupt, states commit suicide, even if, as is highly improbable, any of them could ever actually lob a missile across the Atlantic or Pacific?'[51] Obviously, if the probability of an irrational rogue leader taking over control of a state of concern is 'zero,' we can re-establish the utility of rational deterrence strategies and render obsolete the need for missile defence.

There are at least three problems with this line of argument. First, in order to accept the logic and associated policy prescriptions underlying it, one would have to conclude that the probability of ever having to face an irrational rogue leader is (and will always be) zero. Anything greater than zero, however, should force even the staunchest critics to reconsider the wisdom of rejecting BMD. But let's assume, for the sake of argument, that the relevant probability is indeed zero. A much stronger case in favour of missile defence can be made if one assumes rogue leaders are perfectly rational. Without BMD to worry about, wouldn't it be more rational for leaders in Iran, Libya, North Korea, Syria, and so on to build and deploy ballistic missiles, since they would now have the luxury of avoiding the technological hurdles and financial costs of incorporating decoys and countermeasures into their programs? Put differently, wouldn't the incentives to proliferate be much greater for a rational rogue leader when ballistic missiles are cheaper to deploy and more likely to work if launched? If rogue leaders are rational (and I believe they are), deploying missile defence is the most logical policy choice. In a world of rational rogue leaders, failure to deploy missile defence will provide the very incentives that increase the prospects of proliferation. Now, if rogue leaders are irrational enough to build missiles even when the costs are high and incentives are low, or on the slim chance that even one irrational rogue leader will, at some point in the future, deploy and threaten to use ballistic missiles, BMD, again, is a no-brainer. Whether rogue leaders are rational or irrational, missile defence makes sense.

Second, if rogue leaders are rational enough to be deterred by threats of U.S. conventional or nuclear retaliation, why would these same rogue leaders be any less rational (or any less susceptible to rational deterrence) when facing a U.S. BMD shield? Wouldn't the shield provide an added level of deterrence, rationality, and security, even if only to protect American citizens from inadvertent launches or those anomalous, irrational rogue leaders? If a rogue attack is laughable, how much harder would we laugh imagining the same rogue leaders contemplating a suicidal ballistic missile launch against the U.S. if they were confronted with a U.S. BMD system? Put differently, why would BMD make these rogue leaders more suicidal, especially when their suicidal attack is less likely to work?

Third, critics often ignore the other relevant objective of U.S. BMD – preventing rational rogue leaders from believing they can rationally deter the U.S. (or NATO) from launching a conventional attack. This fact was explicitly acknowledged in the 2002 U.S. National Security Strategy: 'For rogue states these weapons are tools of intimidation and military aggression against their neighbors. These weapons may also allow these states to attempt to black-mail the United States and our allies to prevent us from deterring or repelling the aggressive behavior of rogue states. Such states also see these weapons as their best means of overcoming the conventional superiority of the United States.'[52] This is perhaps the most important point critics repeatedly overlook (or intentionally ignore): BMD has as much to do with protecting the U.S. (Canada, and Europe) from intentional and inadvertent attacks (irrational or otherwise) as it does with protecting and enhancing U.S. (Canadian, and European) bargaining leverage in places like Bosnia, Kosovo, Iraq, Afghanistan, Taiwan, and North Korea. In fact, the reason ballistic missile technology is likely to proliferate in the future has more to do with lessons learned from the past decade of NATO- and UN-sponsored interventions: Coalition victories in these conflicts create added incentive for rational rogue leaders to acquire and quickly deploy ballistic missiles as insurance against being on the losing side in similar conflicts with the United States, NATO, and the UN in the future. These crises also serve as a compelling reminder to American, Canadian, and European officials that defensive technologies are becoming more crucial, not simply for our security but perhaps for the security of hundreds of thousands of people in the midst of a humanitarian catastrophe such as Kosovo. The key problem as we enter the twenty-first century is not how to deter rogue states from launching a nuclear attack, but how best to prevent them from thinking they can deter the U.S. or NATO from launching a human security intervention.

Fourth, from the point of view of crisis management, consider how each stage of a hypothetical foreign policy crisis with rational North Korean leaders will play out: in the midst of a crisis, does the probability of an attack, initiated by Pyongyang, increase or decrease if North Korean leaders are facing U.S. BMD? Now take a step back: are officials in North Korea more/less likely to escalate tensions in the face of U.S. BMD? And in the initial stage of a crisis, are the incentives for resolving the conflict higher or lower for North Korean officials with/without U.S. BMD? At each stage of this hypothetical crisis, U.S. BMD is likely to be more stabilizing – unless, of course, the North Korean leadership is irrational. That is precisely the problem, some critics will argue: North Korean leaders may act irrationally in the midst of a military crisis, thereby upsetting the logic of this simplistic, 'rationalist' interpretation of their preferences. That may be true, but in addition to defending the culturally insensitive assertion that North Korean leaders are less rational than their Western counterparts, critics are now faced with having to explain why North Korea would actually contemplate launching a ballistic missile attack (for whatever irrational reasons) – an act previously assumed to be unlikely and a fabrication of the U.S. military industrial complex.

It is irresponsible (not to mention disingenuous in the extreme) for anyone to characterize as absurd the fear that some state, at some point in time, could conceivably contemplate threatening to launch a missile. Even the most optimistic assessments of the proliferation and terrorism problems should force the staunchest proponents of multilateralism to reconsider their recommendation that we depend exclusively on these approaches and dispense with all efforts to deploy missile defence.

Without doubt, nuclear, biological, and chemical weapons have been acquired by states whose leaders are convinced of the potential for these weapons to increase their security by enhancing their bargaining leverage in potential crises with military rivals. The declaration by North Korean leaders (13 March 2002) that their country has the capacity to retaliate if attacked by the U.S. was a direct response to the most recent U.S. Nuclear Posture Review, which expanded the list of potential targets to include North Korea. Clearly, North Korean officials believe they need a credible nuclear threat to balance U.S. power. American officials believe it is imperative to expand the scope of strategic stability to include deterrence of WMD use by new and aspiring nuclear rivals. North Korean officials, in turn, are compelled by strategic imperatives to respond accordingly. The common thread, of course, is that the perceived benefits associated with proliferation are far greater than those tied to disarmament.

Given decisions by India, Pakistan, and North Korea to develop their respective nuclear programs, and the fact that each one of these states experienced 'constructive engagement' overtures from the international community (following short-lived economic sanctions), Iranian leaders are likely to conclude that the only thing that works is nuclear weapons, because nuclear states seem to be treated differently – consider the obvious comparisons with Iraq.[53] No state has ever dismantled a nuclear weapons program because of some moral imperative to do so; leaders of states who decide to dismantle their programs typically do so because they are convinced that these weapons contributed very little to their security and/or foreign policy interests. Regardless of the international pressure on India, Pakistan, Israel, Iran, Syria, and others, these states are unlikely to reverse or dismantle their respective WMD programs, and critics of BMD have yet to describe a credible diplomatic path that promises that result.

Conclusion: Critical Biases

The term *critic*, used throughout this chapter, refers to individuals (or groups) who put forward any one of the arguments/criticisms evaluated above. Although there is extensive overlap in the positions of those opposed to BMD, all critics do not necessarily share the same views or assign equal weight to the various arguments; there is no homogeneous position, nor is one likely to emerge for such a complex issue. To the extent that a certain degree of uniformity (conformity) exists it has more to do with the tendency to overlook deficiencies with weaker components of the critics' overall case. For example, regardless of evidence to the contrary, few BMD critics are likely to explicitly (publicly) reject the conventional wisdom that BMD will provoke an automatic arms race. Despite the acquiescence by China and Russia to the death of the ABM Treaty and deployment of U.S. BMD, dire predictions by critics of an uncontrollable arms race are likely to continue unabated.

Likewise, few if any critics will ever support the view that BMD is worth $100, $80, or even $60 billion over ten years, regardless of the success level of BMD testing, and regardless of the costs and failures of alternative strategies critics offer as replacements. The prevailing assumption seems to be that a collection of weak arguments, if repeated often enough, combine to produce a persuasive criticism. But the opposite is true. Any critic, but especially those who have impressive credentials, should be prepared to confront the weaknesses in the overall position being defended, because we desperately need a meaningful and substantive debate on these issues. Without it, the discussion (and attendant policy prescriptions) will continue to be driven entirely by political agendas and stifled by efforts to protect the intellectual capital invested in defending simplistic (and largely outdated) assertions about the dangers of defending people against ballistic missiles.

In many ways the BMD controversy represents a microcosm of the old realist-idealist debate in the field of international relations. It's not really about the technology, costs, proliferation risks, or any of the other arguments. Implicit in the arguments against BMD is an unmistakable bias against the underpinnings of U.S. foreign policy. The bias is typically revealed in the form of two distinct fears associated with, on the one hand, American unilateralism, and, on the other, American isolationism. Both fears are derived from predictions about what the U.S. might (or might not) do with the added security of a functioning BMD system.

With respect to the first bias (i.e., fears of American unilateralism), consider the typical responses critics offer when asked the following questions. Why should we be worried about BMD? The answer: because, among other problems, an emerging U.S. strategic advantage would force Russia and China to augment their nuclear arsenals in order to protect their interests and re-establish strategic stability. But why would officials in Moscow and Beijing be worried about maintaining strategic stability? Well, for two reasons. First, the offensive advantage that comes with a fully deployed and functional BMD system would significantly enhance U.S. capabilities to launch a pre-emptive first strike (and withstand a retaliatory second strike). Second, by reinforcing security with this defensive shield, American officials would have added incentive and opportunity to pursue an unfettered, global unilateralist foreign policy agenda. But what specifically would a global unilateralist American foreign policy agenda look like? Why, precisely, should Russian or Chinese officials be worried? And how, exactly, would these unilateral initiatives jeopardize European or Canadian interests? The point to this hypothetical exchange is to emphasize that a position on U.S. unilateralism should depend on whether the specific foreign policy initiatives in question are morally defensible and/or make a meaningful contribution to global peace and security. For example, what's wrong with undermining Russian or Chinese interests in places like Bosnia and Kosovo? Both Moscow and Beijing threatened to veto any Security Council resolution to allow NATO's intervention to proceed, forcing the intervention to be conducted outside the auspices of a UN-sanctioned multilateral operation. Similarly, what's wrong with undermining Russian or Chinese interests in places like Chechnya and Taiwan, respectively? Or Russian and French oil interests in Iraq in 2003, especially if intervention freed Iraq from the grips of Saddam Hussein's brutal regime? Contrary to the foreign policy bias implicit in the critics' case against BMD, unilateralism is not necessarily evil, and exclusive reliance on multilateral legitimacy that comes with a UNSC resolution is not necessarily good.

Concerns about U.S. isolationism represent a more subtle form of the foreign policy bias implicit in the critics' case against BMD. The problem has less to do with what the U.S. might try to accomplish unilaterally with BMD, and more to do with the global obligations U.S. officials will continue to avoid meeting – such as dismantling nuclear stockpiles; investing a larger portion of GNP in aid, development, and debt relief; forcing Israel to make significant concessions on creating a Palestinian state; responding more forcefully to human rights abuses in Chechnya or China; signing the Kyoto Protocol; fighting AIDS in Africa; and so on. The concern here is that BMD would simply be one more impediment to creating a more equitable balance of global power, influence, and vulnerability. A vulnerable U.S. would level the playing field and force American officials to pursue a more balanced foreign policy agenda, one that critics of BMD espouse.

The bias in the critics' case is also expressed through inconsistent predictions about the future. Proponents of arms control and disarmament have attempted for over a decade now to convince everyone that the world has undergone fundamental and irreversible transformations, particularly after the collapse of the Soviet Union and the end of the cold war. According to this view, globalization and constantly expanding levels of economic interdependence have created a worldwide environment in which large-scale conflict involving major powers is becoming increasingly remote and, for a number of dyads (i.e., pairs of states), obsolete. Economic and trade relationships are far better predictions of interstate behaviour today than traditional factors tied to military competition, power, and status. For example, American officials spend more time and diplomatic energy/capital on renewing China's Most Favoured Nation status or getting China admitted into the World Trade Organization than they do managing military crises between China and Taiwan. Liberal internationalism prevails and economic interdependence rules.

Consistent with expectations derived from this ever expanding web of interdependencies, and perhaps the clearest indication that the entire payoff structure has shifted away from viewing the military as an effective tool of statecraft, defence expenditures in practically every major European capital continue to fall. Given these transformations and the incredible improvements in cooperative relations between East and West, there is virtually no justification for large numbers of nuclear weapons. If conventional forces and military power are becoming less relevant in a world dominated by trade and financial markets, large numbers of nuclear weapons are irrelevant and perhaps always were.

But this entire liberal internationalist framework of analysis, not to mention the complex combination of arguments and evidence put forward to establish its relevance in a post–cold war setting, gets discarded by many critics once the subject of BMD comes up. Many of the same persons who point to these changes when accounting for the irrelevance of nuclear weapons now believe that a relatively minor expenditure of 2 per cent of the overall U.S. defence budget can reverse these fundamental transformations in a relatively short period of time. A decision to deploy BMD, they claim, will produce an arms race that is virtually identical to those we experienced during the cold war, and for identical reasons – security, military power, control, and influence. Apparently 'realism' prevails, and power and self-interest rule. If this is indeed a more accurate representation of contemporary international politics, apparently the cold war is far from over, China and Russia have legitimate security concerns about U.S. objectives to develop a first-strike advantage, and the U.S. may at some future point in time actually contemplate launching a pre-emptive first strike against rising powers like China.

Either we live in a post–cold war world in which new incentives combine to produce a new logic guiding international relations (as I argue in chapter 1), or we don't. Growing levels of economic inter-dependence have either changed payoff structures such that nuclear weapons are becoming increasingly obsolete, or payoff structures (and associated demands for access to nuclear weapons) remain the same. If payoff structures have changed for the better, we should not expect a relatively minor investment of 2 per cent of the U.S. defence budget to have such a profound impact on international relations. On the other hand, if BMD can reverse this new (and obviously very fragile) post–cold war system in such a short period of time, critics should acknowl-edge that we live in a world in which military might, nuclear weapons, and ballistic missile technology are still very relevant and useful, both for existing and future nuclear powers. Now, if these weapons retain their utility, BMD should be viewed as prudent. If nuclear weapons have no value, according to liberal internationalist dictates, BMD should not be perceived as threatening by Russia and China. Regardless of one's views about post–cold war transformations, then, the case against BMD is weak. Critics need to establish a cold war mindset to defend claims about the destabilizing effects of BMD, but doing so seriously undermines their case against the utility of nuclear weapons in a post–cold war world.

Chapter 5

WMD Proliferation: The Case against Multilateral Arms Control and Disarmament

Burden(s) of Proof

Some will no doubt argue that the approach I've taken in the previous chapter essentially reverses the burden of proof. As David Mutimer asserts,

> National Missile Defence is a major, costly and quite dramatic alteration of the strategic status quo. Just how costly it will be is at the heart of the debate about its future, but its cost is certain to include billions of dollars and political disruption with American allies and others. Under the terms of almost any understanding of debate, whether logical, formal or political, those proposing have the responsibility to make the case for the proposal. Those who are opposing need to show sufficient problems with the proponents' case that it should not be accepted. To shift metaphors for a moment, not proceeding with the proposal, in this case NMD, must be the default position. By beginning with a response to the array of critical arguments, Harvey [in my article 'The International Politics of National Missile Defese: A Response to the Critics,' 2000] neatly attempts to reverse this burden, or default, as the clear implication of his article is that critics must make the case against NMD, or forward it should go.[1]

There are at least three obvious problems with this line of argument. First, in applying the burden-of-proof analogy Mutimer misrepresents the prosecution and the defence. According to critics, proponents of BMD are guilty of one or more of six crimes: (1) fabricating rogue state and other proliferation threats, (2) creating a dangerous security environment that threatens to destroy decades of arms control agreements,

(3) producing yet another destabilizing arms race, (4) wasting at least $80 billion for an illegal defence system prohibited by the ABM Treaty, (5) lying about whether BMD technology has worked (or will ever work), and (6) doing all this to save an outdated military industrial complex from disappearing after the cold war. If the critics could prove that the evidence they have compiled is true, supporters of BMD would be guilty as charged – indeed, BMD would deserve nothing less than the death penalty. But if proponents are the 'defence' in this case, doesn't the burden of proof lie with the prosecuting critics? Don't they have the obligation to establish the charges underlying their case? My defence of BMD includes a long list of logical errors and empirical flaws in the critics' arguments. If this were a court of law any competent judge and jury would throw the case out of court if the prosecution decided to ignore the problems with their case.

Second, this debate is not taking place in a court of law, and we are not lawyers defending clients in a criminal case. The last thing academics should do in the midst of such an important policy debate is to re-create the problems that plague the legal profession. Getting off on a technicality associated with default positions should never be considered an option. Whose 'understanding of debate' is Mutimer referring to here? How exactly does one arrive at a conclusion that 'the responsibility for making the case for the proposal rests with those proposing it.' Meanwhile, those opposing BMD need not suffer the burden of proving their case against, as if these two sides of the debate are somehow separable. When exactly were critics given the legal or constitutional right to refuse to respond to questions on the grounds that their answers might incriminate them?

How, one might ask, should proponents defend BMD against these charges? Well, by pointing to major weaknesses in the critics' case, by establishing evidence of ballistic missile proliferation (the security problem BMD is designed to address), and by demonstrating that, in comparison to alternatives, BMD is a more secure, effective, and cost-efficient approach to the problem. And how, one might ask, should critics make the case against? By addressing the problems with their arguments, by establishing that ballistic missile proliferation is not a serious problem, and/or by demonstrating how their alternatives to BMD offer more secure, effective, and cost-efficient strategies for dealing with the problem of ballistic missile proliferation.

Everyone involved in this debate should be prepared to acknowledge and defend the validity of any and all arguments they make.

Consider for a moment Mutimer's burden-of-proof argument from the perspective of American or Canadian public officials whose primary responsibility is the security and welfare of citizens. They would expect critics to establish, perhaps even beyond a shadow of a doubt (given the consequences), the following: leaders in North Korea, Iran, Syria, and Libya have no interest in acquiring ballistic missile technology and are not currently developing (or spreading) that technology; the strategy of 'constructive engagement' has helped to address the demand and supply side of proliferation; and, finally, aspiring nuclear powers can be convinced that acquiring nuclear capabilities will not enhance their security. Critics have a responsibility to provide at least some evidence to support their case.

The level of cooperation between the U.S. and certain rogue states may vary over time (recall that it took forty-five years to transform the U.S.-Soviet rivalry), but this is less relevant than the fact that nuclear-equipped rogue states (whoever they may be) are expected to remain a problem in the future. Their status as rogues may change from time to time, for a variety of political or economic reasons, but it is the expectation that there will be rogue threats in the future (thanks to proliferation) that explains the rational push for unilateral approaches to security today. Again, there are typically three reasons why proliferation of WMD to potential rogue states is threatening: (1) because these missiles could at some point in the future be launched at the United States or one of its allies (e.g., Israel, South Korea), (2) because proliferation increases the probability of an accidental launch, and (3) because WMD could conceivably be used by rogue states to prevent (deter) the United States, NATO, and other Western allies from getting involved in the right kind of war – for example, to prevent humanitarian catastrophes. In fact, a fourth reason can now be added to the list in light of recent moves by Pyongyang to use brinkmanship and the threat of horizontal and vertical proliferation to gain additional economic concessions from the U.S., above and beyond those granted to North Korea in 1994 by the Clinton administration.

Third, and most important, Mutimer misses a relatively straightforward point about how to make a case for or against any policy alternative: refuting critics' claims about BMD's technological limitations, economic costs, and proliferation risks is essential to strengthening the case for deployment simply because it diminishes concerns about those political, economic, and military costs that critics keep bringing up. If the financial and security costs of BMD are found to be far less significant than critics claim (or, in comparison with alternatives put forward by critics, far more reasonable), that evidence is directly relevant to making a case for BMD.

Measuring NACD's Successes and Failures

Proponents of the multilateral NACD (non-proliferation, arms control, and disarmament) regime continue to assert that their preferred approach may not be 'sufficient' to achieve every non-proliferation and arms control objective but it is a 'necessary' component of any comprehensive strategy to disarm the planet. This raises the obvious question, What are the other 'necessary' components for success and where exactly do unilateral strategies fit in? More important, what criteria should we use to evaluate the success and failure of competing policies? Assuming that all this information is essential for making definitive claims about 'necessity' and 'sufficiency,' the following explores the issue of success/failure in more detail.

The main challenge for proponents of the NACD regime is the lack of demonstrable proof that multilateral arms control works. As a regime with a very specific and straightforward set of objectives, it has never achieved the kind of success that would warrant giving its proponents the moral or intellectual authority to dismiss unilateral alternatives, such as BMD.[2] Without such evidence there is no logical, empirical, legal, moral, or policy-relevant foundation for embracing multilateral arms control. Several additional points related to measuring the success and failure of the NACD regime should be noted.

Ongoing disagreements over appropriate criteria for measuring success and failure preclude definitive statements about the real (and relevant) contributions of the NACD regime. For instance, should we rejoice in the success of indefinite renewal of the NPT or remain highly sceptical of the treaty's capacity to prevent signatories (including, but not limited to, China, Russia, Iran, North Korea, Iraq, Syria, and Libya) from acquiring and/or selling prohibited WMD technology? Should we focus on the portion of any draft arms control treaty that achieves consensus or the portion that remains contested because of a combination of insurmountable political, financial, or military hurdles? Consider, for example, how much of the 450 pages of text in the most recent draft of the Biological and Toxin Weapons Convention remain highlighted and bracketed – that is, contested. Should we focus on the minutiae of pre-negotiation concessions on the location and timing of the next conference, chairmanship, conference schedules, and so on, or should we acknowledge that the combined efforts of those involved in virtually thousands of similar conferences have failed to stop WMD and ballistic missile technologies from proliferating to states that want them? Examples of NACD successes typically highlight less significant accomplishments in the area of 'process' rather than 'outcome' or minor revisions to the text of draft treaties because these 'successes' are far easier to identify. But this approach simply lowers the bar for measuring progress – indeed, the evaluative criteria for the NACD regime is increasingly removed from straightforward questions about whether WMD technology continues to proliferate and how we can prevent it.[3]

Proponents of multilateralism are quick to offer as clear 'evidence' of success a long list of multilateral treaties, protocols, agreements, and conventions; nuclear-weapon-free zones; and hundreds of multilateral declarations, verification programs, monitoring agreements, protocols, export control guidelines and clarifications / modifications / amendments, and other memoranda of understanding. In addition, multilateralists are likely to list as illustrations of progress hundreds of governmental and non-governmental institutions, organizations, conferences, annual meetings, boards, and agencies with arms control, verification, and monitoring mandates; hundreds of United Nations resolutions and legal opinions designed to address proliferation; hundreds of independent departments, intelligence agencies, and legislative committees established by Western governments (with billions of dollars invested worldwide) to solve one or another part of the proliferation puzzle; and virtually thousands of non-governmental organizations and think-tanks with the same mandate receiving hundreds of millions of dollars in public and private funds. All this activity is held up as concrete evidence of what four decades of multilateral arms control and disarmament activity has accomplished – incontrovertible evidence that multilateralism is alive and well.[4]

But evidence that multilateralism is rampant and spreading does not, in any way, constitute proof of successful multilateralism. Notwithstanding all this activity, there is no demonstrable proof that we have dealt effectively with the proliferation problem or that the planet is any safer today than it was before we engaged in all this activity. Indeed, nuclear, chemical, and biological weapons (and their delivery vehicles) continue to proliferate and pose a more significant global threat today than ever before.[5] While the actual number of WMD may be decreasing (largely because of the reductions in Russia's arsenal alone) the number of ballistic missiles (the primary means of delivery for WMDs) is increasing worldwide, as is their accuracy, range, sophistication, and the number of actors who possess these systems. These trends have been occurring for some time.[6]

Russia continues to sell its expertise and equipment abroad to China, North Korea, Iran, Libya, Syria, and India. Often, states purchasing weapons from Russia offset the expense by selling the technology to others. For example, China purchases equipment and expertise from Russia but pays for the acquisition in part by leasing missiles and operators to Saudi Arabia or by selling slightly older technology to North Korea. North Korea purchases technology and equipment from China but recoups some of the costs by selling equipment and expertise to Syria, Egypt, and Iran. North Korean missiles are funded partly by Iran in return for operating joint production facilities or by providing transportation services to other third parties. Iran can cover the expenses of its weapons programs by sending technicians and experts abroad, to places like the Democratic Republic of the Congo, in exchange for premium consultation and training fees. In essence, states who possess ballistic missiles and WMD are investing more resources to improve and expand their arsenals, and the number of states providing such technology to aspiring nuclear powers is rising. Russia is perhaps the greatest proliferator, seconded only by China.

In addition to the detailed summary of publicly available information on WMD proliferation on the Web (see note 5), table 5.1 provides a few snapshots derived from recent reports that highlight the challenges. Gordon Giffin sums up the issue well: 'One can safely assume that these countries are not going to the vast expense of developing and building such devices simply to store them in a warehouse: They intend to use them, or to threaten to use them ... These weapons are also their means to counter our conventional forces and to break the cohesion of our alliances and coalitions.'[7] Obviously the status quo is not conducive to protecting or enhancing global security. To assume that current multilateral strategies are risk-free because they haven't resulted in 'catastrophic failure' is not a basis for sound arms control and non-proliferation policy.[8]

With respect to other dimensions of the proliferation challenge, consider the problems with trying to control nuclear trash from the former Soviet Union. In the 1960s and 1970s the USSR manufactured close to a thousand small nuclear batteries to generate electricity and to power lighthouses and transmission towers. 'Combining the strontium in these

TABLE 5.1
WMD Proliferation: Demand and Supply

1 Iran is stepping up efforts to build nuclear, chemical, and biological weapons, missile systems with help from Russia, China, North Korea and Western Europe. Iran tested a missile with an 800-mile range – enough to reach Israel – and another (Kosar) has a reach of 2,485 miles, within range of Western Europe.[a]

2 'China increased its missile-related sales to Pakistan last year and is continuing to supply nuclear, chemical and biological weapons and missile goods to North Korea, Libya and Iran. China has ongoing contacts with Pakistani nuclear weapons officials – contrary to a pledge made by Beijing in 1996 to halt aid to nuclear programs in Pakistan that are not under international controls.'[b]

3 U.S. satellite photographs confirmed that North Korea is exporting missile components from the port of Nampo (west coast). 'The exact types of missile components were not disclosed, but the goods are believed to be for foreign production of North Korea's homemade Scud B or Scud C missiles. The Scud B has a range of about 186 miles, and the Scud C can hit targets up to 310 miles away. North Korea also has exported medium-range Nodong missiles, which have a range of 620 miles ... Intelligence officials said North Korea is a major missile supplier to several nations, including Egypt, Pakistan, Iran and Libya.' According to a CIA report to Congress, North Korea exported 'significant ballistic missile related equipment and missile components, materials and technical expertise to countries in the Middle East, South Asia and North Africa.'[c]

4 President Putin announced on 12 March 2001 (after talks with President Khatami of Iran) that Russia 'will sell billions of dollars worth of arms to Iran in breach of an agreement with the United States that has underpinned regional security for the past five years. Moscow is expected to sign deals worth up to $7 billion (£4.8 billion) ... Mr Khatami signed deals to buy spare parts for Russian-made arms, such as MiG fighters and Su24 bombers, plus new air defence systems. Sales could include the S300 PMU2 anti-aircraft missile system, with a range of 200 kilometres (124 miles) and the Yakhont missile, which has a range of 300 kilometres (186 miles) and could be used to block oil shipments in the Gulf.'[d]

---

[a] Information compiled from reports by Bill Gertz, 'Pakistan Gets More Chinese Weapons,' *Washington Times,* 9 August 2000, and Jack Spencer, 'Missile Defense: No Laughing Matter,' *Washington Times,* 27 August 2000.
[b] Gertz, 'Pakistan Gets More Chinese Weapons.'
[c] Portions of the CIA report were cited in Bill Gertz, 'Weapons Sales Concern Seoul, United States,' *Washington Times,* 9 March 2001. According to the CIA report, the sales are 'a major source of hard currency for the Pyongyang government and the money is then used for continued missile development.'
[d] Alice Lagnado, 'Moscow Defies US with Iran Arms Deal,' *Times* (London), 13 March 2001. See also Gertz, 'Pakistan Gets More Chinese Weapons.'

batteries with conventional explosives could make a town highly radio-active,' writes Ian Traynor. In the post-Soviet chaos, the devices were abandoned, frequently without supervision. There are believed to be many more of them scattered in remote areas of Moldova, post-Soviet central Asia and the Russian far east ... The IAEA knows of almost 400 cases of trafficking in nuclear and other radioactive materials since 1993. Of those, 18 involved small volumes of weapons-grade pluto-nium or highly enriched uranium, and most of those cases originated in the former Soviet Union.'[9] This is but one illustration of the nature of the problem that directly relates to terrorism. China continues to spread arms technology to regimes that sponsor terrorists, and *Janes Defence Weekly* reported in February 2002 on U.S. sales of military hardware to India, which included an agreement to hold joint military exercises; the fact that Beijing is suspicious of U.S.-India ties is likely to encourage additional sales of technology to Pakistan and Iran.[10] And the absurdity spreads.

The Myth of Multilateral Alternatives to BMD

To make a much stronger case for the NACD regime its proponents must do more than highlight the costs, risks, and technological impedi-ments of BMD – that tactic will never be sufficient. More compelling evidence is required to demonstrate that multilateral arms control actu-ally works. However, the evidence indicates that when it comes to preventing WMD proliferation, the NACD's 'successes' will never be as positive (or constructive) as its 'failures' are negative. Multilateralists will no doubt respond to this assertion with a list of ways that their approach can be made to work more effectively. But they must also show that proposed solutions are realistic, cost-effective, technologi-cally feasible, and more likely to produce a greater return in security than corresponding solutions to BMD's deficiencies. Proponents of multilateralism have spent almost no time developing this case, for obvious reasons – the impediments to multilateral success are far more significant, more fundamental, more entrenched in domestic and inter-national politics, and therefore less resolvable than any technological, financial, or security hurdles associated with BMD success (including overcoming worries about Russian and Chinese responses).

Consider the following: no other security threat has produced more multilateral institutions, agreements, treaties, organizations, non-gov-ernmental organizations, committees, meetings, expenses, and activity, yet WMD continue to proliferate. If the combined investments in multilateral approaches to non-proliferation and disarmament have failed to stop the spread of WMD, arguably among the most serious security threats of all time, what does that suggest about the capacity of multilateral institutions to protect citizens against threats that do not pose the same kind of urgency or produce the same kind of motivation to succeed? Success with respect to proliferation, in other words, is the crucial test of multilateral approaches to security. If we have failed here, we are not likely to succeed when it comes to less pressing security threats, for obvious reasons – the incentives to acquire and spread WMD are too strong.

The evidence is not persuasive to demonstrate that multilateral alternatives promise a better return for security investments – prevention, pre-emption, conventional deterrence, constructive engagement, diplomacy, import-export controls, economic sanctions, transparency, monitoring, verification, and codification have all failed in the past, and are even less likely in the future, to ensure U.S. or Russian security from WMD threats. Collectively, these approaches have made meaningful progress on arms control and non-proliferation, but (1) they failed throughout the cold war to prevent the spread of nuclear and ballistic missile technology; (2) they remain incapable of providing levels of security to render nuclear weapons obsolete; and (3) they are inappropriate to address current and future security threats. In spite of the need for change, however, the arms control and disarmament community continues to defend status quo policies as if they were the only legitimate options. Several problems with each policy will be addressed in turn.

Prevention

Consider for a moment the following NACD failures (a list that continues to grow): the inability to enter into force of the second Strategic Arms Reduction treaty (START II); the lack of serious negotiations on START III; re-affirmation by major powers of the need for nuclear weapons (such as NATO's New Strategic Doctrine, U.S. Presidential Directive 60, Russia's revised nuclear policy); refusal by Russian, American, and European (NATO) officials to accept a relatively straightforward commitment to 'no first use'; the persistence of tactical nuclear weapons and their inclusion in nuclear force planning doctrine; south Asian nuclear tests and subsequent decisions by Western powers to lift economic sanctions against India and Pakistan; the impending demise of the Comprehensive Test Ban and the death of the Anti-Ballistic Missile Treaties; proliferation of weapons technology to outer space; failure of the Non-Proliferation Treaty (NPT) to stop signatories and non-signatories from spreading and acquiring nuclear material and weapons technology; and so on. If we are indeed on the verge of devaluing the currency of nuclear weapons because of an apparent consensus that such weapons are absurd, unethical, and immoral, why is this absurdity spreading? There are two possible answers: (1) the leaders of new and aspiring nuclear states don't understand how incredibly foolish this is, or (2) members of the arms control and disarmament community can't quite grasp how perfectly rational it is.

Regardless of whether one looks at the supply or the demand side of any dimension of proliferation, the evidence does not favour the optimists.[11] It takes only one nuclear weapon to produce the catastrophe the NPT was designed to prevent. The simple fact is that more states now have the capability to inflict that level of damage and devastation. Despite years of effort by the international community to prevent the spread of ballistic missile technology from the supply and demand side, WMD technology continues apace, and proponents of the status quo provide almost no evidence that the proliferation puzzle can be solved in the future using the same techniques.

Proponent of arms control will argue that the NACD regime has successfully slowed the pace of proliferation and that a more relevant measure of effectiveness is the number of nuclear weapons states in the world today. As noted in the previous chapter, however, most non-nuclear states are not interested in acquiring nuclear weapons, simply because they provide no added security or because security guarantees from allies are more than enough. The success of the NACD regime must be measured by how many aspiring nuclear states are prevented from acquiring the technology to deploy nuclear weapons and their delivery vehicles, and how many states continue to provide the requisite technology to help them along.

Some optimists point to decreased arms expenditures by rogue states as evidence that something must be working: 'Taking Iran, Iraq, Syria, Libya, and North Korea as a group: since the late 1980s their military spending has fallen 70 percent; their arms imports are barely 10 percent of what they once were ... More generally, without the technical support, funds, and arms once provided by superpower patrons, yesterday's rogue giants have lost the capacity to equip, train, sustain, or employ armed forces of the size and quality typical of the 1980s.' But the implication of these trends is that rogue leaders will become more reliant on asymmetric threats, ballistic missiles, and WMD weapons technology, because these alternatives are more affordable than having to sustain large conventional mistakes.[12]

Pre-emption and Conventional Deterrence

Pre-emption and conventional deterrence can be evaluated in one of two ways: (1) as less appealing alternatives to BMD (the approach selected here and addressed below), or (2) as unilateral or multilateral strategies. The following will identify some of the difficulties with effective application of these strategies, whether unilaterally or multi-laterally, and why they are inadequate to justify excluding missile defence as a necessary option.

Mounting a successful pre-emptive attack requires accurate intelligence about the location of enemy missiles, domestic and international support for the attack, locally deployed military capabilities, and a political leadership willing to take the risks and incur the costs. Pre-emption may appear to be a relatively straightforward solution to the problem of proliferation, but such strikes are very difficult to mount for political and operational reasons, as the lead up to the 2003 Iraq war demonstrated. Proponents of this strategy rarely outline the conditions under which a pre-emptive strike would be justified. In the case of pre-emption again North Korea, for example, the costs of such a conflict in lives and devastation would be astronomical, with estimates of one million deaths in simulated war games with South Korea.[13] As Daniel Schorr points out, the preferred doctrine for North Korea is 'tailored containment,' in which diplomatic and economic pressure is directed at Pyongyang, but military threats are virtually non-existent.[14]

Therefore, while pre-emption will always be viewed by any current or future American president as a legitimate option for responding to emerging threats of terrorism and proliferation, and while presidents will view pre-emption as emanating from a moral obligation to protect American citizens, the enormity of the threat alone (whether high probability–low impact; low probability–high impact or some combination) will never be sufficient. In 2003, the Iraq regime represented a pre-emptive target that carried an acceptable level of risk for what was considered at the time to be an important security objective; North Korea, on the other hand, did not, despite the fact that the nuclear threat posed by the leadership in Pyongyang (and its record of prolifer-ating nuclear and ballistic missile technology worldwide) remains con-siderably greater than was the case in Iraq.

With respect to the deterrent value of conventional war, a common assertion among critics of BMD is that the United States is more than capable of retaliating against any attack on its soil with a devastating conventional strike. The U.S. capability to disable or destroy targets quickly without suffering many casualties, or to change a regime in a relatively short period of time (three weeks in the case of Iraq) is so credible and so potent as a deterrent that BMD is unnecessary, redun-dant, and a waste of time and money. Rational rogue leaders would never provoke the kind of devastation the U.S. is capable of inflicting on its enemies, or so the argument goes.

Conventional capabilities of major powers can achieve relatively easy (and major) military victories without having to suffer even small num-bers of casualties, but this growing list of successful interventions in Iraq (1991), Bosnia (1995), Kosovo (1999), Afghanistan (2001), and Iraq (2003) create the very incentives for leaders in Iran, North Korea, Syria, and Libya to acquire and deploy ballistic missile and WMD technology as quickly as possible to avoid being on the wrong side of another U.S., NATO, or UN victory. Paradoxically, the conventional weapons that make BMD obsolete actually create conditions that spread the very WMD technology that makes BMD essential.

Conventional retaliation will never produce potent enough threats to be reliable, because the costs and risks of facing even a large conventional attack are likely to be acceptable under the right conditions (see chapter 3 for a discussion of how contemporary multilateralism undermines the credibility of conventional deterrence). For example, despite suffering a devastating loss against a U.S.-led coalition in 1991, and despite the presence of a perfectly credible threat of military retaliation by a U.S.-U.K. coalition, Saddam Hussein was willing to risk the consequences of yet another devastating war in 2003. The Taliban made the same miscal-culations in the period leading to the 2001 war in Afghanistan.

Many of the same critics of BMD who prefer reliance on conventional military capabilities are also more likely to reject the application of that force when it does take place. U.S. and NATO retaliation in Kosovo, for example, was criticized as unjust by many critics of BMD, because the intervention was disproportionate, unnecessary, and, therefore, a crime against humanity – despite the fact that it was in response to ten years of ethnic war in Bosnia, 250,000 deaths (many from ethnic cleansing), two million refugees, another 3,000 deaths in Kosovo circa 1998–99, and the expulsion of anywhere from 500,000 to 800,000 Kosovar Albanians over a one-month period. The Balkans case offers two key lessons for those who believe that the threat of large-scale conventional retaliation can deter all forms of attack, including a nuclear attack by a rogue state.

First, one should never underestimate the capacity of the public to reject as obscene and unacceptable any military response (including conventional war) even in retaliation for the deaths of thousands of Americans – witness the criticisms, by American citizens, of the U.S. retaliation against Afghanistan for the terrorist attacks on 11 September. Similarly, one should never underestimate the tendency of states of concern to mistakenly assume that their Western counterparts (and Western publics) lack the political will to sustain a long and potentially brutal and costly conventional war, even in retaliation for a ballistic missile attack. Threats of massive conventional war will never be as effective as nuclear threats when deterring ballistic missile attacks, because conventional retaliations (even those as significant as the 1991 gulf war) will never be perceived by opponents as more costly than even a small nuclear retaliation.

Second, leaders face several operational problems when applying conventional deterrence in contemporary crises. A credible deterrent force requires a large 'power projection' capability to intervene to protect U.S. interests, a capability that is expensive to maintain and entails, according to John A. Hopkins and Steven A. Maaranen, 'a complex of naval, air, and ground forces and their support. To operate these forces effectively requires an overseas base network, which [the U.S. is] losing, and a forcible entry capability, which is doubly challenging especially if there are no local bases to rely on.'[15] The point is that localized conventional superiority requires a large standing force to project a deterrent threat, and that projection cannot be sustained for long periods – a lesson that was becoming abundantly clear in the months leading up to the 2003 Iraq war. This level of mobilization and deployment is extremely costly (approximately $87 billion by October, 2003), far more expensive than the current budget for BMD. Ironically, any effort by the U.S. administration to maintain such a deterrent capability will almost certainly be rejected by the same critics of BMD as excessively intrusive and a waste of time and money.

One final point is in order regarding the utility of conventional threats as an effective form of coercive diplomacy. On 20 January 2003, the North Korean leadership threatened to re-establish a nuclear weapons program, to deploy and sell more sophisticated ballistic missiles, and warned of 'pre-emptive war' in response to U.S. threats of economic sanctions. That level of hostility, in response to a mere threat of economic sanctions, does not appear to support the view that Pyongyang is either containable or particularly worried about U.S. conventional retaliation – they appear to be encouraging crises and certainly do not seem deterred from acting in ways that threaten U.S. security interests. The more relevant lesson from the past decade is that in each of the five major wars between 1990 and 2003 involving U.S. and NATO troops, and despite overwhelming superiority of the U.S. military to achieve security, U.S. threats did not produce the level of fear or rationality implied in the conventional-war thesis – the coercive threats failed, the attacks took place, and, as expected, the regimes fell. Milosevic in 1999 obviously did not learn from NATO's successful intervention in Bosnia four years earlier; the Taliban in 2001 did not learn from NATO's intervention in Kosovo two years earlier; Saddam did not learn from any of these previous cases, including his own experience in 1991; and so on.

Diplomacy, Soft Power, and Constructive Engagement: The North Korean Case

As for constructive engagement, there is no persuasive evidence that it has worked in the past and no reason to expect that it will work in the future. The Clinton administration made North Korea the top recipient of U.S. aid in Asia (about $600 million from 1994 to 2000) but was never able to inspect, as Pyongyang promised, the two sites that were suspected of storing plutonium.[16] More than a decade of constructive engagement between the United States and China has done little to stop China's modernization plans or to prevent it from selling advanced nuclear and missile technology to aspiring nuclear powers. More important, proponents of constructive engagement have yet to provide a logical explanation for why their preferred strategy should be expected to work in the first place – improving economic relations with Iran and Iraq in the 1990s (by, for example, lifting sanctions), or North Korea today will simply provide the very capital these states need to augment their nuclear and ballistic weapons facilities. To continuously push for constructive engagement when the record of failure is so clear will only serve to ensure that the strategy will never work, no matter how much we wish it would.

Among the more common policy recommendations put forward by proponents of soft power is for the United States to simply buy out North Korea's nuclear program. According to one version of the argument, from John Steinbruner, 'Various economic measures can give countries powerful incentives to voluntarily abandon or curtail their weapons programs or deployments. Some might consider "incentives" a euphemism for "bribery" but, if the threat is considered grave enough and the opportunity presents itself, there should be no hesitation to "buy out" the nuclear or missile programs of potential adversaries. Overall, economic incentives to induce countries not to develop or transfer missiles or warheads are likely to be vastly cheaper and more effective than building a system to shoot them down after they are launched.'[17] Consider some of the problems with this apparently obvious 'solution.' It is not at all clear how one would go about implementing this in practice, or how much it would cost. How intrusive would the verification process have to be to establish whether constructive engagement is working? What, exactly, is the probability that this strategy will successfully compel North Korean leaders to make all necessary concessions regarding on-site inspections? Who would control the money? What do we do if it fails? What kind of sanctions would we put in place to enforce compliance, and what if the sanctions fail? What incentives do officials in Pyongyang have to use these funds to feed people? In sum, what are the risks and costs, and what are the unintended consequences? With respect to the latter question, what message would this policy send to other developing countries whose leaders may be contemplating acquisition of their own WMD program knowing now that they can sell the program (or even the threat of developing one) to the United States? With such a lucrative product (fear or security) to sell, wouldn't the logic of supply and demand dictate to North Korean officials that ongoing proliferation on their part is rational and profitable? On balance, why would these rational rogue leaders give up the security benefits, international engagement, and diplomatic recognition they seem to be getting by developing, deploying, and selling these weapons?

Canada's decision to establish full diplomatic recognition of North Korea is a case in point. The objective was to use diplomacy to persuade the regime to abandon its plans to develop and sell long-range missiles. But increasing the attention paid to North Korea (or India and Pakistan) has sent the wrong message to aspiring nuclear weapons states: the quickest path to international respect and diplomatic recognition is to acquire WMD technology as soon as possible. North Korean leaders are not likely to change course, because their strategy appears to be working. Indeed, North Korea's decision to restart its nuclear weapons program is the most recent illustration of the emerging collapse of the non-proliferation regime and associated treaties. As with many other multilateral arms control agreements, signatories often ignore both the demand and the supply side of their NPT obligations. If states maintain an official commitment to the treaty yet continue to acquire and sell weapons of mass destruction, how relevant is the NPT to addressing the proliferation problem? That is the true measure of success or failure.

Perhaps the most interesting feature of the re-emergence of North Korea's nuclear weapons program is the fact that officials in Pyongyang are demanding bilateral negotiations with the U.S. and prefer to avoid dialogue with the United Nations (through the International Atomic Energy Agency), the European Union, South Korea, or any other state or international organization. Ironically, the decision in Washington to focus on Iraq in 2003 and to let the international community deal with North Korea through multilateral mechanisms was criticized by multilateralists who demanded greater unilateral initiatives from the Americans.

The most obvious problem is this: leaders of the most dangerous states, those who need to be restrained, are the least likely to be affected by global opinion. As Walter B. Slocombe acknowledged from his experience as a U.S. State Department adviser, 'the prospects of affecting regime action by appeals to the good sense or innate caution of the citizenry are minimal simply because the regime will have been careful to insulate itself very thoroughly from such public pressures. Even the most powerful international instrument of pressure short of military force – economic sanctions – has a very feeble potential for deflecting such states' actions.'[18] Consider the concessions already granted to North Korea through the Framework Agreement signed by Washington and Pyongyang – as of the end of FY 2001, $800 million has already been put into the construction of the reactors; a further $1 billion would be required to finance the power grid; as of July 2000, South Korea has spent $2.9 billion, the United States $1.9 billion, Japan has invested $1.3 billion, and the EU has spent $82 million towards the agreement.[19] In the '90s, the Clinton administration provided the regime with two light-water nuclear reactors and offered an aid package that included 500,000 tons of oil, yet throughout the same period U.S. intelligence confirmed the existence of an advanced uranium enrichment and ballistic missile program and tracked sales of this technology to Iran, Syria, and Pakistan.[20] As expected, the 'success' record for constructive engagement in North Korea over the past decade is anything but impressive. Despite the removal of nuclear weapons from South Korea in 1991 and the major concession to cancel U.S.–South Korea military exercises, the Bush administration was unable to gain anything approaching serious reciprocal concessions from Pyongyang.[21] On 18 April 2003, North Korean officials responded to events in Iraq by stating that 'the Iraqi war teaches a lesson that in order to prevent a war and defend the security of a country and the sovereignty of a nation it is necessary to have a powerful physical deterrent.'[22] All this occurred during the same period in which an estimated two million people died of starvation.

The evidence from the North Korean case does not support the claim that constructive engagement works, or that diplomacy is the best alternative for dealing with aspiring nuclear states committed (for whatever reason) to acquire WMD. Constructive diplomatic overtures to North Korea will continue to fail, most obviously because there are no substantial costs to the leadership in Pyongyang if sanctions are imposed on the country, nor is there any domestic political pressure to worry about.

Effective diplomatic pressure on North Korea also requires some level of engagement with partners, including and especially China. But China has its own interests and will prefer to play a diplomatic role that befits and benefits its status as a major power. That the U.S. needs China puts Beijing in an excellent bargaining position. Despite the fact that it would be relatively easy for China to put pressure on their North Korean client to produce the concessions the Americans seek, a final resolution to the North Korean impasse is not as appealing to Chinese officials as prolonged engagement. By managing the crisis but avoiding both escalation and a clear resolution, China emerges as an important, if not essential, regional ally. This leverage is always valuable in a unipolar setting in which the U.S. hegemon is worried about its security. North Korea represents an extension of China's own nuclear rivalry with the U.S. The unfortunate reality is that this approach is perhaps the safest for managing the crisis and avoiding escalation, but the diplomatic dance playing out over North Korea's nuclear program does nothing to relieve the terrible suffering of North Koreans.

Despite the problems with constructive engagement, and the benefits to North Korea and China of maintaining the status quo, proponents continue to make the case that diplomacy can work to remove nuclear weapons from North Korea. Consider the following examples from James E. Goodby:

> It is likely that only a comprehensive peace settlement on the Korean Peninsula would provide the diplomatic context for Pyongyang to give up its nuclear ambitions once and for all. Although Kim's government has publicly rejected the idea of third-party mediation, some kind of multilateral effort, perhaps a Northeast Asian security conference, is urgently needed to derail North Korea's drive to build a nuclear arsenal. The agenda must be a broad one ... As a first step, a peace settlement would include a verified denuclearization of the Korean Peninsula, as agreed by the two Koreas in 1991. But ultimately it must address other military issues in Korea and must include economic and political cooperation ... Lasting peace in the region will never happen in the absence of societal change in North Korea. The best hope for that is to open the country to outside influences – in short, through a policy of engagement.[23]

The arguments in favour of constructive engagement, such as those put forward by Goodby, often begin and end with these sorts of pronouncements. But with respect to North Korea, the real problem is not that we have too little constructive engagement but too much of it. Any argument that proposes more of the same should at least provide the details for how to achieve any of these laudable goals and objectives. How, precisely, can the country be opened to 'outside influence' through 'engagement'? How exactly is a comprehensive peace settlement attained by tying those negotiations to nuclear ambitions that remain non-negotiable? How can the process become multilateral when the leadership in Pyongyang demands bilateral talks with only the U.S. administration? It would also help to provide more precise answers to questions such as these: why is a Northeast Asia security conference urgent, how exactly would such a conference achieve anything substantial (especially when there is no interest in Pyongyang to multilateralize the negotiations), and how would such a conference succeed in derailing North Korea's nuclear program? Proponents of constructive engagement rarely provide the details regarding the mechanisms through which these 'denuclearization' objectives are achieved. What threats would have to be issued (from economic sanctions to war) that would be conducive to pressuring Pyongyang into making these concessions, and at what costs and risks? How should the U.S. deal with explicit warnings from officials in North Korea of a catastrophic

war if the U.S. or UN (IAEA) imposes economic sanctions? And, perhaps most important, what's in it for China and why would they want
to help the U.S. achieve those objectives?

The case against constructive engagement can be strengthened by
exploring other applications of the strategy. Consider, for example, the
investments made by Canada to constructively engage Cuba over
the past decade (e.g., $34 million from 1994 to 1999 in grants from the
Canadian International Development Agency; $30 million [2000–2001]
to cover Canadian exports to Cuba). Aside from crafting a 'distinct'
foreign policy, what has Canada accomplished to secure Canadian values, interests, and objectives, such as human rights, economic development, and/or political and economic reforms in Cuba?[24]

Andrew Mack's excellent report on the arms control implications of
North Korea's decision to withdraw from the NPT makes several noteworthy observations. He notes, for instance, that through the International Atomic Energy Agency's mandate, the NPT promises to provide
countries who forgo nuclear weapons the opportunity to acquire safe
nuclear technology and power.[25] But states often join the NPT 'in order
to gain civilian expertise to use in a clandestine nuclear weapons program.' The assumptions that states sign on to the NPT for the sake of
peace and stability, or global efforts to rein in nuclear arms, or for the
good of their citizens, are all highly suspect. Yet these assumptions are
the cornerstones of the treaty and of optimistic expectations about its
potential success. But North Korea achieves more security, bargaining
leverage, and economic and political concessions from the international
community by defecting from the treaty, not by supporting it. Staunch
supporters of multilateralism are often more than willing to grant North
Korea's demands, which in turn fuels further abuse.

The current crisis between the U.S. and North Korea, more than any
other in the past half century, illustrates the problem with multilateral
arms control and signals to aspiring nuclear states that these weapons
have become the 'strategic equalizer.'

Codification

Perhaps the most significant problem with the current NACD regime is
the assumption that codification is a necessary condition for success.
Proponents assume that the only way to control the spread of WMD, or
to establish meaningful cuts in the nuclear arsenals of major powers, is
to tie those states rigidly to a set of specific rules, regulations, and

guidelines for the rest of time. In this view, the only path to collective security is to ensure that states recognize the supremacy of international law and acquiesce to its tenets, principles, and treaties, regardless of their defects. The assumption is that until and unless nuclear weapons states remain absolutely committed to specific codified limits stipulated in standardized arms control agreements we will never experience 'real' security.

There are several problems with codification that advocates rarely acknowledge. Codification closes options. That may appeal to those who continue to rely exclusively on bilateral approaches to strategic stability, but it is not likely to appeal to those responsible for identifying security strategies for new and emerging WMD threats.

Irreversible codification diminishes incentives to enter into good agreements, even if potential signatories accept the principles and objectives underlying the treaty in question.[26] The current impasse in the U.S. over the Comprehensive Test Ban treaty is a case in point: the U.S. continues to abide by the ban but refuses to be tied indefinitely to its limits. When major threats are assumed to be static and unchanging (as they were throughout the cold war), codified agreements are perceived as less risky. But as security threats change and become more complex (as they are today), codified agreements that limit options carry far more security risks.

Codification makes it more difficult to withdraw from potentially dangerous agreements that maintain force numbers at artificially high levels, as in the ABM Treaty.[27] That treaty compelled both sides to field forces large enough to maintain mutually assured destruction, not because the relationship required this balance but because the logic of bipolar strategic stability demanded these numbers. Indeed, the parity principle perpetuates an obsolete code of conduct that maintains force numbers for mutual annihilation that Washington does not want, Moscow cannot afford, and the relationship does not demand.

With these problems in mind, what evidence is left for critics to use in support of their view that ballistic missile proliferation will somehow be controlled in the future or, better yet, cease to pose a security threat if the U.S. scraps BMD in favour of focusing more of its attention on multilateral alternatives?

American Investments in Multilateral Arms Control

None of the preceding analysis should be taken to suggest that the U.S. government has decided to rely exclusively on unilateral BMD as an approach to these or any other security threats. In fact, multilateral approaches remain central to the American strategy and were explicitly acknowledged as important in the 2002 U.S. National Security Strategy. One of the objectives to strengthen non-proliferation efforts is

> to prevent rogue states and terrorists from acquiring the materials, technologies, and expertise necessary for weapons of mass destruction. We will enhance diplomacy, arms control, multilateral export controls, and threat reduction assistance that impede states and terrorists seeking WMD, and when necessary, interdict enabling technologies and materials. We will continue to build coalitions to support these efforts, encouraging their increased political and financial support for nonproliferation and threat reduction programs. The recent G-8 agreement to commit up to $20 billion to a global partnership against proliferation marks a major step forward.[28]

Consider also the many U.S. concessions tied to transparency and surprise inspections offered to appease Moscow's concerns about BMD deployment (information exchanges, numbers and locations of ABM interceptor missiles, notification of key events pertaining to the ABM system, inspections to verify raw data, short-notice inspections, and so on). These are designed to enhance predictability and, together, represent the very ingredients of credible arms control agreements for which the disarmament community has been fighting for years.

Washington tends to receive far more criticism for its unilateral initiatives than praise for its contributions to multilateralism. This often creates an exaggerated impression that Washington prefers nothing but unilateralism even when the record is more balanced. For example, the International Science and Technology Centre is a U.S.-government–sponsored organization that provides former Soviet Union weapon scientists alternative research opportunities to limit temptations to sell their expertise to states attempting to develop WMDs. These initiatives, part of the Nunn-Lugar program, are funded through the Departments of Defense, Energy, and State. Since 1991 they have spent almost $3 billion on counter-proliferation programs that have been extended to 2006. Over the next five years, the administration hopes to spend $4.5 billion on the programs in Russia, a 64 per cent increase from 1999.[29]

Spending on the verification of WMDs, based on specific programs under the Strategic Arms Reduction Treaty (START), will cost the U.S. government about $8 billion annually. The cost of one satellite and its launch to monitor such agreements is about $1.25 billion.[30] The National Nuclear Security Administration, created in FY 2000 to rationalize three other programs, is an agency of the Department of Energy and an integral part of the U.S. non-proliferation mission. One of its five goals is to 'detect, prevent, and reverse the proliferation of weapons of mass destruction while promoting nuclear safety worldwide.' [31] Budget allocations for FY 2001 were $6.8 million, $7.6 million for FY 2002, and a request of $8.03 million in FY 2003, including a discretionary budget of $1.3 billion to handle crises as they emerge. In addition, for FY 2003, the State Department requested $725 million for UN Peacekeeping operations, $618 million for international organizations with security and arms control mandates, and another $108 million for programs that enhance peacekeeping capabilities. The administration also requested $4.4 million in FY 2003 for UN operations in Iraq. Consider other expenditures (and increases) from FY 2001 to FY 2003, outlined in Table 5.2.[32]

Other well-known examples of American initiatives include: (1) the Agreed Framework with North Korea – in return for 'promising' to cease development of nuclear and ballistic missile technology, the United States agreed to provide 500,000 barrels of heavy fuel oil (approximately $30 million/yr), and in concert with South Korea, Japan, and several European states, provided two light-water nuclear reactors for power generation and additional infrastructure to transport the power; (2) the decommissioning and disposal of Russia's obsolete nuclear stockpiles, and continuing work through international institutions, such as the World Bank and the IMF, to provide substantial loans and credit to Russia; and (3) U.S. diplomatic initiatives to decrease tensions between nuclear rivals such as India and Pakistan.

Yet, notwithstanding all these efforts and investments, the unfortunate truth is that proliferation and development of WMD and ballistic missiles continues without any tangible demonstrations of appropriate levels of success to warrant exclusive reliance on these multilateral efforts. Despite the funding, fuel, equipment, and expertise donated to North Korea, and the substantial quantities of food provided by the international community to avoid mass starvation, Pyongyang continues to behave aggressively towards South Korea and Japan, continues to develop nuclear weapons and longer-range ballistic missiles such as the Taepo Dong series, and continues to produce and export ballistic missiles to third parties in the Middle East. Similarly, while accepting the funding provided under the Nunn-Lugar plan, Russian scientists continue to double-dip by selling their expertise for profit. This support will significantly reduce the time and costs required for aspiring nuclear

TABLE 5.2
Non-Proliferation Budget Requests (in millions)[a]

Department	FY01	FY02	FY03
Department of Energy			
Nonproliferation and verification R&D	238	322	283
Nonproliferation and international security	149	76	93
Nonproliferation programs with Russia	476	686	802
Total	$914	$1,030	$1,114
Department of State			
Nonproliferation and Disarmament Fund	15	14	15
Export Control and Border Sector	19	59	36
Bioweapons Redirection	35	67	52
IAEA Voluntary Contribution	50	50	50
Comprehensive Test Ban Treaty International Monitoring System	18	20	18
Korean Peninsula Energy Development Organization	75	91	75
Total	$212	$301	$246
Department of Defense			
Cooperative Threat Reduction	$442	$403	$417
Total Nonproliferation	$1,568	$1,734	$1,777

Composition (% of total) for FY03 – Energy 63%, State 14%, Defense 23%.

[a] Http://www.stimson.org/fopo/pdf/FollowingtheMoney.pdf.

powers to develop this technology. Finally, despite the assurances, guarantees, and financial support provided by the United States to India and Pakistan, border clashes continue unabated, as do anti-American demonstrations and attacks on U.S. assets within these countries.

In addition to underplaying the U.S. contribution to multilateral strategies, critics also tend to exaggerate the extent of European multilateralism.

European Unilateral Self-Interests and Priorities

Shirley Williams provides perhaps the best summary of the European position, one that uncovers fundamental differences over BMD and its impact on strategic stability.

Europeans are constantly seeking a firmer basis for the security for which they yearn. But they do not seek invulnerability. Shaped by a history during which their countries were invaded, defeated, occupied and despoiled, they know that to be a chimera. So they believe instead in political and institutional alternatives to conflict: the European Union, UN peace-keeping, the international criminal court, the European defence initiative, the Balkan stability pact and so on. Most of all, they want to keep the cat's cradle of disarmament treaties, UN conventions, barriers against nuclear proliferation and so on that constitute the frayed blueprint of a world of international law and order. To many Europeans, NMD presents itself as a threat to that objective, likely to encourage the proliferation of nuclear warheads in an angry, unsettled Russia, a China going through the throes of an uncertain succession, and an India determined to keep its newly claimed place on the ladder of nuclear powers.[33]

Although several high-ranking European officials continue to express concerns about Russian and Chinese proliferation in the aftermath of BMD deployment, their positions should be viewed in the context of two distinct security concerns.

Some European officials are worried that BMD will push the U.S. into the isolationist camp – as Americans become independently secure through their own defences, they are more likely to shift foreign policy priorities away from traditional commitments to European stability. While European critics of the U.S.-led NATO intervention into Kosovo should welcome the prospects of such a shift in American priorities, most are worried and for obvious reason: recent crises in Bosnia and Kosovo gave European leaders the clearest illustration to date that they have neither the will nor the capability to manage these sorts of conflicts on their own, or prevent escalation and spill-over if and when they erupt. Political shocks from the 1998–99 Kosovo crisis renewed European interests in creating their own security and defence identity, but these efforts have more to do with appearing less dependent on the U.S. while at the same time avoiding complete independence from the U.S. That few European leaders are pushing to reverse the downward spiral in defence spending is indicative of the true objectives underlying European Security and Defence Identity (ESDI): Europeans want control over decisions regarding deployment of the planned sixty-thousand-strong rapid reaction force but also want access to NATO assets (read 'American assets') should they need them. The jury is still out on whether the U.S. will make any significant concessions in that regard, particularly in light of recent tensions with European allies over Iraq.

It is important to note that European concerns about decoupling are entirely distinct from Russian and Chinese worries, which have more to do with the probability of the U.S. coupling itself more assertively to European security interests. The latter assumes the additional protection from BMD will give American officials the assurances they need when contemplating future interventions like those throughout the nineties. It is not clear which of these two mutually exclusive positions, decoupling or coupling, represents the more accurate assessment of BMD's implications for U.S. foreign policy, but the fact that both interpretations are accepted is informative. The decoupling option is more apparent today as American leaders continue to push for rapid NATO expansion, and as NATO becomes less central to U.S. security priorities in the aftermath of 11 September and the second gulf war. New alliances with Afghanistan, Pakistan, and other states in the Middle East are emerging as central to American security priorities.

Other Europeans are concerned that BMD could propel Europe into the position of surrogate target for missiles that otherwise would have been aimed at an unshielded North America. Although European views vary from one capital to another, those who are critical of BMD are more likely to be concerned about second-order vulnerabilities. That is, over time the threats associated with WMD proliferation may become a more serious concern, not because of any foreign policy initiative they might defend but because BMD may create incentives for terrorist and rogue leaders to use European assets and cities as proxy targets. From the point of view of prospective terrorists, European proxy targets are appealing for three reasons: (1) geographic proximity makes it easier to hit with ballistic missiles, (2) the lack of a defensive shield means that it is cheaper for rogue states and terrorist leaders to build and deploy ballistic missiles without having to worry about countermeasures, and (3) a ballistic missile directed at a European capital is cheaper to build and more likely to work if launched. Not only does this particular fear explain why European officials are opposed to BMD, but it also explains why they appear to be more favourably disposed to President Putin's alternative – an ABM-compliant, theatre missile defence system based on boost-phase technology. This alternative is, for Europeans at least, far more conducive to addressing their fears about BMD, proliferation, and surrogate targeting.

Despite European reservations about BMD, criticisms from Paris,

Berlin, London, and Moscow have become increasingly muted since 11 September and will become even less pronounced as France and Germany work to re-establish their ties with the Americans by addressing the diplomatic fallout from divisions in NATO and the UN over Iraq. The terrorist attacks on 9/11, the Iraq war, and future terrorist attacks will essentially kill any criticism from Europe regarding fabricated threats from rogue states – even in the context of the difficulties experienced by inspectors, after the Iraq war, to find the WMD that were identified in hundreds of UN (IAEA, UNSCOM, UNMOVIC), U.K., German, French, and American intelligence reports over the previous decade. Europeans have all but conceded the point that the United States is facing serious security threats and have come to understand that the U.S. does not need permission from allies to proceed with BMD. David Warren writes, 'No person and no nation has ever required the slightest permission to act in self-defence. No moral principle or law can possibly require this.'[34] The muted reactions by Russia and China to the U.S. withdrawal from the ABM Treaty will circumvent whatever remaining European criticisms exist, and Europe's corporate and business communities see potentially lucrative contracts emanating from a full-blown BMD program.

The clearest evidence of a European change of heart was witnessed in the closing communiqué issued by NATO foreign ministers after a one-day meeting on 29 May 2001: all references to the ABM Treaty were dropped, marking a major reversal by European allies. As Ben Barber points out, 'at the same meeting a year ago, the ministers described the treaty, which stands in the way of U.S. plans for a missile defense system, as "a cornerstone of strategic stability." NATO Secretary-General George Robertson said support for the ABM was dropped, in part, because last year "there was a different American administration."'[35] It is more likely that the new European consensus is derived from the view that BMD is essentially a *fait accompli* – a justifiable reaction to the end of the cold war and 11 September. Javier Solana, the former NATO secretary-general, issued a statement that the 1972 ABM Treaty 'is not the Bible. For us Europeans, what we would like is for the major powers to [reach a deal] by consensus if possible.' He added that the U.S. has 'a right to deploy' BMD.[36] In fact, one European diplomat recently indicated that the problems with 'national' missile defence 'may be mainly semantic.'[37] He implied that if the name is changed the concept might be more palatable to European allies – which explains the shift in language from NMD to BMD over the past two years.

Conclusion: A Fait Accompli

The arguments outlined in the previous two chapters combine to explain the current U.S. position on BMD. The effect of 11 September was to speed up the process, because it provided the clearest and most vivid illustration of the enormity of the security threats facing the U.S. and its allies.

The main preoccupation for American and Russian force planners in the future will be how to accommodate bipolar strategic stability with the American need to address emerging ballistic missile threats. How do you reconcile nuclear deterrence in a bipolar context with discriminate deterrence in a multipolar setting? The 'how' of BMD will be resolved in favour of a limited, multiple platform (land, sea, and air), layered (boost, mid-course, and entry level) missile defence system. Following another successful fully integrated test in July 2002, the Bush administration is moving quickly to construct radar and interceptor sites in Alaska and California and to speed up the BMD testing schedule.

Bipolar strategic stability and deterrence will be redefined in the context of contemporary international politics or it will cease to be relevant. While the logic of mutual nuclear deterrence is impeccable, its relevance will continue to vary from context to context, depending on the health of the relationship in question. In a post–cold war world, stability must be expanded to accommodate the realities of a complex international system with expanding sets of interlinking and interdependent nuclear relationships.[38]

The critical policy puzzle is to find the right balance of *mutual* and *discriminate* deterrence to enhance both bilateral and multilateral stability. The challenge for U.S. policy makers is to develop a defensive system with adequate transparency and predictability to convey clearly to Russia and China that it is indeed directed against third parties.[39]

It is important to understand that BMD is not just another initiative on the foreign policy agenda of a naïve, unilateralist American president with little understanding of the complexities of international politics. Nor is it simply a matter of waiting for cooler minds to prevail, another U.S. election, or the right mix of advisers to guide the president along a saner path. The U.S. effort to develop an alternative approach to strategic stability by constructing a new balance of offensive and defensive systems is not a temporary phenomenon – it represents the beginning of a long-term strategy to reshape and update strategic stability to address contemporary (and emerging) threats in a more complex international environment. We are experiencing a paradigm shift in nuclear deterrence, arms control, and disarmament thinking that will remain a permanent fixture of all future force-structure debates in the U.S. and Russia.[40]

Current and future U.S. efforts to deploy a BMD go well beyond ideological divisions in the United States. The difference between the Democrats and Republicans on BMD deployment is minor and superficial. During the 2001 campaign, Clinton and Gore understood the political costs/risks of prematurely breaking ground in Alaska, which explains why they decided, during the 2000 election, to pass the buck to the next administration. At the time, the Pentagon adjusted their timeline for development of a more advanced (and lighter) booster rocket to 2006/7. With that new date in mind, it made very little sense to begin breaking ground for construction of BMD radar sites on Shemya Island (in Alaska's Aleutian chain) to meet a 2005 timeline, since the interceptors wouldn't be ready for another year. The diplomatic and domestic political costs associated with undermining the spirit (if not the letter) of the ABM Treaty, at that point in time, were considered too high, with no measurable benefits in security to compensate. The decision to postpone deployment also made sense from the perspective of a legacy-conscious president hoping to avoid responsibility for killing the ABM Treaty.

To the extent that differences exist at all they have more to do with the tactics of opposition party politics and legislative bargaining strategies. Few members of the U.S. Senate or House of Representatives are seriously opposed to any form of BMD. Those who are critical are more likely to voice concerns about specific approaches to deployment rather than deployment itself. Very few (if any) legislators have made BMD the core item on their legislative agenda, so even if they are critical they are very likely to support the president's program in exchange for White House concessions on issues that are more important to their constituents. And with respect to those constituents, public support for BMD remains very high throughout the United States; it is likely to remain strong enough to persuade current and future U.S. administrations to continue with testing and deployment plans, especially given the prospects of constantly improving interceptor technology.

The paradigm shift towards a new nuclear age will also affect the NACD regime and associated strategies. Coordination and consultation, not codification, will become cornerstones. Rather than codifying numeric goals and expecting them to be valid over time, complexity dictates that states will increasingly choose to retain the prerogative to adapt to changing circumstances and will avoid irreversible restriction by artificially rigid rules and regulations.[41] Artificial competition, adversarial negotiations, never-ending bargaining periods that typically are drawn out for years (often for purely domestic political reasons), detailed verification procedures, and so on are all part of an outdated approach to arms control that will soon reach the point of irrelevance. The assumption that there are no realistic alternatives to the status quo is simply wrong. Moreover, rejection of U.S. offers to unilaterally disarm, simply because unilateral disarmament takes place outside the confines of traditional arms control, is profoundly shortsighted, counterproductive to NACD goals, and dangerous.

This has become a moral imperative for any American president: 'No U.S. president can responsibly say that his defence policy is calculated and designed to leave the American people undefended against threats that are known to exist,' writes Gordon Giffon.[42] With respect to European concerns, former secretary of state Henry Kissinger sums up the American reply: 'Total vulnerability should not be the price the United States is asked to pay' for transatlantic solidarity. He went on to point out that BMD is significantly safer and saner than ongoing dependence on mutually assured destruction.[43] And David Warren notes, 'If it is possible to devise an anti-missile defence that could intercept surprise launches, you do it. You would be crazy not to. It is worth a very large amount of impounded tax money; and worth running risks that the technology can work in practice as well as theory. Why is it worth tens and hundreds of billions? Because tens and hundreds of millions of people stand in the way of being gassed or incinerated. That is not an abstract idea: we know precisely what the consequences would be of missile attacks on major cities.'[44]

Finally, deployment makes sense because of increasing probabilities of success, the inevitability of further improvements through trial and error, and the risks and benefits to overall security when compared with alternative strategies. The fact that BMD has not yet reached 100 per cent accuracy is irrelevant and always will be. What is relevant to policy makers is whether the pace of innovation is such that BMD will, at some future point, produce a high-enough probability of success to warrant development, testing, and deployment. The current program provides more than sufficient promise of enhanced security to offset the costs and potential risks, and the status quo approach to bipolar strategic stability grounded in MAD is obsolete. It raises far too many barriers to the changes that need to be made to deal with the proliferation problems multilateral approaches cannot solve.

Chapter 6

The Inevitability of Terrorism, and American Unilateralism: Security Trumps Economics

Sceptics in more traditional fields of scientific inquiry often claim that unlike their areas of expertise (e.g., physics, chemistry, and biology), forecasting human social, political, and international behaviour is virtually impossible. Actions of inanimate objects and chemicals under controlled experimental conditions, they assert, are much easier to predict than the behaviour of political officials interacting within and across complex social settings, bureaucracies, states, and international systems. But, in reality, accurate social scientific predictions are relatively easy to make, because some trend lines are considerably more obvious and informative than others. The hard part is convincing sceptics that the future is clearer than they might otherwise imagine, and that the evidence to support certain predictions is compelling enough to serve as a strong basis for informed policy guidance. Rejecting any and all claims to predictability is perhaps one of the more serious impediments to developing sound foreign and defence policies, because if one starts with that assumption, any speculation about the future – and any policy recommendation that flows from such speculation – can be accepted or rejected for reasons other than sound logical and empirical analysis. That is precisely the problem with conventional wisdom on the benefits of multilateralism in a world of globalized terrorism and proliferation (reviewed in chapter 1). Following a more systematic evaluation of the flawed logic and contradictory evidence underlying the conventional wisdom (chapter 2), in-depth case studies of the Iraq war (chapter 3), and missile defence/proliferation (chapters 4 and 5) were used to defend the mutually exclusive prediction (and prescription) that unilateralism will remain an important feature of U.S. foreign, security, and defence policy for decades.

If one measures the relative severity of a security threat in terms of the amount of attention and resources invested by a government to address the problem, American officials are clearly convinced that the nexus between terrorism and WMD proliferation represents 'the greatest existential challenge of our age,' according to Charles Krauthammer.[1] And, as noted in the U.S. National Security strategy, 'New deadly challenges have emerged from rogue states and terrorists. None of these contemporary threats rival the sheer destructive power that was arrayed against us by the Soviet Union. However, the nature and motivations of these new adversaries, their determination to obtain destructive powers hitherto available only to the world's strongest states, and the greater likelihood that they will use weapons of mass destruction against us, make today's security environment more complex and dangerous.'[2] This challenge is the most recent manifestation of a subset of historical patterns, strategic imperatives, and associated behaviours that have always been fundamental to international politics – the unilateral economic, political, and military self-interests of states and peoples, the primacy of security and defence, the primordial competition to acquire the status and influence that comes from hard power, and, of course, state or regime survival. But unlike many other security challenges, this particular confluence of emerging threats is severe enough to be transformative – it is changing the way the U.S. hegemon deals with the world and in the process changing the world.

The domestic and international forces at play today are creating a security environment in which American officials will become increasingly committed to (and politically dependent on) a set of priorities in which *security trumps everything*. Critics of U.S. foreign policy must move beyond the superficial assumption that these American security imperatives are only temporal in nature, simply a product of a conservative-hawkish cabinet led by a Republican president who managed a victory in the 2000 U.S. national election. That assumption is dangerously superficial – terrorism and proliferation have become facts of life for American citizens and, by extension, any current and future U.S. government. The U.S. will continue to be threatened by, and experience the devastating physical and psychological consequences of, terrorism. This emerging security reality, essentially the Israelization of the United States, will create ongoing pressures on American officials to institutionalize a patterned response that will continue to mould and shape the U.S. foreign and security policy paradigm. And this new paradigm will prioritize a multiplicity of approaches, strategic alliances, and policies that will inevitably include unilateral initiatives. Barring any fundamental transformations in human nature, exclusive reliance on multilateral approaches to security, guided by process-related legitimacy derived from multilateral consensus, is not likely to become a serious policy option for the U.S. government, or any government, for that matter. Contrary to the conventional wisdom described in chapter 2, the transaction costs and security risks of exclusive reliance on multilateralism are increasing, not decreasing, and will never overcome the imperative, on occasion, to act alone.

The following chapter offers several additional explanations for the inevitability of terrorism and unilateralism.

The Inevitability of High-Impact Terrorism: Globalism and the Privatization of War

There are three reasons to expect an increase in the scope of terrorist activity in the future – the 'democratization of technology' (Thomas Friedman), the 'privatization of war' (Joseph Nye), and the 'miniaturization of weaponry' (Anthony Cordesman and Tara O'Toole). Building on Friedman's analogy of democratizing technology, Nye explains the effects of low costs and increasing accessibility: '[T]errorists can be more agile and more lethal than before. In the 20th century, a malevolent individual like Hitler or Stalin needed the power of a government to kill millions of people; if 21st-century terrorists get hold of weapons of mass destruction, that power will for the first time be available to deviant groups and individuals. This *"privatization of war"* is not only a major change in world politics, but the potential impact on our cities could drastically alter the nature of our civilization [emphasis added].'[3] The lethality of biological agents, for example, is many times that of either chemical or nuclear weapons.[4] Approximately thirty kilograms (66 pounds) of anthrax could kill five hundred times as many people as three hundred kilograms (660 pounds) of deadly Sarin nerve gas.[5] Well-dispersed anthrax could kill 20 per cent more people than a 12.5 kiloton nuclear bomb (approximately the size of the bomb dropped on Hiroshima).[6] Two hundred fifty pounds of anthrax could kill three million people – the equivalent effect of one million tons of TNT.[7] In the conclusion to her sobering report on the subject, O'Toole outlines several factors that will increase the probability of biological use in the future:

First, in the near term, non-state actors may be helped by rogue states, removing technological obstacles to the efficient use of any kind of CBRN [chemical, biological, radiological, nuclear] weapons. Second, proliferation in the developing world or the insecurity of the Russian or [former] Iraqi CBRN arsenal may provide a ready source of agent to terrorists. Third, future advances in genetics may make it easier to create more potent and more easily useable agents. Fourth, technological barriers to the effective dispersal of biological agents are disappearing, and knowledge of these advances will inevitably spread. Finally, our medical and public health systems do not have the capacity to handle a large-scale, CBRN terrorist attack. In the future, terrorists using biological weapons are likely to have a much greater capacity for mass destruction.[8]

The democratization of technology, privatization of war, and miniaturization of weaponry will, among other effects, increase the number of individuals, groups, terrorist regimes, and rogue nations that, if committed, could acquire the capacity to produce, transfer, and use more lethal weapons. The capacity to inflict unacceptable levels of pain and suffering on a larger number of targets will spread, by definition. The nexus between an increasing number of potential terrorists with various motivations to terrorize, accessibility as a consequence of the democratization of technology and knowledge, and the evolving capacity to transfer increasingly lethal weapons encompass only a portion of the threat profile governing U.S. foreign and security policies today.

The privatization of war also means that conflicts will become more intractable, because there are likely to be a larger number of committed spoilers on both sides of future disputes with the power to quash consensus by escalating the violence and making concessions more difficult. The protracted nature of conflict in the Middle East is perhaps the best illustration of the problem: although the U.S. managed, in 2003, to exclude Yasser Arafat from the negotiation process and obtain commitments from Israeli officials and the new Palestinian leadership to refrain from any violence and hostilities, attacks continued. As predicted, terrorist attacks have been relentless, despite agreements. There are thousands of individuals committed to at least four terrorist groups (Hamas, Islamic Jihad, Hezbollah, and the Palestinian al Aqsa Martyrs Brigade) with the power to control the success and failure of negotiations. Even if the Hamas leadership decides, for whatever reason, to negotiate a cease-fire, persons within Hamas who are more committed to extremist demands retain the power to shape perceptions of progress and the commitment to stay the course. Now consider the implications of greater accessibility to more lethal weapons in the future.

In addition to the problem of existing conflicts becoming more intractable, the number of intractable conflicts are likely to proliferate as a result of the privatization of war. In the future, persons committed to their own causes will emerge as relevant players and potential enemies with an unending list of grievances to resolve. The 2001–2002 anthrax attacks in the U.S. could conceivably have been initiated by a single person; it took only nineteen men to transform the priorities of a superpower and, as a result of its response, the priorities of an entire international system.

Fighting an Unwinnable War on Terror

Conventional wars have a beginning, middle, and an end, but the war on terror has no temporal reference points, and certainly no end is in sight. This difficult challenge explains why some critics were so concerned about assigning the title 'war' to the struggle against Al Qaeda and other terrorist groups. Victory will remain elusive because the only available solutions tend to produce unintended consequences that create additional (and sometimes worse) problems. The military intervention in Afghanistan, for example, succeeded in dismantling Al Qaeda's terrorist infrastructure and changed a very oppressive Taliban regime, but these successes created additional hurdles for counter-terrorism. The pre-11 September Al Qaeda may no longer exist, as Dennis Pluchinsky points out, because it has 'metamorphosed into a new structure with different characteristics, tendencies, procedures, communication codes, false documents, membership, command and control, travel methods and financial channels.' Pluchinsky offers the following analogy to illustrate the problems intelligence communities now face as a result of the 'success' in Afghanistan:

Imagine your boss coming to your desk, placing a lunch-size brown bag twisted at the top on your desk and asking you to tell him what the contents mean. The contents are some 120 disparate pieces of a puzzle. As you look over the puzzle pieces, you immediately notice that about one-third of the pieces are blank, and another one-third appear to have edges that have been cut off. As you look at those remaining pieces that are intact and have some part of a picture on them, you sense that this is really a mixture of about four different puzzles. Now keep in mind that you have no box top to tell you what the puzzle(s) should look like and you do not know how many total pieces are in the puzzle(s). Welcome to the art of terrorist threat analysis![9]

Perhaps most disturbing of all is the likelihood that Al Qaeda will 'eventually transform itself into a "leaderless resistance" movement like the Earth Liberation Front or the defunct Revolutionary Cells in Germany ... [and] could become amorphous, with no structure, hierarchy, central headquarters, or group dynamics.'[10]

Consider another paradoxical outcome of the war on terror: successful efforts to track and impede terrorist funding will force these groups to become even more competitive in recruitment, funding, and sponsorship efforts. To use a corporate analogy, the 9/11 terrorist attacks had a very positive impact on Al Qaeda's stock, recruitment, and support base. The attacks became the benchmark for successful terrorism and the new corporate model for a profitable and efficient terrorist organization. This will very likely motivate other groups to outperform 9/11 in an increasingly competitive security environment. Ironically, the more successful the global effort to control terrorist financing, the more ruthless the competition to demonstrate to potential sponsors the value of their product and organization – all of which gives added meaning to the phrase Terrorist Inc.

Costs and Benefits Are Identical to Terrorists

Recall from chapter 3 that successful coercive diplomacy requires effective communication (e.g., an explicit threat of serious consequences), a clear commitment to the issue at stake, the capability to impose painful military, economic, or political costs on the opponent, and the resolve to carry through on these threats. The theory stipulates that if these conditions are met, the coercive threat should deter the specified act or compel the opponent to act in ways that are consistent with the demands stipulated in the ultimatum. Successful coercion relies on an opponent's rational calculation that the costs of defection (attacking) will be significantly higher than any benefits to be gained – if the costs are too high, or the benefits too low, the coercive threats should prevent the attack.

Conventional methods of coercion are not likely to work against terrorists, for several reasons. The privatization of war will make it very difficult to identify prospective attackers or pinpoint the location of targets, so communicating threats of retaliation will become more difficult. CIA Director Louis Freeh predicts that the next attack will be in cyberspace, citing the following example: 'A very young man in Stockholm, Sweden, hacked his way into the 911 system in southern Florida and began to shut them down as a challenge. He knocked out police, fire and rescue services, without leaving the comfort of his own home.'[11] It is impossible to issue threats of retaliation when someone is sitting in front of a computer thousands of miles away. It is comparatively easy to apply coercive threats against states whose leaders are easily identifiable and whose potential acts of aggression are more apparent. Tom Ridge, head of the Department of Homeland Security, acknowledges the challenge to protect oil and gas refineries, power plants, and electrical substations, water treatment plants and reservoirs, dams, pipelines, schools and hospitals, banks and financial institutions, our airports and seaports, bridges, and highways.[12] These attacks, all potentially devastating, are almost impossible to predict and increasingly difficult to deter.

Perhaps the most serious impediment to successful coercion is the unconventional nature of a terrorists cost-benefit calculations. Mobilizing a powerful military for a massive retaliation against an opponent's impending attack would convey, in most cases, that the costs (e.g., death or regime change) are likely to be too high. But such cost-benefit calculations are less relevant when the opponent is willing (indeed, determined) to die for the cause – the costs, in other words, represent the benefits. If terrorists rationally seek to provoke a disproportionate response by major powers, as David Lake warns in his study of rational extremism, the more indiscriminate the retaliation, the more likely it is that innocent civilians will die. This will alter the preferences of moderates in these societies and produce greater support for the extremist cause. Paradoxically, the more successful a military operation to address the terrorist threat, the more likely it is to provoke the very behaviour it was designed to prevent.[13] When one combines the increasing inability of major powers to employ coercive compliance through traditional threats of punishment with the privatization of war, the democratization of technology, and the miniaturization of weaponry, the enormity of the security challenge becomes clear.

Despite the willingness of the American public to provide the government with the tools it needs to succeed in the war on terror, the trends described above will make it very difficult for Washington to demonstrate success. The most difficult challenge was identified by Joseph Nye, who argued that the very features of America's greatness – its economic wealth, technological prowess, critical infrastructure, skyscrapers, political freedoms, and civil liberties – also represent its central weaknesses. The larger, freer, and more technologically and economically interdependent a society, the more numerous, interlinked, and accessible its potential targets, and the more significant the ripple effects of relatively small attacks.

The Inevitability of Unilateralism: Domestic Determinants and Pressures

Blurring the Probability-Impact Calculus

A related consequence of the privatization of war is the absence of a clear distinction between *high probability–low impact* threats (for example, a single attack by an individual against a single target) and *low probability–high impact* threats (such as a ballistic missile attack on American soil). The American reaction to the 9/11 attacks revealed that impact can no longer be based (if it ever was) on an objective and dispassionate assessment of lives lost (about 3,000) and the amount of damage inflicted. Impact is a product of the public's perception of the attack, the corresponding political demands the public places on the government to prevent another attack, and the pressure the government subsequently exerts on individuals, groups, organizations, companies, and other governments to accommodate the need to maximize security. The nature of probability-impact calculations was changed by 9/11 because it demonstrated that a relatively small and inexpensive attack by nineteen men could have profound effects on the entire system – including a complete overhaul of U.S. foreign and security priorities, extensive revisions to refugee and immigration policies and practices, two major wars (one without the consent of key allies), serious divisions within NATO and the United Nations that have arguably transformed international alliance structures, and system-wide disruptions in domestic and global economies resulting in the loss of hundreds of thousands of jobs worldwide, and dozens of bankruptcies of major companies in the airline and travel industries. The distinction, therefore, between high- vs low-*impact* threats is becoming increasingly blurred, because the threshold for what Americans consider to be unacceptable damage has proven to be quite low in the aftermath of 9/11. Moreover, the democratization of technology, miniaturization of weaponry, and privatization of war will make it very difficult in the future to distinguish high- from low-*probability* threats. Both trends combine to create great pressure on political officials to demonstrate successes in a security environment in which success will become increasingly difficult to demonstrate.

Public Expectations of Success and the Growing Impact of Smaller Failures

Any discrepancy between the level of security the public expects (or feels it deserves) and the actual level of security the government appears to be delivering will increase the public's threat perceptions and produce corresponding political pressure to fix the problem. The larger the discrepancy between these two perceptions, the more pressure the government will experience from the public and business communities to do something about it, and the larger the impact when the response kicks in. The reason 9/11 had such a profound impact was that public expectations of security were much higher in the aftermath of the cold war – the attacks were so unusual and unexpected and more devastating as a result. The more unexpected the attacks, the more disturbing the effects on the public's sense of security when they happen, and the more significant the pressure to fix the problem (see table 2.2 and appendix B for examples of Washington's response to 9/11). While the pressure may not be direct, overt, and relentless, elected officials understand that the public's sense of security is paramount to their own political survival.

Herein lies the crucial security paradox facing the U.S. government today – the more money invested in homeland security and other counter-terrorist and counter-proliferation programs, the higher the public's sense of security and corresponding standards for measuring success and failure, and the more significant (and negative) the political and economic impact when the next failure occurs. To illustrate the point: if another attack like 9/11 occurred today, say, on Chicago's Sears Tower, notwithstanding the billions of dollars the U.S. government has invested to prevent a repeat of 9/11 (and whatever successes these investments produced), the corresponding loss in the public's sense of security is likely to be even more significant and disruptive than the 9/11 attack. Public standards for measuring success and failure are directly related to the sacrifices and costs incurred for security, so even small attacks today are likely to be perceived as significant failures.

America's Diminishing Pain Threshold and Its Addiction to Security

As the public's expectation for security rises, any indication that the government has wasted billions of dollars on homeland security and risked the lives of thousands of American troops in foreign interventions will produce a disproportionate response demanding that political officials do more. This dynamic will become apparent following the first (and subsequent) suicide bombings on U.S. soil, because these attacks will be perceived by the public, the media, and opposition parties as confirmation that all the unilateral and multilateral efforts have failed to protect American citizens. As expectations for higher levels of security increase, and as corresponding thresholds for acceptable levels of pain decrease, billions of dollars will continue to be spent by both parties in a never-ending competition to convince the American public that their party's programs will be more successful.

The U.S. government will become addicted to security for two reasons: spending will never be sufficient to achieve absolute success in the war on terror, and it will take that much more effort, after each attack, to demonstrate a clear commitment to stop the pain. Consider some of the early signs of addiction: spending more on the addiction despite diminishing returns; convincing critics that the dangers and unintended consequences of the obsession are exaggerated; an increased willingness to accept the negative legal, moral, social, and financial ramifications of the practice; becoming defensive when others question your behaviour and priorities. The analogy can perhaps be pushed a little too far, but it is possible to identify a similar dynamic at play in the United States today as it relates to public demands for security, the government's willingness to answer the call, and concerns expressed by both opponents and allies in the international community. Both political parties will try to outdo each other in an effort to exhibit (and distinguish) their respective commitments. Even small attacks will provoke a disproportionate reaction to prove to both domestic and international audiences (friends and foes) that Washington will not relent in its effort to secure the nation. Consider the millions on critical infrastructure protection: by most accounts, these investments (in both United States and Canada) are considerably out of proportion to the potential impact of most critical infrastructure attacks, but government officials today are motivated to prevent any indication that they have been hit. In fact, so high are the worries about a public backlash that Washington has established colour-coded categories for threat warnings to apportion some of the blame to the public for not heeding Washington's warnings to be more careful.

The spiral in security spending will inevitably produce diminishing returns as the public continues to demand a level of protection the government is increasingly incapable of providing. Threats and related attacks will continue to motivate Washington to win an unwinnable war, a war in which losses (failures) will always loom larger than gains (successes). The positive effects of arresting five terrorists known to have been planning an attack that could conceivably have killed thousands will never be as significant as the negative impact from a single car bombing that kills ten people. The U.S. government faces enormous challenges, therefore, to establish some balance in expectations and assessments of success and failure, because the public will invariably perceive a greater loss in security from a minor attack than a corresponding gain in security from stopping a potentially major attack.[14]

The side effects of the addiction (to push the analogy a little further) – for example, security legislation that places limits on civil liberties, rigid enforcement of refugee and immigration regulations, enhanced intelligence and surveillance capabilities for the CIA and FBI, increased trade and related transactions costs at the border – will become less acceptable to the public in the absence of another attack. Obviously as concerns over security subside, the American public will become less willing to suffer the side effects or incur the economic costs of added protection. But if high-impact terrorism is inevitable, for reasons outlined in the preceding section, U.S. citizens will become increasingly willing to pay for (and demand) more security. For instance, U.S. Attorney General John Ashcroft's successes at obtaining from Congress additional search, surveillance, and intelligence capabilities were related to the proximity to various terrorist attacks – he managed to obtain additional powers regardless of whether deficiencies with existing security and surveillance measures were responsible. As noted in a recent report on the legacy of the 9/11 attack, 'war on terrorism affects people's daily lives, they become frightened; they look to government to provide law and order for protection. They become less concerned with justice (i.e., fair and equal treatment of people accused of crimes) and more concerned with preventing harm ... Consequently, citizens are more likely to give wider latitude to government to provide order. And governments in Europe, as well as the U.S. government, have acted accordingly.'[15]

TABLE 6.1
America's Security Challenges

1 Inevitability of high-impact terrorism
2 Privatization of war
3 Democratization of technology
4 Miniaturization of weaponry and lethality
5 Blurring of probability-impact calculus
6 Israelization of the United States
7 Fighting an unwinnable war on terror
8 Paradoxes of power and weakness
9 Terrorist cost-benefit calculations and the diminishing utility of rational coercion
10 The public's expectations of success and disproportionate impact of smaller failures
11 America's declining pain threshold and its addiction to security
12 Diminishing trust in traditional alliances
13 Failure of multilateralism in practice
14 Relative promise/success/utility of unilateralism in practice

Security Trumps Economics

The domestic and international factors described above, and listed in table 6.1, have created an environment for American officials in which security has become the prime directive, the priority that will continue to trump almost everything else – including trade and economics. To revisit Bill Clinton's 1992 campaign slogan, when it comes to the debate over whether sound security and defence policies are more important than sound economic policies, it's *not* 'the economy, stupid!'

The *economics of security* trump the *security of economics*, because terrorist attacks will have an immediate and negative impact on the economy anyway, but weak or misguided economic policies have little to no impact on security. Security failures often strip control over the economy from the hands of government officials, and in a security-conscious society faced with these challenges the loss of 3,000 lives will invariably be perceived by the American public as a more significant tragedy than the loss of 300,000 jobs. Conversely, the potential to save 3,000 lives will be perceived as more important than the potential to create 300,000 new jobs. Consider the significant economic downturn in the U.S. economy, from 2001–2003, as a result of the crash in technology stocks, the 9/11 terrorist attacks, corporate scandals at Enron and WorldCom and a corresponding decline in overall consumer confidence in the stock market, the wars in Afghanistan and Iraq, a growing U.S. deficit, and, notwithstanding these economic pressures, President

Bush's commitment to follow through on a substantial tax cut. Despite all this, George W. Bush's popularity remained solid, while support for the Democrats, generally assumed to be more competent on economic matters, was much lower throughout this period. The fact that the Republicans did considerably better than the Democrats in the 2002 midterm election is particularly relevant to the point. President Bush may not have had a very popular economic plan or record, but the Democrats had no security plan.

The Inevitability of Unilateralism: External Determinants and Pressures

Diminishing Confidence and Trust in Traditional Allies

With respect to corresponding effects at the international level, the more effort by other states (including traditional allies) to balance American power (to protect their own economic and political interest) by demanding American compliance with multilateral consensus, the more incentive Washington will have to demonstrate a credible commitment to protect its own security by acting alone, if necessary. Conversely, the more effort Washington puts into unilateral initiatives to protect core security interests, the more pressure European powers will place on the U.S. to remain committed to multilateral consensus.

Among the characteristics of post-9/11 international politics is the speed at which strategic relationships are changing. Consider some recent events that a few years ago would have been highly improbable: Pakistan and Uzbekistan emerging as more important to American security priorities than France or Germany; a French, German, Russian, and Chinese coalition to thwart U.S. efforts in Iraq; NATO being prevented from providing Article V security guarantees to a member, Turkey, because of French and German opposition to the U.S.-U.K.–led war on Iraq; a decision by the Turkish government to withhold support for deploying American troops through Turkey into northern Iraq as a way of pressuring American officials for a better financial package (see chapter 3); and Canada's decision to oppose the war in Iraq despite being one of the biggest contributors to the military campaign. The consequences of this for American perceptions of (and reliance on) traditional allies will be considerable. Sharing security priorities is the hallmark of an effective alliance, but as security priorities continue to diverge, U.S. officials will feel more threatened by the institutional constraints imposed on their actions by other major powers that do not share motivations to respond to the same threats. Tensions are likely to mount as other powers become more dependent on institutions that constrain U.S. hegemony. The features of American power that enhance its capacity to act alone are the very elements of American power that France, Russia, and China would like to control.

In an unpredictable world in which alliance support is less trustworthy, and one in which constantly shifting strategic coalitions are more efficient for addressing emerging threats, unilateralism is becoming more rational at a time when exclusive reliance on multilateralism is becoming less prudent. The more significant the threat, the less likely the U.S. government will comply with constraints imposed on it by multilateral institutions that cannot demonstrate an impressive enough record of achievement to warrant serious consideration. These tensions are expected to become more pronounced as the threats from terrorism and proliferation increase over time.

Diminishing Reliance on Multilateralism

In theory, multilateralism has features that make it particularly well suited to addressing contemporary threats – working together in harmony for a common cause is always better than working alone. But the more relevant question is whether the kind of multilateralism we have managed to create is capable, with all its deficiencies, of addressing the problems of globalized terrorism and proliferation. Does exclusive reliance on multilateralism hold the most promise for managing and resolving these problems? Listing the potential benefits of cooperation is often held up as proof that multilateralism is becoming more essential, but proponents of this view need to distinguish the multilateralism of our dreams from the multilateralism of contemporary international politics. As the Iraq case study shows, international politics encompasses interactions among multiple unilateralisms and competing multilateralisms, and the latter are typically a function of the capacity of major powers to generate support for their respective coalitions by accommodating the unilateral self-interests of other states (see chapter 3). The propensity of leaders to use, or bypass, multilateral institutions for strategic reasons places contemporary multilateralism in its proper context. This is all we, as a community of states, have managed to accomplish to date; perhaps it is the best we can do, or the best we can hope to accomplish in a post-9/11 world.

While the theory may look good on paper, multilateralists have an obligation to explain how their ideal can be made to work in practice. When forced to confront evidence of the many impediments to effective multilateralism (outlined in chapters 3 through 5), the record is anything but impressive. As such, it would be irresponsible for any leader, including the U.S. president, to develop policies based on the illusion of shared values. Multilateral approaches will be rejected by any state if they preclude options considered essential for achieving core security objectives, or if collective action cannot achieve the same objective with fewer risks and costs. If France threatens to veto a U.S.-U.K. led Security Council resolution, or if Turkey denies the U.S. military access through its territory, American officials will invariably move on to plan B. Going to the UN Security Council to obtain multilateral consensus is not inconsistent with a preference for maintaining unilateral options if requesting multilateral support carries few risks and is relatively cost-free. No leader would rationally prefer to solve a problem on his own if working with others would accomplish the same outcome. However, it is the outcome (and associated costs and risks) that matters, not the process. The fact that the U.S. may gain something from process-related legitimacy will not make Washington dependent on that support.

Efforts to obtain multilateral support do not constitute evidence of a new liberal internationalist order. Multilateralism is not unexpected in a state-centric, realist world if multilateral cooperation serves the self-interests of the states involved. On the other hand, multilateralism would be anomalous if compliance with institutional constraints jeopardized the core security interests of the state in question, yet despite these costs compliance remained strong. A decision by South Africa, Brazil, or Chile to dismantle their respective nuclear weapons programs, for example, is not as unexpected as, say, a decision by India, Pakistan, and/or Israel to dismantle theirs. The strategic relationship and security context of the rivalry, not the values these states have for multilateral arms control, accounts for their behaviour. States do not support multilateralism because it serves the interest of global stability and world peace; they support these institutions because they offer the best hope of checking and balancing other powers, including the U.S. But American power is not something they feel compelled to check because it is abusive to Bosnians, Afghans, Kosovars, Iraqis, or any state in which the U.S. military has intervened; France, Russia, and China feel compelled to check American power because, despite the fact that American interventions have helped to reverse the effects of genocide, massive human rights abuses, and other crimes against humanity, these operations undermine the capacity of officials in Paris, Moscow, and Beijing to use their power and influence to protect their interests.

American presidents will always prefer operations that obtain multilateral legitimacy, because these actions are more likely to be supported by the public. But they will not be constrained by that priority alone when the risks of exclusive reliance on multilateral consensus are too great. The fact that asymmetric threats are spreading and that terrorist attacks will become more likely and more lethal in the future will compel any U.S. president to assign less weight to the public's preference for multilateralism (and the corresponding losses in public support if multilateral consensus is not obtained) and comparatively more weight to the public's demand for security. As the margin for acceptable error in the war on terror decreases, the public backlash following the next terrorist attack will be considerably more damaging (and politically costly) than the failure to obtain multilateral consensus for an intervention that is designed to prevent those attacks.

Prioritizing Unilateral Arms Control and 'Discriminate Deterrence'

With respect to predicting how American officials will modify the old concept of strategic stability' to address the problem of stopping terrorists and other potential proliferators, it is clear that deterrence will be redefined or it will cease to be relevant. While the logic of nuclear deterrence (and mutually assured destruction) is impeccable, its utility will continue to vary from context to context. As stated in the U.S. National Security Strategy, 'In the Cold War, especially following the Cuban missile crisis, we faced a generally status quo, risk-averse adversary. Deterrence was an effective defense. But deterrence based only upon the threat of retaliation is less likely to work against leaders of rogue states more willing to take risks, gambling with the lives of their people, and the wealth of their nations.'[16] The logic of cold war strategic stability implied that responsible, rational leaders would refrain from hostilities that would threaten the survival of their nation and their leadership or their capacity to fight and win a war.[17] Deterrence in a bipolar nuclear rivalry, therefore, was relatively straightforward because the dominant relationship was simple – one enemy, one threat, one strategy. But if deterrence is primarily about relationships, as old threats diminish, as new ones emerge, and as bipolarity collapses under the weight of multipolar pressures, a complex mix of strategies will be needed.

There are at least three important implications that follow from these observations. First, as Paul Nitze and J.H. McCall argue, '[W]e can no longer construct a security strategy and policy around the belief that sheer numbers and firepower will deter aggression generally; we must create better, more specific, focused policies and strategies with better technology for the job ... Post–Cold War deterrence will require creating forces that can offer a credible deterrent on these new terms.'[18] This does not mean that the logic of bipolar strategic stability is obsolete, only that the relevance of policies derived from that logic is diminishing as new (and multiple) friendships and rivalries develop.[19] Deterrence is becoming a complex problem and will require flexible approaches – there is no longer one but multiple versions of strategic stability.[20]

Second, not only will the number of deterrence relationships multiply but they will also become more interdependent and interlinked, so that decisions and strategies in one setting will increasingly affect the stability of other relationships (positively and/or negatively). Third, future deterrence strategies must be able to discriminate between opponents and tailor deterrent threats in more selective, context-specific ways. Perhaps the best statement of how deterrence will change in a post–cold war setting can be found in a 1988 U.S. Defense Department report.[21] 'Discriminate deterrence' focuses on the realities of asymmetric threats and emphasizes a range of contingencies that go beyond the two extreme threats that have always dominated force planning, namely, a Warsaw Pact attack on Central Europe and a pre-emptive nuclear attack by Russia. The point about discriminate deterrence is that the conditions required for success (credibility, commitment, resolve, and the capability to inflict unacceptable damage) are less likely to work for future threats involving states, regimes, and transnational terrorist groups with no proven channels of communication, no shared assumptions about crisis management, few cultural similarities, and no economic incentives to guide preferences. In the future, threats will have to be tailored to specific opponents and acts.

Of course, if deterrence success in the future depends on the capacity to become actor and situation specific, maintaining a deterrence system based solely on the logic of mutual vulnerability and bipolar strategic stability is likely to become exceedingly dangerous. This is the predicament the U.S. would have faced if it continued to be confined by the

ABM Treaty and MAD. The challenge for Washington is to develop a defensive system with sufficient transparency and predictability to prove to Moscow and Beijing that it is indeed directed against third parties, rogue states, and terrorist groups committed to acquiring this technology.

As for how much is enough, future force requirements will be more difficult to estimate as the system becomes more complex. In the past, a truly meaningful assessment of the numbers required for strategic stability focused on several interdependent variables.[22] These considerations will remain central to nuclear force planning in the future, but it will become more difficult to make accurate estimates of each variable for each opponent across a complex set of scenarios. A definitive account of appropriate numbers applicable for all relationships will be virtually impossible. The propensity to err on the side of caution, therefore, will result in deployment of more rather than fewer weapons – another reason why nuclear abolition will never be a serious policy option.

This dynamic also explains why arms control and disarmament in the future will be very different from the rigid and legalistic approach used in the past. The main determinant of all future negotiations will be to provide maximum flexibility and freedom of manoeuvre. This is certainly apparent in the most recent agreement signed by Presidents Bush and Putin in May 2002. The three-page document marks a significant trend-setting departure from the voluminous arms control treaties that set the standard throughout the cold war.[23] The other important feature of the new agreement is the unilateral commitment to reduce numbers of warheads from around six thousand to about seventeen hundred.[24]

Prioritizing Pre-emption

Critics will claim that the application of America's pre-emption doctrine in the Iraq war went well beyond efforts to protect the U.S. against an imminent attack, as stipulated in the U.S. National Security Strategy [USNSS] 2002. Instead, it involved a strategy of preventive war 'in the sense of using military force where the only threat is a vague and uncertain one of possible conflict at some indefinite point in the future ... ' writes Walter B. Slocombe. '[This is] an unbounded invitation to use of force on mere suspicion of the ambitions or intent of another nation.'[25] But current and future American administrations will continue to defend this position with reference to international law and the right of self-defence. Slocombe provides perhaps the best explication of the prevailing view in Washington, a view that is unlikely to be affected by the political ideology of the party in the White House or Congress:

> The argument begins with the proposition that international law unquestionably recognizes a right of self-defense and moreover acknowledges that exercising that right of self-defense does not require absorbing the first blow. As the NSS puts it, under long-recognized international law principles, 'nations need not suffer an attack; they can lawfully take action to defend themselves against forces that present an obvious danger of attack ...' The problem is not so much that WMD will be used with little warning – attacks with conventional weapons have all too often achieved tactical surprise – but that surprise use could be decisive and that the capability can be so successfully concealed that pre-emption is operationally impossible even if warning were available. On this basis, a strong case exists that the right of 'self-defense' includes a right to move against WMD programs with high potential danger to the United States (and others) while it is still feasible to do so.[26]

The case for preventive war against a rogue regime in Baghdad, with a twelve-year record of proliferation, genocide, and defiance of international law, is obviously stronger than many critics are willing to admit. This is the security framework that will continue to inform and guide the U.S. government for the foreseeable future, since any official who does not subscribe to this interpretation of America's priorities, or that questions the logic and evidence to support them, will not survive in politics very long. This will continue to apply even in the face of serious difficulties with unilateral applications of force, or in light of the problems with, for example, post-war reconstruction efforts in Iraq. If American and British troops in Iraq continue to die from attacks funded by supporters of the previous regime, those losses are likely to be viewed as acceptable in the war on terror and proliferation. The costs and risks of most alternatives, from pulling out to handing the entire project over to the UN, are simply too high.

Other states are not likely to share this view, and tensions will no doubt recur. However, it is also conceivable that other major powers, including France and Russia, will begin to see their interest as interdependent enough to give the U.S. a little more leeway in its responses. According to Kenneth Janda, the potential effects of interdependence were clearly relevant to Europe's responses to 9/11:

> About the same time that President Bush announced his plans to create military tribunals, France expanded its police powers to search private property without warrants, Spain curbed organizations associated with a Basque guerrilla group (E.T.A.), Germany loosened restraints on telephone taps, Britain gave prosecutors the right to detain indefinitely and without trial foreigners suspected of terrorist links, and the European Union formulated a common arrest warrant and a common definition of a terrorist act. Daniel Valliant, France's Interior Minister, said, 'The scale of the attacks on the U.S. and the way they were carried out has made us aware that no one is safe from such terrorist acts. We now speak in terms of before and after September 11.'[27]

In part, these responses highlight the effects of globalized terrorism in a post-9/11 world: '[G]lobalization has come to function like a broad collective security agreement which regards an attack on one member as an attack on all. In this interpretation, economically advanced nations with global connections might imagine that the September 11 terrorist attack on the United States was, or might be, an attack on them. In this view, the terrorist attack on America was indeed an attack on civilization.'[28] The devastating impact of 9/11 on the global airline and travel industries, including dozens of bankruptcies and the loss of billions of dollars and hundreds of thousands of jobs worldwide, illustrate the extent to which the economic pain was globalized. If future attacks on the U.S. produce the security spiral described earlier, and if the challenges outlined in table 6.1 affect American priorities and responses as predicted, the ripple effects on the global economy will be significant.

If security trumps economics south of the border, and if Washington continues to define the strength of its relations with other nations not in terms of their capacity to enhance bilateral trade but through the prism of a new security environment, the implications for Canada are obvious. The health and prosperity of Canada-U.S. relations will ultimately depend on very clear signals from Ottawa that it not only understands the pressures and imperatives that determine U.S. security priorities, but that it is willing to make the necessary concessions to accommodate a set of priorities that both Canada and the U.S. fundamentally share. Canada's decision to stay out of the Iraq war, the inability of the Liberal government to defend that decision on 'principled' grounds (chapter 7), and the anti-Americanism that prevailed in the Canadian Parliament throughout this period sent the opposite message to Washington and to the American public. As a result, Canada will be forced over time to work that much harder to maintain (let alone enhance) the privileges that come from a strong economic relationship with the United States. The final two chapters address these issues in more detail.

Chapter 7

The Moral Foundations of Canadian Multilateralism: Distinction Trumps Security

SECTION I: Canada's Dishonest Multilateralism

Regardless of how commendable are a nation's goals of establishing a truly multilateral global order, the refusal to acknowledge the many deficiencies of multilateralism outlined in the previous six chapters is morally suspect. This is particularly true given the lack of any compelling evidence that multilateral approaches to terrorism and proliferation have succeeded in enhancing global security, or that similar commitments in the future will be any more successful. Yet despite these failures, an almost religious commitment to multilateralism has emerged as the only game in town for Canada, perhaps a consequence of Canada's status as a declining middle power in a world of many new and emerging influential states. There is nothing wrong with this – Canada has done a good (although not error-free) job as the world's favourite multilateralist (at least in the eyes of most Department of Foreign Affairs and International Trade [DFAIT] officials). But Ottawa should not assume that the priorities Canada is forced to accept by virtue of our declining position in the world are the only priorities that should be imposed on the rest of the world. If Canadian security is in fact a priority, honest multilateralists would at least acknowledge the deficiencies that plague the approach and would accept, at least occasionally, the potential contributions of alternatives. It bears repeating that a little less multilateral apathy in Rwanda in 1994 and a little more unilateral (independent) initiative would have saved hundreds of thousands of lives.

Moreover, Canada's 'distinctive' emphasis on soft power does not provide Canada with anything distinctive – every state on the planet has some level of soft power. The assertion that Canadian officials somehow corner the market on the intellectual and diplomatic tools required to apply this kind of diplomacy more effectively than others is a tad insulting to every other state that lacks hard power. In reality, as Mark Proudman correctly points out, '[S]oft power is useless – it is not power at all [and] can result in a kind of ideologically induced blindness to the existence of real and intractable conflict ... International actors emphasize those types of power that they have the ability to exercise, and simultaneously deny those problems with which they are ill-equipped, materially or ideologically, to deal.'[1] Proudman's insightful observation explains the backwardness of Canadian foreign and defence policies, which tend to be derived not from a balanced assessment of our values, interests, and strengths but from our weaknesses. If Canada cannot sustain a meaningful contribution to, say, a war to disarm Iraq, it makes sense to emphasize the benefits of letting diplomacy work through multilateral channels. At least that approach postpones the inevitable demonstration that Canada can no longer afford to sustain its contributions to multilateral security. The result is a kind of dishonest multilateralism that is vigorously supported by Canadian officials not because the approach maximizes Canadian security but because it creates the impression that Canada is making a meaningful contribution to global security through soft power and multilateral diplomacy.

An honest commitment to Canadian security would acknowledge that all options available for dealing with terrorism and proliferation, including unilateral (or non-UN sanctioned) options, should be evaluated on the basis of their capacity to solve these problems. Canadian officials should be willing to confront the mounting evidence that exclusive reliance on multilateralism has failed and should address the following questions: if U.S. unilateralism is not the answer, what is? How exactly will multilateralism work to address the specific problem in question, and what precisely are the costs and security risks to Canadians and Americans if that approach fails? Canadian officials have an obligation to provide answers to these questions but rarely do. This dishonesty plays out in much of what passes as Canadian foreign policy today. Several recent illustrations are outlined below.

Dishonest Commitments to Multilateral Security and Defence

In response to the U.S. president's plea to NATO ministers (at the 2002 NATO Meeting in Prague) to increase defence expenditures, Defence

Minister John McCallum pointed out that this was 'a Canadian matter' and that 'a number of Canadians were a little bit ticked off.' Consider the irony here: we have the world's favourite multilateralist state rejecting a suggestion from the world's least favourite unilateralist that Canada (and other NATO members) should spend more on defence in order to become a more effective contributor to multilateral security, peacekeeping, and peace support operations.

On the surface, Bush's request was entirely appropriate, considering that many Canadians are in favour of, for example, a more sustained deployment of Canadian troops in Afghanistan (beyond the eight hundred troops that were forced to return after their first six-month deployment); new helicopters to replace the aging Sea Kings; increasing Canada's ranking among the world's leading peacekeepers, now around thirty-fourth; and meeting the NATO average in percentage of GDP spent on defence. Despite these preferences, Bush's very reasonable suggestion was dismissed by McCallum because, as he correctly notes, Canada reserves the right to make unilateral (sovereign) decisions about how to make Canadians secure. In other words, no state or collection of states has the right to define Canada's security priorities and defence imperatives. Yet Canadian officials, namely Prime Minister Jean Chrétien and Foreign Affairs Minister Bill Graham, reserve the right to demand from U.S. officials that they avoid unilateral approaches to protecting American citizens from terrorism and WMD proliferation in favour of a multilateral consensus. This very selective application of multilateralism by Canada is dishonest. As Richard Gwyn observes: 'Multilateralism serves us because, like all relatively weak powers, we benefit from the international rule of law that ties down powerful nations. Multilateralism provides the perfect platform for the helpful fixer/peacekeeper policies that Canadians support so strongly. *We are also defence unilateralists* ... It's reckless irresponsibility to do this when the conditions that made it possible to maintain the hypocrisy no longer exist. The change agent is, of course, 9/11 [emphasis added].'[2] Gwyn's point is that by proclaiming a multilateralist agenda, particularly one that is not derived from any principles to which most states could at least respect (see Canada's Unprincipled (Hyper-) Materialism in Iraq, 2003, below), while at the same time refusing to pay for it, Canada is on the verge of becoming 'invisible' and 'irrelevant.' This, as he argues, is particularly damaging 'for a nation that depends significantly on foreign policy activism as a means of expressing our national identity.' Robert Kagan makes a similar point about Europe's dishonest commitment to developing a truly multipolar international system in which multilateral approaches to security would be enhanced:

> If Europeans genuinely sought multipolarity, they would increase their defense budgets considerably, instead of slashing them. They would take the lead in the Balkans, instead of insisting that their participation depends on America's participation. But neither the French, other Europeans, nor even the Russians are prepared to pay the price for a genuinely multipolar world. What France, Russia, and some others really seek today is not genuine multipolarity but a false multipolarity, an honorary multipolarity. They want the pretense of equal partnership in a multipolar world without the price or responsibility that equal partnership requires. They want equal say on the major decisions in global crises (as with Iraq and Kosovo) without having to possess or wield anything like equal power.[3]

The effect is potentially damaging to global security, because while Canadian and European officials want to increase their status, prestige, and influence in the world, they are trying to do this, says Kagan, 'at the expense of American power but without the strain of having to fill the gap left by a diminution of the American role.' Without an honest commitment to fill the security gap produced by the decline of a relatively benign hegemonic power, such as the United States, or worse, without a commitment to balance the rising influence of less benign (and certainly less democratic) global powers, the international system could become significantly less stable.

Dishonest Commitments to Multilateral Aid and Development

The dishonesty with which Canadian officials practice multilateralism can also be seen in pronouncements by Prime Minister Chrétien that the international community work towards resolving global poverty in part because of the presumed linkages to terrorism. These statements are dishonest for two reasons. First, there is no empirical evidence to support the linkage between poverty and terrorism. In their systematic study, Alan Krueger and Jitka Maleckova conclude the exact opposite is true. According to the authors, 'The evidence we have assembled and reviewed suggests there is little direct connection between poverty, education and participation in terrorism and politically motivated violence. Indeed, the available evidence indicates that, compared with the relevant population, participation in Hezbollah's militant wing in the late 1980s and early 1990s [was] at least as likely to come from economically advantaged families and have a relatively high level of education as they were to come from impoverished families without education opportunities.'[4] These findings corroborated similar results in studies by Charles Russell and Bowman Miller: '[T]he vast majority of those individuals involved in terrorist activities as cadres or leaders is quite well educated. In fact, approximately two-thirds of those identified terrorists are persons with some university training, university graduates or postgraduate students.'[5] Since education is related to social status and class, the implication for the negative (not positive) correlation between poverty and terrorism is obvious. As Krueger and Maleckova conclude, '[E]ducated, middle or upper class individuals are better suited to carry out acts of international terrorism than are impoverished illiterates because the terrorists must fit into a foreign environment to be successful.' Several other studies point to similar flaws with conventional wisdom.[6]

Thomas Friedman provides a more compelling explanation for terrorism that goes beyond the simplistic assertions about poverty. The real problem has to do with the indignities associated with a failure to modernize and democratize: 'Muslim states are failing at modernity' and exist primarily in environments 'dominated by anti-democratic regimes and anti-modernist religious leaders and educators.'[7] These are the forces that are producing the 'undeterrables,' the 'young men who are full of rage' and who, Friedman observes, are raised to view Islam 'as the most perfect form of monotheism ... but they look around their home countries to see widespread poverty, ignorance and (political) repression. And they are humiliated by it, humiliated by the contrast with the west and how it makes them feel, and it is this humiliation – this poverty of dignity – that drives them to suicidal revenge.' It is the 'deficits in freedom, modern education, women's empowerment,' all of which thrive in the absence of liberal political institutions, not a deficit in income and wealth that explains the central challenge of our time.

The second problem with Chrétien's disingenuous recommendations is that he promised in 1994 to raise the level of Canadian aid and development assistance to 0.7 per cent of GNP, yet Canada's level of aid fell during his tenure as prime minister to 0.25 per cent of GNP. Canada's efforts under Chrétien to combat poverty fell from eighth among OECD nations to seventeenth out of twenty-two (recent figures place Canada's ranking at nineteenth of twenty-two).[8] Canada is eleventh in the OECD for overall dollar amounts invested in aid and development assistance, tends to distribute foreign aid in smaller amount to more states (thus diminishing the overall impact of the aid that is distributed), and, in comparison with other OECD members, allocates a larger portion of our foreign aid budget to central administration in Ottawa.[9] As Jeffrey Simpson commented, 'The façade suggests that Canada is a respected, robust player in the world, whereas Canada's influence has been in steady decline. The reasons for that decline are many, including the rise of other powers, but one explanation is entirely Canadian – the yawning gap between Canada's ambitions and capabilities. This country has a self-image of an international do-gooder; the reality is Canada doesn't pay its way. The gap between pretence and reality is everywhere apparent: in foreign aid, defence and foreign policy. Never has the world meant more to Canada; never has Canada meant less to the world.'[10]

Perhaps most disconcerting of all is that this record was compiled at a time when Canada successfully raised its own standard of living (and associated quality-of-life indicators) to be ranked by the UN as the best country in the world in which to live – a statistic the prime minster repeatedly and enthusiastically flaunted at every opportunity. Consider the irony (and dishonesty): Canada's quality of life increased at a time when Canada's contribution to the quality of life of the planet's less fortunate declined. If one follows the logical implications of these two trends, Canada is arguably more responsible than other OECD members (including the U.S.) for the poverty that, according to Chrétien, induces the terrorism inflicted on American citizens. As Norman Hilmer and Maureen Appel Molot note, 'Chrétien has put considerable effort into what will be the last summit [Kananaskis] in Canada under his leadership ... Helping [Africa] pull itself up by the bootstraps is a laudable enough goal, if one can get past the understandable scepticism that Africa is merely the flavour of the month and that good money will be thrown after bad ... Good works are usually the sign of a leader winding down.'[11] Perhaps the saddest irony of Chrétien's Kananaskis plan for Africa is that Africa, the poorest continent on the planet, doesn't produce many terrorists. Comparatively wealthy countries in the Middle East, such as Saudi Arabia, do. I doubt very much that the Canadian International Development Agency has any plans to shift a significant portion of Canada's development assistance to the Middle East, for very good reasons – the nineteen Saudi, Emirati, Lebanese, and Egyptian-born middle-class terrorists who flew planes into the World Trade Center and the Pentagon (and their millionaire sponsor, Osama bin Laden) hate liberal democracies for what they are, not for what they have.

Dishonest Evaluations of BMD and Multilateral Arms Control

It is reasonable to expect that a position on whether any policy is morally justifiable should depend, in large measure, on the answers to at least four questions: (1) are the objectives important (i.e., is there a moral imperative to seek these goals), (2) do the policy prescriptions stand a reasonable chance of succeeding, (3) do the strategies run the risk of producing unintended consequences that are morally unaccept-able, and (4) does the balance between morally acceptable and unac-ceptable outcomes favour the status quo (doing nothing) or action (implementing the strategy in view of the risks and potential costs and benefits)?

Critics of BMD who claim to be representing the higher moral ground on the issue rarely work through the moral implications of their position. Without question, arms control and non-proliferation satisfy the first condition – few would disagree with the importance of disar-mament goals. But traditional multilateral approaches have failed to accomplish meaningful progress, and threats from proliferation of WMD are mounting. To reject any and all alternatives because of the intellec-tual capital and energy invested in defending the regime is itself mor-ally suspect.

Bosnia, Kosovo, and Afghanistan are all human security crises that compelled the international community (and Canada) to respond to end large-scale human suffering – that is precisely how the intervention was defended at the time by Lloyd Axworthy, Canadian foreign minis-ter from 1995 to 2000. But the decision to intervene, and the underlying moral justifications offered to explain our actions cannot be evaluated in isolation. These actions must be viewed as part of a complex set of decisions and choices that have consequences for other policies. As noted earlier, among the many effects of these interventions is that they create strong incentives for some leaders (especially those in the so-called axis of evil who are being targeted by the U.S. for terrorism and proliferation) to acquire ballistic missile technology as quickly as pos-sible, largely to prevent the U.S. and NATO from achieving relatively quick (and morally defensible) victories in the future.

Those within the NACD regime who defend multilateral arms con-trol continue do so not because there is compelling evidence to demonstrate that these strategies have succeeded, or evidence that these policies offer the only real hope for success in the future, but because policy alternatives, like BMD, are defined from the outset as morally reprehensible and destined to fail. Proponents of multilateralism, however, rarely acknowledge the double standards they apply when they demand verifiable evidence that unilateralism actually works. The criteria they recommend when assessing BMD success/failure (that is, financial costs, security risks, political and technological impediments, and so on) are never applied with equal vigour to their multilateral accomplishments.

Unlike the NACD regime, BMD can be assessed in straightforward and easily verifiable terms. Do the interceptors hit their targets? How often? And can we reasonably expect to overcome existing technological hurdles with a sufficiently high degree of success to warrant deployment? Using any reasonable criteria for measuring success, the BMD program makes a far less ambiguous contribution to U.S. and Canadian security than the NACD regime simply because we can actually see successes and failures. While no policy initiative will ever make an entirely unambiguous contribution to security, some programs are easier than others to evaluate. With this in mind, the current rate of success for fully integrated BMD tests from 1999 to 2002 is five successes out of seven, or roughly 72 per cent. Using any reasonable measure of proliferation (vertical or horizontal), the NACD has done much worse, notwithstanding the typical arguments put forward to defend its record.

Consider the following: if a single BMD interceptor fails to hit its target, critics immediately conclude that the technology is unworkable and the entire system should be scrapped. But if we apply the same standards to the NACD, what should we conclude from the thousands of failures to intercept a variety of WMD technologies from spreading to state and non-state actors who remain committed to obtaining these weapons?

Now, consider these failed intercepts in relation to the billions of dollars, pounds, francs, and yen invested in non-proliferation and disarmament over the past forty years by hundreds of government agencies, international institutions, and non-government organizations. Our collective inability to prevent North Korea from acquiring nuclear weapons constitutes a far more significant and profoundly disturbing failure than two of seven BMD interceptors missing their target, especially considering its status as a signatory to the non-proliferation treaty. In fact, BMD's 72 per cent success rate is pretty good considering the alternative – a zero per cent probability of stopping any incoming missile if BMD is not deployed. More important, the U.S. can reasonably expect to overcome BMD's existing technological hurdles with a high-enough degree of success to warrant the investment of a mere 2 per cent of the U.S. defence budget. In contrast, consider the diminishing probability (and increasing risks and costs) of dismantling (by constructive engagement or force) North Korea's nuclear program; please see chapter 5 for a more detailed discussion of the impediments to constructive engagement.

Again, the purpose here is not to indict the multilateral NACD regime for its failures but to recommend opening up the space for alternatives that typically get dismissed by proponents of multilateral arms control. Unless the regime can demonstrate success, unilateral approaches like BMD will continue to be perceived as absolutely essential.[12]

SECTION II: Canada's Unprincipled (Hyper-) Multilateralism in Iraq, 2003

Canada's 'principled' decision to stay out of the Iraq war encompassed three core tenets: (1) no support for regime change; (2) support for interventions that are sanctioned exclusively by multilateral consensus in the UN Security Council; and (3) independent control over our own foreign policy. Two secondary rationales for the prime minister's decision included deference to Canadian public opinion and domestic political imperatives – namely, the election in Quebec. A more considered assessment of the actions taken by the government highlights several factual and moral deficiencies with the Chrétien doctrine and Canada's actions throughout the war.[13] The case provides an excellent illustration of the confused and disjointed nature of Canadian foreign policy at a crucial point in time when Canadian officials needed to convey focus and clarity. It is also a study of Canada's fading status as a relevant power.

Selective Opposition to Regime Change

A principled position against regime change in Iraq would presumably have continued to oppose that policy especially when it succeeded – perhaps even more forcefully. Instead, the prime minister supported the U.S.-U.K.–led coalition, declared at the time that he wanted the U.S. to win, and defended the policy of replacing Saddam Hussein with a more democratic leader and regime. Together, this amounted to a policy in favour of regime change. The unanimous motion passed in Parliament (April 2003) to 'bring to justice Saddam Hussein and all other Iraqi officials responsible for genocide and crimes against humanity' did very little to clarify where Canada stood on the issue. Steven Hogue, a spokesman for the prime minister, was quoted as saying that 'we welcome efforts to prosecute Saddam Hussein and other regime members responsible for genocide.'[14]

It was the confusion over the orders given to the Canadian commander of the naval task group in the gulf, to avoid turning over to Americans any Iraqis fleeing the war, that prompted U.S. Ambassador Paul Cellucci to sum up Canadian policy as 'incomprehensible.'[15] As John Ibbitson correctly observed, '[I]f a country that Canada had no particularly strong relations with invaded another country for the sole purpose of installing a friendly government, and then asked Canada to help repair the damage of the invasion, would we be there? Of course not. We wouldn't legitimize the invasion by helping to repair the damage it caused.'[16] But the Canadian government was among the first to offer support for the reconstruction of Iraq, despite the fact that the UN was given a relatively minor role in post-Saddam reconstruction efforts – United Nations Security Council Resolutions 1483 and 1511 establish the United States and United Kingdom as the 'Authority' in Iraq.

That Canada took advantage of its sovereign right to establish an independent position against regime change, one that was obviously distinct from that of the U.S., Britain, Australia, and sixteen of nineteen NATO members, was not the problem. The issue was that Ottawa never clearly articulated a principled position against regime change; it was simply opposed to the process by which this particular regime was changed. As Foreign Affairs Minister Bill Graham said at the time, 'The government and, I believe, the large majority of the Canadian population do not believe that regime change is something that we should engage in lightly in the international world without having the international approval of the United Nations Security Council. That issue is past us. Of course we wish our American allies a speedy end to this. Of course we wish them well in their endeavours.'[17] On another occasion, Graham noted that '[m]any Canadians are supportive of the United States, and we as a government are supportive of the United States' desire to get rid of Saddam Hussein, to deal with the weapons of mass destruction issue around the world ... We are obviously not indifferent to its outcome ... We recognized that Saddam Hussein is dangerous, had used weapons of mass destruction before and now that the conflict is engaged ... [W]e wish our American friends and British friends Godspeed.'[18] Graham was drawing an obvious and important distinction between 'principled' opposition to regime change (regardless of process) on the one hand, and the view that regime change should not be engaged in lightly without UN approval. But regime change in this case was given considered thought by the United States and United Kingdom, received the support of sixteen of nineteen NATO members, several new and future members of the European Union, and over forty-five other states, including Japan, Poland, and Australia. Canadian officials ultimately favoured regime change and wished those endeavouring to change the regime 'Godspeed,' all in the absence of a United Nations Security Council resolution. In the end, what 'principle' was Canada defending?

When Washington appeared to shift the rationale for the war from WMD disarmament to regime change, Prime Minister Chrétien issued the following statement at a press conference in Mexico: 'If we start to go and change every government that we do not like in the world, where do we stop? Who is next? Give me the list.' But consider the benchmarks established in the Iraq case – the prerequisites for intervention that must be met by subsequent coalitions of the willing include genocide; chemical attacks against Kurdish (or other) minorities; two unprovoked invasions of Iran and Kuwait within ten years; refusal to comply with seventeen UN resolutions over a twelve-year period; links to (and financial support for) suicide bombing attacks against Israel; concerted efforts to develop WMD as outlined in David Kay's interim report after the war (chapter 4); embezzlement of billions of dollars in illegal oil revenues accumulated throughout the UN sanctions regime; a collapsed economy; refusal to negotiate the oil-for-food program until 1995, after four years of widespread poverty, malnutrition and related health problems, and increasing child mortality rates; systematic political repression and executions; pariah status in the international community, and the list goes on. It was obviously this record that prompted the prime minister to state in the House of Commons that, in the end, the United States had a 'privilege and right' to attack Iraq.[19] The truly moral question to ask in this case is not, Who is next? or Where do we stop? but If not Saddam, who?; If not Iraq, where do we start?

A more thoughtful position on regime change would uncover serious flaws in the logic underlying Canada's position. To demand only that process-related legitimacy is satisfied implies granting to fewer states the power to quash the requisite consensus to act in the interests of those who are suffering under brutal regimes. Ironically, an exclusive commitment to process-related legitimacy increases the prospects for unilateralism, because it grants to any single member of the Security Council the power to thwart any consensus for its own interests, whatever they may be. The Chrétien doctrine implies that process-related consensus and legitimacy are satisfied when five permanent members of the UNSC are in agreement but not satisfied when close to fifty states offer explicit support to a U.S.-U.K.–led coalition and dozens of other states offer implicit support (please see appendix A). When one considers the fact that the non-permanent members of the UN Security Council would likely have followed the lead of the permanent five (with the possible exception of Syria), process-related legitimacy really comes down to the wishes of two states – France and Russia. This is a recipe for hyper-unilateralism, not multilateralism.

Periodic (and Selective) Support for Multilateral Consensus

Prime Minister Chrétien's unwavering support for a UN process in which multilateralism guided Canadian foreign policy suffered from similar ethical deficiencies that were clearly apparent in view of several recent cases, the most noteworthy being Rwanda (1994); the United Nations Security Council failed to generate the multilateral consensus that would have saved close to one million lives. A policy that demands that level of collective agreement for the sake of legitimacy implies that the United States, France, Canada, and dozens of other countries were right to avoid military intervention in Rwanda to stop the genocide. Critics were right to ask whether 'a blind, unthinking devotion to multilateralism' was a principle Canada should be defending.[20]

The lack of UN Security Council harmony on rules of engagement in the Balkans between 1990 and 1995 failed to prevent 250,000 deaths, systematic rape and ethnic cleansing, one million refugees, and the massacre of six thousand Muslims in the UN-controlled 'safe haven' of Srebrenica. According to the Chrétien doctrine, non-intervention was a 'principled' policy because multilateral consensus to stop the war was missing.

More recently, UN Secretary-General Kofi Annan was 'deeply disturbed' by the twenty fresh mass graves found in the spring of 2003 in the Democratic Republic of Congo, and demanded that 'all concerned unconditionally respect the basic human rights of innocent civilians.'[21] In a subsequent letter to the Security Council, Annan requested deployment of a UN peacekeeping force to the town of Bunia (Congo): 'In view of the likely further worsening of the situation, and its serious humanitarian consequences, I should like to request that the Council urgently consider my proposal for the rapid deployment to Bunia of a highly-trained and well-equipped multi-national force, under the lead of a member state, to provide security at the airport as well as to other vital installations in this town to protect the civilian population.'[22] But the UN Security Council had not yet reached the multilateral consensus required to stop the killing. The amount of time, energy, and political and financial capital required to move the UN Security Council towards either unanimous or majority support for any resolution, let alone one dealing with humanitarian intervention, is prohibitive.[23] Reports by prominent human rights organizations estimate that approximately three million people have been killed in the DRC over the past four years. France sent about a thousand troops to the region in the summer of 2003, the same summer in which France hosted the G8 summit and deployed close to five thousand troops throughout France to guard against protesters and possible terrorist attacks. Yet, according to the Chrétien doctrine, non-intervention in the DRC amounted to a principled position, because, like Rwanda, the automatic legitimacy that comes from multilateral consensus was absent. Despite the obvious moral and ethical deficiencies with this position, the prime minister continued to defend the policy: 'The Canadian position is that on matters of peace and security, the international community must speak and act through the UN Security Council. The position of Canada is that we were insisting right at the beginning, you remember, that Canada act through the United Nations, through international institutions. We believe in multilateralism very strongly.'[24] But truly moral and principled foreign policy would try to prevent these human rights atrocities from happening despite the absence of that consensus. Unfortunately, although Ottawa was asked to provide peacekeepers to the DRC it was forced to decline for lack of resources.[25]

The application of military power against Milosevic in Kosovo in 1999 helped to stop and reverse the effects of ethnic cleansing, facilitated the return of hundreds of thousands of Kosovar refugees, and changed the governing regime in Kosovo and, ultimately, in Serbia by sending Milosevic to the Hague to be tried for war crimes. All this was supported by Ottawa and accomplished despite the absence of UN

Security Council agreement and the threat of Russian and Chinese vetoes. When asked in an interview to address the inconsistency between the 1999 Kosovo decision and the 2003 Iraq policy, the prime minister stated, 'When you went to Kosovo, for example, there we did have a resolution, but I mean, the Russians opposed, and being, you know, because it was Serbia and question of the same religion and so on, people considered that as special veto, everybody was one in agreement, but we went there anyway under the NATO umbrella and it worked because we went there to stop the genocide and we succeeded.'[26] There are three problems with this explanation. It is not at all clear what 'people' the prime minister was referring to, or where in the UN Charter could be found a reference to anything like a 'special veto' for a shared religion. With respect to justifying the intervention on the grounds of genocide, the case has been made that the devastation and aggregate scope of human suffering inflicted on the Iraqi people under Saddam Hussein exceeded 'by huge orders of magnitude' the atrocities committed in Kosovo under the Milosevic regime.[27]

Chrétien's statement is also a dishonest depiction of his (and Canada's) foreign policy record over the past ten years as it relates to UN multilateralism. When comparing Canada's position in the 1991 and 2003 gulf wars, the prime minister accurately noted that '[i]t was the same position that we had during the first Gulf War, in 1990–1991 when the Government at the time said to George Bush Sr. that if you have a war we will be with you, but only with the support of the United Nations, with a resolution of the United Nations. And that's exactly what George Bush Sr. got. So we did participate.'[28] Although the prime minister took credit for Canada's consistency, the actual position he defended in 1991, as leader of the official opposition, was the exact opposite one, as he stated in the House of Commons in January 1991: 'We strongly reject the use of force now because the embargo has not been given enough time to work.'[29] Chrétien believed at the time that his alternative approach (sanctions) was more principled, even though the war received two UN Security Council resolutions to 'use all necessary means' to force Iraq to comply to the wishes of the international community. Obviously on matters of peace and security in 1991, Jean Chrétien believed that Canada did not necessarily have to 'speak and act through the UN Security Council,' as he asserted in 2003.

In 1995, without a mandate from a UN Security Council Resolution, Prime Minister Chrétien supported the U.S.-led NATO bombing campaign in Bosnia-Herzegovina that ultimately led to the Dayton peace accords and an end to ten years of Balkan wars, ethnic cleansing, and genocide. In 1998, when Iraq failed to comply with a series of post–gulf war UN resolutions on disarmament and UNSCOM inspections, the prime minister supported the U.S.-U.K.–led non-UN-sanctioned bombing campaign against suspected weapons sites in Iraq. On 9 February 1998, addressing Parliament, Chrétien stated, 'If we do not act, if we do not stand up to Saddam, that will encourage him to commit other atrocities ... [T]he choice is clear. It is a choice dictated by the responsibilities of international citizenship, by the demands of international security and an understanding of the history of the world in this century.'[30] He went on to say,

Saddam's determination to develop and use weapons of mass destruction, chemical warfare in particular, is well documented. Anyone doubting the serious character of the threat this man represents has only to recall how he turned these weapons against his own people. Equally well documented are his ongoing efforts to block the work of UNSCOM, the United Nations Special Commission created to ensure compliance with Security Council Resolution 687 ... That is why, if it comes to that, we believe a military strike against Iraq would be justified to secure compliance with security council resolution 687 and all other security council resolutions concerning Iraq.[31]

Consider also the statements in 1998 by Lloyd Axworthy (minister of foreign affairs) and Art Eggleton (minister of defence), both staunch critics of the 2003 Iraq war but supporters of Chrétien's party line on the 1998 bombing: 'This whole thing could be solved in 10 seconds if Saddam Hussein lives up to the commitments made in 1991,' said Axworthy, at the United Nations, 11 February 1998; 'Saddam Hussein has shown in the past that we cannot trust his word,' noted Eggleton. If more countries joined the U.S. coalition, the defence minister argued, then 'this might be enough for Saddam Hussein to back off and to comply with UN resolutions.'[32] And with respect to the search for principles worthy of defending, it is important to note that the 1998 bombing of Iraq, which Chrétien, Axworthy, and Eggleton supported, took place after seven years of UNSCOM inspections. By contrast, the 2003 gulf war, which they collectively opposed, occurred after four years without inspections, a higher probability of WMD deployment, and continued intransigence and defiance by Saddam of several additional UN resolutions, all culminating in a unanimously endorsed UNSC

1441 that called for immediate, full and unconditional compliance under the pressure of 'serious consequences.'

In 1999, the prime minister supported and participated in the U.S.-U.K.–led NATO bombing campaign in Kosovo and Belgrade without a UN Security Council resolution. The Canadian air force dropped bombs on sites in Belgrade when Canada was not at all threatened by Slobodon Milosevic (a similar argument was made about Saddam Hussein when defending Canada's decision to stay out of the 2003 Iraq war). The arguments put forward by the Liberal government to defend the 1999 decision are virtually indistinguishable from those made by U.S. and U.K. officials prior to the 2003 gulf war. And, most recently, the Chrétien government in 2002 sent Canadian troops to support the non-UN-sanctioned NATO-led deployment in Afghanistan, under U.S. command. The Canadian record over the past ten years is not one marked by a blind commitment to multilateralism through the UN.

The most obvious problem with the decision to stay out of the 2003 gulf war is not that it was anomalous in light of Canada's foreign policy record over the previous decade, but that it did not appear to serve any identifiable Canadian interest, while the other decisions did. Allan Gotlieb, a former Canadian ambassador to the U.S., lists the most relevant criteria for judging the utility of any foreign policy initiative: 'Is it good for the country? Is it in our national interest? Is it a moral foreign policy? Is it genuinely internationalist? Or is it crypto-isolationalist? Not least in importance, is it compatible with our historic partnership with the U.S. in the cause of freedom and peace? In assessing its implications for Canada-U.S. relations, do Canadians understand the profound psychological changes that have taken place south of the border since the U.S. was, in effect, invaded?'[33] Those who defended (and continue to defend) Canada's decision to stay out of the war would have a very hard time providing honest answers to these questions.

Canadian officials could easily have demonstrated a strong commitment to multilateralism with reference to UNSC 1441 and the coalition of more than forty-five countries supporting the invasion (as described in appendix A). The 2003 gulf war was really a debate between two versions of multilateralism, a principled argument Canadian officials could easily have made to defend their decision. Critics might respond by arguing that measuring multilateral consensus with reference to what the U.S. managed to cobble together really stretches the meaning of multilateralism, but that view implies than some states are more relevant that others when measuring legitimate consensus.

Russia, for example, should perhaps carry more weight than, say, Rwanda (which joined the U.S. coalition). But the slippery slope here implies that the U.S. position in any crisis should carry more weight than, say, Iraq, Canada, or perhaps even France. If all states are equal under the very UN charter that critics of American foreign policy are fond of defending, claiming the U.S.-U.K.–led coalition did not constitute a multilateral response to Iraq is a considerably more difficult position to defend.

Canada's Subordination and Subservience to Hyper-Multilateralism

This leaves the third tenet of the Chrétien doctrine, 'independence' and related assertions that Canada 'exercised independent judgement, and that independence has both supported and differed from the U.S. judgement of world events, especially military conflicts. That pattern continues today,' notes Jeffrey Simpson.[34] But few if any critics of Canada's position ever questioned the right (indeed, obligation) of Canadian officials to exercise independent judgment and to select policies based on the best interests of the country. The issue was not that Canada took a position independent from that of the United States; the problem was that the independent position Canada defended in this case did not protect our right and obligation to act independently.

In the weeks prior to the unanimous passage of UNSC 1441 (designed to create a more robust inspections regime in Iraq), DFAIT officials were constantly repeating Canada's five main concerns about a U.S. unilateral intervention in Iraq: (1) there are many other rogue states that are just as bad as Iraq, and we can't attack them all; (2) there is no clear proof of a link between Iraq and 9/11; (3) there is no clear proof of WMD proliferation by Iraq; (4) there is no clear proof of WMD deployment by Iraq, and (5) there is no clear proof of an intention to use WMD. Put differently, unless the U.S. produced a picture of Saddam Hussein with his finger a few inches from a big red button, the U.S. had no right to attack Iraq. But apparently the same Canadian officials did not require the same evidentiary prerequisites to establish a just and moral cause for intervention in Iraq *if* the intervention was sanctioned by a UN Security Council resolution. In that case, none of the five conditions applied, as if a multilateral operation sponsored by a UN Security Council resolution (such as UNSC 1441) automatically established the moral clarity required to attack another country without cause.

Other critics of the U.S.-U.K.–led war in Iraq tried to clarify Canada's stance by arguing against the principle of pre-emption embedded in the 2002 United States National Security Strategy. Carol Goar defended Ottawa's position by outlining what she viewed as the underlying 'principles' guiding Canada's 'independent' policy:

1 Canada does not endorse the principle of launching pre-emptive attacks on dictators who might, at some time in the future, pose a security threat;
2 Canada does not support the use of military force when diplomacy is producing visible results;
3 Canada does not believe that thousands of Iraqi citizens should be killed in order to secure 'regime change';
4 Canada does not see how bombing Iraq will reduce the risk of terrorism, stabilize the Middle East or make the world safer;
5 Canada does not want to be part of an invading force that 'does all of this' but that has not been sanctioned by the United Nations.[35]

But these principles were not the ones Canadian officials were defending. In fact, none of the impediments to Canadian participation outlined in Goar's first four conditions would have been relevant if the UN Security Council passed a second resolution beyond UNSC 1441. As long as the Security Council gives the go-ahead, Canada would have no problem (1) endorsing the principle of launching pre-emptive attacks on dictators who may or may not pose a security threat; (2) using military force when diplomacy is producing results; (3) risking the deaths of thousands of citizens to secure regime change; and (4) bombing Iraq in the hopes of reducing terrorism and stabilizing the Middle East. The fifth condition is particularly interesting because it illustrates the previous point – it excuses Canadian participation in an invading force (with all the above-mentioned costs, injustices, and risks) if it is sanctioned by the United Nations. But there is no clear explanation why a UN resolution is sufficient to either avoid or excuse the costs outlined in items 1 to 4. How, for example, would a UN-sanctioned bombing campaign be any less likely to avoid the deaths of thousands of Iraqi citizens and anger hundreds of thousands of Muslims? In fact, the fifth 'Canadian principle' noted in Goar's list is perhaps the least accurate in view of Canada's foreign policy record in similar international crises over the past decade (outlined earlier).

A commitment by Ottawa to rely exclusively on multilateral consensus does not establish autonomous control over our foreign policy. In fact, it accomplishes the exact opposite – subservience and subordination to any single member of the Security Council who decides to veto any consensus that does not support their own unilateral economic, political, or military self-interests. It is not at all clear how any of this enhances Canada's capacity to act independently or in the interest of Canadians.

In an effort to defend the governments position, Bill Graham stated, 'So I think we should be where we are, have our policy, which is to support the multilateral system with good arguments and to say to them, "We're staying put. And you're better to have an ally and friend that debates frankly than to have someone who says yes, sir."'[36] Leaving aside the absence of 'principles' underlying this unwavering support for multilateralism, the government continued to assume the policy was derived from good arguments that the U.S. will come to understand and respect. But that assumption reveals a complete misunderstanding of why American officials were angry – it was not that Ottawa took an independent line, it was that the independent line Canada took did not support the principles it claimed to be defending. Indeed, the position espoused by the Chrétien government, if successful, would establish hyper-unilateralism as the authoritative process through which decisions are made on the UN Security Council, because it assigns to any single member of the Security Council the ultimate authority to kill any multilateral consensus. It was the fact that the arguments made so little sense that upset the U.S. administration and some Canadians. They did not amount to anything more than blind faith in the power of reason and good will to produce a just and moral consensus. Canada's position was essentially a commitment to an ideal of multilateralism that simply doesn't exist. As Walter Slocombe, a former American security adviser, argued when evaluating the position espoused by Canada and some European leaders, 'to require United Nations approval as an absolute condition of legitimate use of military force is to say that no military action of which Russia or China (or, in principle, France, Britain, or, indeed, the US) strongly disapproves is legitimate, no matter how broadly the action is otherwise supported, or how well justified in other international legal or political terms ... [This] amounts to saying that permanent members or alternatively, a majority-blocking group of the non-permanent members are the absolute custodians of the legitimacy of international force.'[37] This does not enhance the prospects for multilateralism on important issues but makes consensus more difficult to achieve. It does not make unilateralism less likely but increases the probability of unilateralism and assigns to that unilateral preference the authority, influence, and legitimacy no state deserves.

Canada's position in the war shifted our historical defence of values espoused by the United States, Great Britain, Australia, new and aspiring members of the European Union (many of which have recently emerged from decades of subjugation under the old Soviet regime) and, as George Jonas describes, 'lead us firmly into the camp of such historic authorities on morality as Russia, such great champions of world peace as Germany, such fearless resisters of tyranny as France, not to mention a nation of such impeccable track record on human rights as China. Or Syria, we mustn't forget Syria.'[38] The refusal to acknowledge the deficiencies and dangers of contemporary multilateralism, and to repeatedly claim that this process represents a vital part of our 'distinctive international personality,' to quote Chrétien,[39] as opposed to a 'vital crutch to our national identity,' to quote a *National Post* editorial,[40] amounts to a policy that is morally suspect and decidedly unprincipled. The moral legitimacy of any policy, including military intervention, should be measured by its outcome. The process, whether unilateral or multilateral should play a minor, if not irrelevant, role.

Sporadic Reliance on Public Opinion

Given the challenges of finding a clear set of moral principles to explain Canada's position on the war, several observers focused instead on second-order explanations derived from domestic politics: (1) Canadian public opinion, and (2) a Liberal victory in the 2003 Quebec election. Both explanations are seriously flawed.

With respect to opinion polls, in the months during and after the war public support in Canada (and in Britain and Australia) continued to rise in favour of supporting the war. Canadian results from polls in March registered the level of support at 56 per cent; interestingly, this is on par with the percentage registered by the same Liberal polling organization (Pollara) for Canadian support of NATO's intervention in Kosovo in 1999, an intervention that did not receive a security council mandate.[41] In April 2003, a COMPAS poll found that close to 72 per cent of Canadians believed Ottawa should have provided some support to the United States, with a full 31 per cent supporting troop deployment and 41 per cent believing that at least verbal support should have been forthcoming. Fifty-six per cent agreed with Washington's decision 'to launch a land invasion to bring down Saddam,' as the poll put it. Among the reasons respondents gave for their position, close to 57 per cent listed either 'Saddam and his terrorist allies' (42 per cent) or 'Islamic extremism' (15 per cent) as legitimate reasons for the intervention.[42] Of those who opposed the war, 31 per cent selected 'all war is bad' as an explanation for their position – a view that would not have changed even with a unanimous and unambiguous UN resolution support an attack on Iraq. A similar poll by *The Globe and Mail* registered 51 per cent in favour of Canada offering help to the U.S., and a Léger poll tracked shifts in Canadian support from August through May – August 22 per cent, March 37 per cent, and May 46 per cent. Despite increases in support for the war over time, there was no effort by the government to revisit the wisdom of their original decision to emphasize public opinion to defend their policy.

Polls in Britain and Australia registered similar majorities in support of the war. As Andrew Coyne points out, these percentages were achieved despite the absence of 'weeks of the most eloquent and passionate exhortation' that British citizens received from their Prime Minister Tony Blair. Indeed, despite efforts by the Liberal government to develop and articulate a 'principled' case *against* the war, Canadians arrived at the same conclusion as the British and Australians, and most Americans.

The Myth of Domestic Political Constraints

With respect to the Quebec election playing a significant role in Chrétien's calculations,[43] there is no clear evidence that the issue was central to either the Parti Québécois (Bernard Landry) or Liberal (Jean Charest) campaigns. Nor is there much evidence from polling data to indicate that even a small majority of the Quebec electorate would vote on the basis of foreign policy, or that this particular foreign policy crisis was central to their concerns in 2003. Typically, foreign policy is ranked quite low in Canadian elections, even lower in provincial elections, and is almost insignificant in Quebec politics. To the extent that the Iraq war was relevant, the prime minister could easily have controlled the level of opposition in Quebec by supporting the U.S.-U.K.–led coalition on the basis of multilateralism, international law (UNSC 1441), human rights, and public opinion but could still have refused to send Canadian troops to the region given commitment to Afghanistan. There is no evidence to suggest that this policy would have seriously undermined Charest's election victory. Given the mixed messages sent by the Chrétien government, particularly the evidence put forward in Parliament by the Bloc Québécois that Canada provided more military assistance to the U.S.-U.K. war effort than most of the countries who supported the U.S., the position taken by Ottawa could conceivably have had a net negative effect on the federalist candidate in the Quebec election.

Dishonest Denials of Canada's Contribution to Multilateral Intervention in Iraq

The dishonesty with which the prime minister attempted to steer Canada's Iraq policy also applied to the 'distinction' between Canadian troops and other soldiers who were actually 'fighting' in the war. According to the official line, Canadian soldiers 'have received instructions from their army to the effect that they could only use their weapons in self-defence. They are not fighting.'[44] On 25 March Defence Minister John McCallum claimed that the troops were in support roles and were instructed to use force only to defend themselves. He went on to make the obvious point that pulling them out at this late date would risk 'putting the lives of allied soldiers at risk.'[45] If Canadian troops had actually received orders to use their weapons only in self-defence, they would have been asked by allies to leave, precisely because that order would have put the lives of American and British troops at risk.

On 31 March 2003, McCallum did little to clarify the distinction: 'We are behind them 100 per cent. We thank them for putting their lives on the line.' However, he went on to explain, once again, that Canadian troops are not in combat, 'they are in a situation of combat around them.'[46] But there is in fact no class system in the military 'where some soldiers in a combat zone are in "direct combat" and others are not,' as Gen. Lewis MacKenzie points out.[47] The distinction simply does not exist, and never has. There is nothing in the just-war tradition, or Geneva Convention, that corresponds to the distinction noted in the prime minister's or defence minister's statements. Despite their best efforts to contort logic and evidence to support the questionable 'principles' underlying the government's policy, Ottawa failed to establish a clear explanation for (or defence of) that policy. With respect to the history of Canada-U.S. military exchange programs – in the Falklands, Northern Ireland, Panama, and in other cases – the Canadian government decided against committing exchange troops, so there is a precedent for taking a different 'principled' stance on exchange program.

The principles were particularly difficult to defend in light of Canada's overall military contribution to the war effort – an impressive level of support that was considered by American officials (including the U.S. ambassador in Canada, Paul Cellucci) to be greater in scope than the military assistance provided by most of the states that fully backed the U.S. campaign. According to a statement by McCallum at the time, however, 'their mandate is to try to find al-Qaida or other terrorists. Certainly they are not authorized in their mandate to intercept or detain or transfer suspected members of the Iraqi regime.'[48] But U.S. officials repeatedly confirmed that the opposite was true – as part of their responsibilities in tracking terrorist and other illegal activity, Canadian sailors were in fact tracking and hunting Iraqis as well. Kelly Toughill's report on the naval deployment was based on interviews with U.S. officials and confirmed that Canadian ships were in fact tasked with 'screening of travellers in the Persian Gulf for Iraqi military officials and government leaders.'[49] U.S. Air Force Lt.-Col. Martin Compton (media relations officer at Central Command in Tampa) verified that 'anyone connected with the Iraqi regime is on the list.' As Toughill explains, 'Canadian sailors who board ships in the Persian Gulf run passengers' identities through a U.S.-controlled database that includes Al Qaeda terrorists and Iraqi officials. Canadian sailors have boarded hundreds of vessels since the operation began. When they board, they sort through passengers and take aside anyone who looks suspicious. Pictures of those suspects are then transmitted to a central database that determines if they are wanted terrorists.'[50] According to Canadian army Maj. Richard Saint-Louis, 'if they have a hit, then those people are turned over to our coalition partners.' Toughill points out that 'the Canadian forces have no control over or specific knowledge about who is included in the database run by the United States, or how that country determines who should be arrested and who should not,' and some of the most senior Iraqi officials from Saddam's regime were included at the top of the official list.[51] Adm. Ken Summers (who commanded Canada's ships in the 1991 gulf war) also confirmed that 'they are providing direct support to the war on terrorism ... and indirect support to the war on Iraq, in that they are escorting through the Strait of Hormuz any allied ship that needs to get in or out of the gulf.'[52] The Canadian commander, Cmdr. Roger Girouard, reported directly to U.S. Vice-Adm. Timothy Keating; the vice-admiral is both the head of the U.S. 5th Fleet (fighting the Iraq war) and the top naval officer in the war on terrorism.[53]

There are at least two points to be derived from the facts presented in these reports. The Canadian naval task group did not and could not clearly separate the roles between terrorism and the war in Iraq. Indeed, the Canadian government really had no control over, nor could they establish definitive operational rules of engagement to separate, those two responsibilities. As General MacKenzie noted, '[A]nyone who thinks our Navy would abort an intercept of a threat to any ship they are escorting because they discover that it's an Iraqi threat doesn't know our Navy.'[54] If Ottawa passed along an order that, for practical operational and tactical reasons, was subsequently ignored in favour of maintaining a commitment to Canada's multilateral obligations, that speaks volumes about the capacity of defence officials to affect operational policy. In reality, the fact that the Canadian navy followed previously established rules of engagement for tracking and arresting suspected terrorists, derived from a list that included Iraqi officials provided by the U.S. military, implies that the Canadian government had a lot less authority and influence than it implied in statements about the distinction between Canadian and American operations – there really was no distinction at all. In the end, as Andrew Coyne correctly observes, 'Other countries may support the war without participating in it, but only Canada is participating without supporting it.'[55]

Canada's Uncompromising Compromise and the Illusion of Middle-Power Mediation

Canada's fence-sitting policy on Iraq, along with efforts by Canadian representatives at the UN to find a solution to the impasse, did not stake out a meaningful compromise, nor was it representative of a distinct middle-ground position. Efforts to carve out a Canadian fence-sitting position on Iraq to highlight Canada's traditional middle-power-mediator role was doomed from the start, primarily because it miscalculated the motivations that drove the three (not two) dominant positions held, respectively, by the American-British coalition, the French-German coalition, and the UN's pre-eminent fence sitter, Hans Blix.

American and British officials were convinced that the current inspection regime had failed and could not be saved. That position was etched in stone, and American officials were given no reason to expect anything else from Saddam Hussein or Hans Blix. French and German officials were compelled to prevent any war, not because the U.S.-U.K. coalition might lose a long and ugly battle, or because Iraqis would suffer, but because the war was likely to be short and successful, consistent with every other just war fought and won by U.S.-led coalitions of the willing over the past decade. French officials know full well that most Bosnians, Kosovar Albanians, Serbs (including those in post-Milosevic Belgrade), and Afghans are all better off today than they were.

Officials in Paris, Berlin, and Moscow also knew that a U.S.-U.K.–controlled post-Saddam Iraq would be far better for Iraqi citizens but less friendly to French, German, and Russian oil interests, which are substantial (please see chapter 3).

Contrary to the spin from Ottawa that Canada developed a unique position in the debate, Hans Blix had already staked out the UN's fence-sitting terrain. Blix continued to produce reports that were mixed and balanced, not because this reflected the situation but in order to avoid war at all costs, to demonstrate the utility of arms inspections and the competence of arms inspectors to do their important job, and to sustain the false impression that the UN's version of multilateralism is more relevant to global security than American multilateralism. That was his job. Immediate, complete, and unconditional disarmament of Iraq (UNSC 1441) was never successfully met by Saddam's regime, and none of UNMOVIC's reports suggested otherwise.

Canadian officials did not fully realize at the time that Hans Blix would never produce a report that declared Iraq in 'material breach' of UNSC 1441. In fact, no career UN diplomat worth his weight in multilateral gold would ever intentionally embrace any part of the responsibility for launching a U.S.-led war that risks the deaths of innocent civilians.

To demonstrate Blix's reluctance to establish material breach, consider the UNSC 1441's call for 'immediate, unconditional, unimpeded and complete' disarmament in light of Blix's own conclusions (outlined in table 7.1).[56] These are not statements issued by a conservative Republican administration in the U.S. determined to attack Iraq; these are the conclusions of the chief weapons inspector. According to Blix:

Unlike South Africa, which decided on its own to eliminate its nuclear weapons and welcomed the inspection as a means of creating confidence in its disarmament, Iraq appears not to have come to a genuine accep-

TABLE 7.1
Material Breach?
Remarks by Hans Blix
Official Report to UNSC, March 2003

Immediate, Unconditional?
'In this updating, I'm bound, however, to register some problems. The first are related to two kinds of air operations. While we now have the technical capability to send a U-2 plane placed at our disposal for aerial imagery and for surveillance during inspections and have informed Iraq that we plan to do so, Iraq has refused to guarantee its safety unless a number of *conditions* are fulfilled ... As these conditions went beyond what is stipulated in Resolution 1441 and what was practiced by UNSCOM and Iraq in the past, we note that Iraq is not so far complying with our requests. I hope this attitude will change ... Another air operation problem, which was so during our recent talks in Baghdad, concerned the use of helicopters flying into the no-fly zones. Iraq had insisted on sending helicopters of their own to accompany ours.'

Unimpeded?
'On a number of occasions, demonstrations have taken place in front of our offices and at inspection sites ... Demonstrations and outbursts of this kind are unlikely to occur in Iraq with initiative or encouragement from the authorities. We must ask ourselves what the motives may be for these events. They do not facilitate an already difficult job, in which we try to be effective, professional, and at the same time correct.'

Complete?
'Regrettably, the 12,000-page declaration, most of which is a reprint of earlier documents, does not seem to contain any new evidence that will eliminate the questions or reduce their number ... Even Iraq's letter sent in response to our recent discussions in Baghdad to the president of the Security Council on 24th of January does not lead us to the resolution of these issues.'

'Iraq has provided little evidence for this production and no convincing evidence for its destruction ... There are strong indications that Iraq produced more anthrax than it declared and that at least some of this was retained over the declared destruction date. It might still exist. Either it should be found and be destroyed under UNMOVIC supervision or else convincing evidence should be produced to show that it was indeed destroyed in 1991.'

'I turn, Mr. President, now to the missile sector. There remain significant questions as to whether Iraq retained Scud-type missiles after the Gulf War. Iraq declared the consumption of a number of Scud missiles as targets in the development of an anti-ballistic missile defense system during the 1980s, yet no technical information has been produced about that program or data on the consumption of the missiles.'

'When we have urged our Iraqi counterparts to present more evidence, we have all too often met the response that there are no more documents. All existing relevant documents have been presented, we are told. All documents relating to the biological weapons program were destroyed together with the weapons.' 'However, Iraq has all the archives of the government and its various departments, institutions and mechanisms. It should have budgetary documents, requests for funds and reports and how

TABLE 7.1
Material Breach? (concluded)

they have been used. They should also have letters of credit and bills of lading, reports and production and losses of material.'

'The recent inspection find in the private home of a scientist of a box of some 3,000 pages of documents, much of it relating to the lacing enrichment of uranium, support a concern that has long existed that documents might be distributed to the homes of private individuals. This interpretation is refuted by the Iraqi side which claims that research staff sometimes may bring papers from their work places.'

'On our side, we cannot help but think that the case might not be isolated and that such placements of documents is deliberate to make discovery difficult and to seek to shield documents by placing them in private homes.'

tance, not even today, of the disarmament which was demanded of it and which it needs to carry out to win the confidence of the world and to live in peace ... Against this background, the question is now asked whether Iraq has cooperated, 'immediately, unconditionally and actively,' with UNMOVIC, as is required under Paragraph 9 of Resolution 1441. The answers can be seen from the factor descriptions that I have provided. However, if more direct answers are desired, I would say the following: The Iraqi side has tried on occasion to attach conditions, as it did regarding helicopters and U-2 planes ... It is obvious that while the numerous initiatives which are now taken by the Iraqi side with a view to resolving some longstanding, open disarmament issues can be seen as active or even proactive, these initiatives three to four months into the new resolution cannot be said to constitute immediate cooperation. Nor do they necessarily cover all areas of relevance.[57]

It is perfectly reasonable for an informed observer to conclude from Blix's report that Iraq was in material breach of 1441. It would be even more difficult to refute this conclusion in light of the findings in David Kay's post-war report on Iraq's clandestine WMD program. The real question at the centre of the debate, then, is whether this breach rises to a level of threat justifying a military intervention. To avoid that conclusion, Blix's reports often buried information on, for example, prescribed weapons systems (such as the drone with an illegal twenty-five-foot wing span), or included evidence in the written report that was excluded from his oral summaries. Consider Blix's efforts to reinterpret some of the evidence to create the illusion of meaningful progress, as noted by Andrew Coyne:[58] Iraqi minders trailing inspectors were re-

TABLE 7.2
Canadian Proposal and Disarmament Requirements

February 28:	Inspectors bring forward their 'clusters' report early and present it to the Council, setting out the key remaining disarmament issues/questions ... Inspectors present a prioritization with timeframes for the Council of the key substantive tasks for Iraq to accomplish, including those involving missiles/delivery systems, chemical weapons/precursors, biological weapons/material, and nuclear weapons.
March 7:	Inspectors update the Council on Iraqi co-operation on substance.
March 14:	Further update.
March 21:	Further update.
March 28:	Final report to the Council by inspectors.
March 31:	Meeting of the Council at ministerial level: if the inspectors have reported substantial Iraqi compliance, a robust ongoing verification and monitoring system, including increased numbers of inspectors/monitors, investigations, etc., would be implemented. If the inspectors have reported continued Iraqi evasion, all necessary means to force them to comply could be used.[a]

[a] www.cbc.ca/news, 'Canada's Proposal Titled "Ideas on Bridging the Divide,"' February 2003.

duced from five to one down to one to one (parity); Iraq 'gave us a number of papers' but, despite the fact that there was 'no new evidence' (according to Blix) in these papers Iraq's efforts 'could be indicative of a more active attitude'; Iraq agreed to allow U-2 reconnaissance flights but with conditions (including information on timing, flight path, and call signs); and Iraq established a new Iraqi commission to search for WMD and documents, but this is something they should have been doing for the past twelve years.

With all this in mind, Canada's proposal (included in table 7.2) was destined to fail, because nowhere in the proposal was there any reference to an explicit threat of serious consequences, or a series of concrete yardsticks that would unequivocally establish either compliance or material breach, one way or the other.

Anything less than a clearly articulated set of rigid benchmarks would have achieved nothing beyond 1441. The problem with the Canadian 'compromise' was the lack of specific standards, which were left up to Hans Blix. But this recommendation offers nothing new other than what had already been included in resolutions over twelve years, in UNSC 1441, and in Blix's existing mandate as UNMOVIC's chief. If Blix could produce mixed and balanced reports on Iraqi 'progress' and 'compliance' to UNSC 1441, a resolution that called for 'immediate,'

TABLE 7.3
British Proposal and Disarmament Requirements[a]

---

1 A public statement in Arabic by Saddam Hussein, to be broadcast on television and radio in Iraq, admitting to the possession and concealment of weapons of mass destruction and declaring his regime's intention to give them up without delay.

2 At least 30 Iraqi scientists selected by UNMOVIC and the IAEA to be allowed to go abroad, together with their families, for interview by UN inspectors. They must cooperate fully with their interviewers.

3 The surrender of all remaining anthrax and other chemical/biological weapons or explanation of their previous destruction.

4 An explanation of the unmanned drone aircraft found by the inspectors, together with the numbers and location of any others.

5 A commitment that the so-called mobile laboratories will be surrendered for destruction.

6 A commitment to the destruction of 'proscribed missiles,' including the remaining Samoud 2 rockets but possibly others also.

---

[a] Brian Whitaker, 'UK proposes six weapons tests for Saddam,' *Guardian* (London), 13 March 2003.

'complete,' and 'unconditional' disarmament that was never achieved, he would certainly have no problem producing equally mixed and balanced reports based on benchmarks he himself is responsible for selecting. Without some explicit threat of war, or a clear set of conditions that go beyond Blix's existing mandate, Canada's compromise really amounted to little more than a shallow offer designed to push the multilateral process along. That is why the proposal was dismissed by the U.S.-U.K. coalition – it had nothing to do with an American refusal to wait any longer. In fact, the U.K. proposal that the U.S. supported also included an extended deadline of a few weeks, additional time the Pentagon needed anyway to prepare for the war.

While the Canadian proposal was designed to avoid an unavoidable war, the U.K. proposal (described in table 7.3) was designed to demonstrate why war was unavoidable, which is why it was accepted by the U.S., and perhaps why Canadian officials did not endorse it.

The U.K. proposal recommended that the UNSC render any final judgments on compliance, not Hans Blix, and included several explicit and demanding targets. The benchmarks explain why the proposal was rejected by France, Germany, and Russia. Similarly, any deadline – a few weeks (U.K.) to a month (Canada) – was equally unappealing to France and Germany, because Saddam's refusal to meet (or only partially comply with) those demands would risk tipping the balance in favour of material breach and, ultimately, a short and successful U.S.-U.K. intervention. This also explains why France adopted the position that a second resolution was not required, and that it would veto any resolution threatening any form of war.

Even if the Canadian compromise had miraculously prevented the war, the more relevant question is whether a disarmed Iraq in which the Baath regime survived and Saddam Hussein (and his sons) retained power constitutes a more 'just' or 'moral' outcome. As Michael Ignatieff observes,

> Iraq is an issue, unfortunately, where multilateralism meets its moment of truth. If we actually believe in international law – and that is the crux and heart of Canadian foreign policy – we don't want to have Iraq defying UN Security Council resolutions on a vital issue. Why? Because this is a regime that has just about the worst human rights record on Earth and is in possession of weapons of mass destruction. It is not just the weapons, lots of other people have the weapons, it is the combination of a rights-violating regime that has an expansionist record in possession of deadly weapons. You can't believe in multilateralism, international law, unless you are prepared also to believe that occasionally you have to step up to the plate and defend it – and by force if necessary. So I am as multilateral as any Canadian, but you can't talk the talk unless you are also prepared to walk the walk.[59]

Given Canada's mixed, disjointed, and confusing messages throughout the Iraq war, Canadian policy ultimately lacked the very principles Canadian officials claimed to be defending – the same principles Iraqis were celebrating when statues of Saddam toppled. The Chrétien doctrine, if it succeeded, would have prolonged their suffering, and, if followed, will prolong the suffering of many others. That position is not at all distinctive; it is shared by dozens of countries whose leaders refuse to address any threat or resolve any conflict that has no meaningful connection to their political, economic, or financial self-interests.

Conclusion

There was no intellectually compelling, logical, or morally defensible middle ground in this crisis, no multilateral fence on which to sit, no middle-power-mediator role to play. As a result, the real message delivered by Canada's diplomats (at least the message other states were hearing) was that only Canadians have the intellectual capacity and moral superiority to figure out a distinctly Canadian soft-power solution that had apparently eluded everyone else on the planet. The more interesting question is why Canada decided not to embrace and defend the U.K. proposal – aside from the more explicit U.K. benchmarks, the difference was essentially two weeks. The answer, I suspect, is that support for the U.K. proposal would have detracted from Canada's image as helpful fixer and mediator, an image the government (and Canadian media) tried very hard to nurture throughout the crisis, but one the American (CNN) and European (BBC) media ignored. For Canadian officials it was never really about the quality of the proposal (there wasn't very much in it) – it was about providing another illustration of Canadian diplomacy that a legacy-building prime minister (and current and former DFAIT officials) can now include in their speeches. Although it failed, the 'Canadian compromise' is an achievement that will be listed along with Canada's contributions to the International Criminal Court, Kyoto Accord, and the Land Mines Treaty. Of course, these speeches will never include even a brief account of the historical details surrounding the Canadian compromise, because the term itself serves an important role in propagating the illusion of Canada's middle-power status and its relevance in the world. Maintaining those myths, and related myths about Canada's distinctive role in the universe, has become an important preoccupation for officials in Ottawa. Unfortunately for Canadians, that preoccupation is becoming increasingly costly to Canadian interests, values, and aspirations.

Chapter 8

Recalibrating Canada's Moral and Diplomatic Compass

The Mounting Costs of Privileging 'Distinction' over 'Security'

Mark Kingwell's defence of 'distinction' as a guiding principle for Canadian foreign policy serves well as a starting point to the final chapter: 'The old cliché has it that Canadians, lacking an identity of their own, construct one out of not being American. I have never understood why this is considered inadequate or feeble. If you were the only dissenter in a room holding a dozen people, standing up and saying "I'm not the same as you" would be a clear mark of moral courage.'[1] But surely moral courage is not something that emerges automatically from an imperative to establish distinction at every opportunity. The mere process of distinguishing yourself as a dissenter is secondary to the underlying issue, policy, value, or interest that motivates the need for distinction. These values and interests, not the act of distinction itself, speak to the moral courage of those who claim to be different. A single dissenter in a room of a dozen people, all of whom are in favour of an immediate international intervention into Rwanda to save close to one million people from genocide, is not morally courageous, but a single dissenter in the same room with twelve opponents to intervention is. Morality is derived from the values one is attempting to defend, and a policy of distinction for the sake of distinction is often blind to this moral clarity.

In an effort to highlight some of the 'obvious' differences between Canadians and Americans (ostensibly to defend his argument) Kingwell goes on to write, 'For generations, we [Canadians] have been busy creating, in your [American] shadow, a model of citizenship that is inclusive, diverse, open-ended and transnational. It is dedicated to far-reaching social justice and the rule of international law. And we're successfully exporting it around the world not by bucking the UN, but by seeing it for the flawed but necessary agency it is.' The most serious problem with this statement is that the 'differences' Kingwell identifies are merely asserted, as if the distinctions are so obvious and apparent that simply listing them establishes the point. But there is absolutely nothing in his list of 'Canadian' values that is in any way distinct from those values that Americans espouse, institutionalize, defend, and export. The priorities and preferences he lists are virtually identical, with the possible exception that the U.S. has been far more successful than Canada at exporting those values. This is certainly true if one compares investments in foreign aid, defence, and diplomacy – the core prerequisites for transferring those values abroad.

Canadians are often subjected to repeated pronouncements by DFAIT officials that Canada was instrumental in developing the International Criminal Court and the Land Mines Treaty, but those achievements should be juxtaposed against the essential role the United States played in ending the cold war and facilitating the democratization and development of Eastern Europe, ending a ten-year war in the Balkans, stopping and reversing ethnic cleansing in Kosovo, changing three brutal and repressive regimes in Serbia, Afghanistan, and Iraq, playing an indispensable role over the years in the Middle East peace process, contributing $15 billion to fight AIDS in Africa, and so on. The balance sheet of accomplishments produced by hard and soft power to protect the values and interests both Canadians and Americans cherish should humble any Canadian official. In reality, writes Jeffrey Simpson, 'Never has the world meant more to Canada; never has Canada meant less to the world ... If indeed Canadians want a somewhat distinctive foreign policy vis-à-vis Washington, then they have to pay for it. Having failed to make that investment, they fill the vacuum with moral superiority that acts as a false substitute for having real assets, tangible commitments and the ability to make real choices.'[2]

Whatever difference one can identify with respect to our approaches to international politics, they have nothing to do with our distinctive cultures, values, or intellect and everything to do with differences in power.[3] Canadian officials prefer multilateralism not because they are better suited with soft-power skills to be more productive international citizens, but because Canada is a self-interested middle power engaged in a never-ending competition with other states for economic and political relevance. Canada is more relevant if powerful states are constrained by international institutions. But constraining the United States by demanding that it comply with multilateral consensus is not necessarily conducive to producing an international system that is more peaceful, stable, or more capable of protecting humans rights and spreading democracy. Indeed, many of those same weak states who demand compliance to international institutions do not share those values.

In a post-9/11 environment the imperative to be confident and proactive when crafting Canadian foreign and security policy has never been greater, and the dangers of a blind commitment to the default 'weak-state' strategy of 'distinction over security' have never been more apparent. As Christopher Sands points out, 'A weak-state strategy for Canada would consider the threat of international terrorism largely a US concern, and seek to placate US pressures within minimum efforts while husbanding Canadian sovereignty and avoiding commitments to undertake new responsibilities with regard to the defence of North America ... If [Ottawa] continues to adopt a weak country strategy it will fade in its ability to represent and defend Canadian interests in the United States, while fading in its attractiveness as a partner for Washington in the management of cross-border issues.'[4] Perhaps Canada's weak-state status defines its destiny – as U.S. power continues to expand and as the status and influence of a growing number of very active middle powers continues to develop, Canada will, by definition, become increasingly marginalized. Consider Canada's declining status in the international system as measured by club memberships, described by Bill Dymond and Michael Hart:

> During the Cold War, Canada was a member of a small select club of functioning democracies ... In the post–Cold War years, democratic politics and market economics have been adopted by most of the countries of the former Soviet Empire and the vast majority of developing countries. WTO membership now numbers 145 countries, with another two dozen in the accession queue. In NATO, the putative enemy has observer status and many of its former satellites are full members. These remarkable developments (many inspired by American initiatives) have created a more prosperous and safer world, but they have deprived members of hitherto exclusive clubs an important part of their identity. Membership in the global *directoire* of the G-7/8 continues to convey a benefit of exclusiveness, but Canada's ability to influence that club to serve Canadian interests is at best modest.[5]

In this context perhaps we should expect Canadian offi/
that much harder at maintaining the illusion that Canada ᵢ.
that Ottawa actually has some choice left, that these choices are ᵥ
relevant, and that Canada has retained control over its priorities and
destiny. Perhaps the weak-state strategy is all Canada really has left.
Unfortunately, that strategy is becoming increasingly costly in the context of terrorism, proliferation, and U.S. unilateralism.

Terrorism and Hezbollah

One of the clearest manifestations of Canada's weak-state policy of
distinction over security was illustrated by Foreign Affairs Minister Bill
Graham's insistence that the social and military wings of Hezbollah
be considered distinct. This position was maintained and defended
despite extensive evidence to the contrary compiled by U.S. and Canadian intelligence sources.[6] Notwithstanding this evidence, and the overwhelming consensus in the terrorism literature that recruitment and
funding are rapidly spreading in liberal democracies, Ottawa continued to portray and defend the distinction between Hezbollah's social
and military wings with repeated references to Canadian 'values.'
    There are several reasons why this policy seriously undermined the
security interests of Canadian citizens. First, the message received by
American officials was that Ottawa is not seriously committed to the
war on terror. Second, Graham's insistence that the distinction was not
only fair but worthy of defending conveyed to prospective terrorists
that their funding networks could function more efficiently in Canada –
what other message could they possibly derive from Canada's policy?
Third, from Washington's perspective Ottawa was perceived as willing
to accept the costs and security risks of being wrong about Hezbollah
(perhaps because those costs were more likely to be paid out in American lives) when there were no corresponding risks from supporting a
very straightforward and rational U.S. security policy – a rationality
Ottawa was forced to finally accept after comments by a prominent
Hezbollah leader calling for renewed suicide attacks against Israel.
Fourth, and perhaps most disconcerting for U.S. officials, was the fact
that there was no compelling political rationale for the Liberal government to maintain this position – there was no election to lose, no
domestic political backlash to worry about, and the official opposition
would have supported an early decision to outlaw both the social and

military wings of Hezbollah. Without any resistance whatsoever to worry about it would have been easy for the Liberal Party to do the right thing for Canadian security, yet for some reason DFAIT officials decided the wrong policy made more sense.

This decision was continuously defended despite evidence compiled by U.S. and Canadian intelligence detailing obvious linkages between Canada and terrorism. As the director of the Canadian Security Intelligence Service (CSIS), Ward Elcock, points out, '[A]bout all of the world's terrorist groups have some presence here.'[7] In 1998, CSIS investigated fifty organizations and more than three hundred persons through its counter-terrorism program and concluded that 'most of the world's terrorist groups, including Osama bin Laden's Al Qaeda, have adherents in Canada ... Islamic terrorists from Algerian, Egyptian, Libyan and Somali groups also have sympathizers in Canada and we have to deal with that reality.' The decision by DFAIT to reconsider the wisdom of Canada's 'distinct' position on Hezbollah was not enough to address serious questions about whether Ottawa's commitment to the war on terror is dependable enough to be taken seriously by Washington.[8] After all, there is another interpretation of Canada's position on Hezbollah that is perhaps even more distressing for American officials – it may have been a policy driven by Canada's unilateral self-interest to avoid the wrath of terrorist groups by demonstrating to them that Canadians are not Americans (a post-9/11 variation of appeasement). This weak-state/appeasement explanation, while perhaps less accurate, is particularly dangerous even if it represents the views of a small minority in Ottawa (or Canada). The most important question, of course, is whether this interpretation is considered in Washington to be more accurate, and what are the implications? Perhaps Mark Steyn's warning is worthy of consideration: '[T]he US is coming to regard Canada the way Australia regards Indonesia. Yes, it's geographically close, an important trading partner, a cheap vacation destination and a nominal ally [but] Bali was a soft target for the terrorists ... The big story since September 11th is that they [Americans] finally see us [Canadians] for what we are: foreigners.'[9] That impression has not been helped by Canada's position in the 2003 Iraq war.

Terrorism, Unipolarity, and Canadian Values and Interests

When it comes to American security after 9/11, current and future U.S. administrations will respond to terrorism and proliferation with unilateral initiatives, for all the reasons outlined in chapters 1 through 7. In essence, multilateralism has become a liability, a security threat. It is perceived by Washington today, writes Joseph Nye, as 'a strategy by smaller states to tie the U.S. down like Gulliver among the Lilliputians.'[10] But none of these states, including American allies, are driven by some higher moral imperative to create a truly global order based on justice and international law; they are motivated by the same fundamental imperatives that drive American foreign policy: power, security, self-interest, and survival. The difference for Canada is that American unilateralism will have a more direct and profound impact on Canadian economic and security interests.

Several important policy implications follow from the evidence and observations compiled in this book. First, Canadian officials should begin to develop planning scenarios to help prepare for a variety of U.S. responses to terrorist attacks. The objective should be to go beyond emergency preparedness (an obviously important component of any immediate response to terrorist attacks in Canada or the United States) and to begin thinking about how Canada should respond more effectively (and immediately) to a range of potential U.S. reactions. These responses should be coordinated in ways that avoid the negative consequences of being caught off guard, and to ensure Canadian interests are not jeopardized in the wake of U.S. unilateralism. A coordinated political, legal, and diplomatic plan will help to avoid the tendency in Ottawa to be reflexive when dealing with U.S. actions or when faced with a major international crisis. Waiting for the United States to act/respond may be appropriate when the policy in question affects some other region or state, but reflexive responses are entirely inappropriate when U.S. actions have a direct (and sometimes instantaneous) impact on Canadian economic and security interests. In a post-9/11 environment, the imperative to be confident and proactive when crafting Canadian responses has never been greater, because 'a nation that merely reacts to events,' as Wendy Dobson warns, 'is likely to see its sovereignty erode and its future determined by others. A nation that exercises its sovereignty anticipates change, prepares options that promote the key interests of its partner, but channels action in ways that best serve its own interests.'[11]

Canada's reaction to 9/11 was coloured, says Jeffrey Simpson, 'by a fear of being seen to have agreed with Washington, and being accused of having "caved," "sold out" or not adequately protecting Canadian sovereignty ... A confident country, whose identity is rooted in its sense of self rather than a determination to highlight differences, would not have worried, as the Chrétien government did, about criticism of being too close to the United States.'[12] Simpson's observations go well beyond the prime minister's tentative response to 9/11 – the same pattern was repeated over the past ten years in Bosnia circa 1990–95, in Iraq 1998, in Kosovo 1999, throughout the ballistic missile debate, in Canada's initial response to the U.S. war in Afghanistan, in Ottawa's subsequent hesitation to deploy the Princess Patricia's Canadian Light Infantry to Afghanistan, when reacting to the U.S. NORCOM announcement, and, most recently, in Canada's confusing position on the U.S.-U.K.– led war in Iraq. Canadian policy is often guided not by a serious and meaningful debate about the implications of change for Canada's security and defence priorities, but by the need to remain distinct, as if our sovereignty depended on our capacity to create differences with the U.S. But being different for the sake of difference is as damaging to Canadian sovereignty as simply following the American lead on every issue.

With limited resources Canadian officials should avoid the tendency to implement (and pay for) quick fixes. This will become increasingly difficult as Canada gets swept along by U.S. unilateralist pressures, but Ottawa should be prepared to handle these pressures in ways that steer U.S. unilateralism in more productive, cost-effective, security-maximizing directions. The objective here is to avoid unintended consequences and to prevent what Malcolm Gladwell refers to as the 'paradox of law enforcement':

> The way in which those four planes were commandeered ... did not simply reflect a failure of our security measures; *it reflected their success.* When you get very good at cracking down on ordinary hijacking ... what you are left with is extraordinary hijacking ... The history of attacks on aviation is the chronicle of a cat-and-mouse game, where the cat is busy blocking old holes and the mouse always succeeds in finding new ones ... During the nineties, in fact, the number of civil aviation 'incidents' world- wide – hijackings, bombings, shootings, attacks, and so forth – dropped by more than seventy percent. But this is where the law enforcement paradox comes in: Even as the number of terrorist acts has diminished, the number of people killed in hijackings and bombings has steadily increased ... Airport-security measures have simply chased out the amateurs and left the clever and the audacious [emphasis added].[13]

Similarly, Barry Rubin has shown that almost all American security measures put in place since 9/11 are designed to prevent a repeat of 9/11, and they will be largely successful. But preventing the same attacks from occurring again is a very small part of what needs to be done: 'Counterterrorist planners need to have some imagination in figuring out the more likely threat and not just a rote repetition of the previous assault.'[14] Solutions should avoid exclusive reliance on inventing new technologies and should focus on making existing technologies work properly. Rubin cites Israel's airport security system as an example – it is among the most efficient and effective in the world, yet has remained virtually unchanged since the 1960s.

Although the gap between Canadian and American objectives in the war on terror is arguably quite narrow, there are specific priorities on which Canadians officials should focus. For example, port security is a high-risk area for future terrorist activity that demands proactive Canadian planning.[15] If Canadian security and sovereignty are priorities, Ottawa should accelerate Canada-U.S. joint planning under Northern Command (NORCOM) and establish additional integrated enforcement mechanisms with the U.S. for homeland security. Another area of immediate concern for Canada is the threat posed by biological weapons proliferation. Terrorists are likely to target food and water supplies with viruses, bacteria, fungi, and other toxins (anthrax, plague, botulism, and salmonella). Ronald Atlas, an adviser to the U.S. government on emerging biological threats, recommends joint Canada-U.S. programs to enhance surveillance and protection of vaccine stockpiles, develop new vaccines to cope with attacks and outbreaks, and establish national centres for preparedness training.[16] According to a recent report by Ed Struzik, there are 750 labs in Canada alone that produce pathogens criminals could use, and there is 'very little control over inventories of biological agents and few licensing procedures that would make it difficult to obtain them.'[17]

Proliferation, Missile Defence, and Canadian Values and Interests

The U.S. will proceed with a layered missile defence system for North America encompassing sea-based, land-based, and air-based components. The layered defence will include testing and development of boost-phase, mid-course, and terminal-phase architecture. That is the only logical and defensible approach that any U.S. administration can take. On balance, the technological, financial, and strategic arguments in favour of deployment far outweigh in logic, rationality, common sense, and empirical validity the arguments against deployment (chapters 4 and 5). It is the strength of the arguments that explains why proponents have been so successful in making their case, why critics have failed to make and successfully defend theirs, and why the U.S. government (regardless of political party) and the American public favour deployment. Debates in Canada regarding the contemporary relevance of bipolar strategic stability, the validity of mutual nuclear deterrence, the future of mutually assured destruction, the obsolescence of the ABM Treaty, and the stability of BMD are only just beginning, but they have been resolved in the U.S. (chapter 1).[18] Ultimately, Canada will support BMD, not because the United States favours it but because, all things considered, BMD makes sense for Canadian security and defence. It is in Canada's interest to find a place under this new umbrella. As David Warren asks, '[W]hy should only the Americans have some defence against missile attack? What makes them so especially worth preserving?'[19]

Critics of BMD continue to warn that it will be costly, but little if any of the costs will be placed on the shoulders of Canadian taxpayers. Ottawa has not been asked to contribute a red cent to the project, and, unfortunately, Canadian companies are unlikely to compete for the larger defence contracts. Some critics will continue to mention costs at every opportunity, but they should admit that the United States is more than willing to pay for (and develop) the technology in-house – for all the reasons outlined above.

Other critics in Canada claim that a decision by Ottawa to reject BMD would have little if any effect on Canada–United States relations. But those who make that argument never explain why the U.S. government would accept such a decision without responding, or why Canada's relations with the United States in North American Aerospace Defence (NORAD) or NORCOM would be unaffected. There are few historical precedents to defend that position. Some have argued that we don't need a cold war relic like NORAD. But just what is the argument here: that there will be no consequences if Canada rejects BMD, or that the consequences of Canada's exclusion from a key element of NORAD/NORCOM will be irrelevant? Few details are offered to defend either position. Most experts agree, however, that a break with NORAD would have a direct effect on Canada's access to the very satellite surveillance technology Canada requires to maintain at least some control over its own sovereignty and security. As Don MacNamara correctly concludes in his report on NORAD and NORCOM, 'It is in Canada's vital interests to survive as an integrated political and economic sovereign state. Sovereignty includes the capability to be aware of and to control activities within the sovereign jurisdiction of the state, its airspace and maritime approaches. The capacity to maintain surveillance over Canadian territory is currently integrated with the NORAD system. Loss of that capacity would indeed affect Canada's sovereign capability.'[20] Some critics have expressed concerns that Canada-U.S. military integration will jeopardize Canadian sovereignty, because it will force Canadian officials to do things differently.[21] But that assertion misses an important point about the relationship between integration, security, and sovereignty: integration will not undermine Canadian sovereignty simply because it forces Ottawa to do things differently (that isn't always a bad thing); integration will undermine Canadian sovereignty if it forces Canada to do things that are inconsistent with the values and institutions it claims to be defending. But when it comes to U.S. foreign and security policies on humanitarian intervention, the war on terrorism, international trade, NATO expansion, the Middle East, proliferation of weapons of mass destruction, homeland security and defence, gulf wars I and II (the latter in terms of Canada's contribution to the war effort), the bombing campaign against Iraq in 1998, Bosnia, Kosovo, Afghanistan, and so on, critics of integration provide no evidence to support the claim that Canada and the U.S. are fighting different battles. Those who are worried about Canada-U.S. integration may not like the values underlying American foreign policy, but many Canadians (and successive Canadian governments) do, and for very rational, morally defensible, ideologically balanced, and cost-effective reasons. In the few cases in which Canadian and American policies diverge, as in the 2003 Iraq war, the government has had a very difficult time defending its position with reference to the values and principles it claimed to be supporting.

With this in mind, Ottawa needs to develop and clearly articulate a national policy on missile defence based on the government's assessment of WMD proliferation and its impact on Canadian defence and security.[22] Policy options should be evaluated on the basis of comparative risks and costs – what are the risks to Canada of proliferation? What are the risks to Canada of supporting (or failing to support) BMD? And what are the immediate and long-term risks and costs of sitting on the fence or relying exclusively on a seriously deficient multi-

lateral NACD regime? If critics of BMD have a better approach to
achieving these objectives, they should put these alternatives on the
table; the previous chapters explain why they will have a hard time
doing so.

The decision by Prime Minister Jean Chrétien in May 2003 to begin
formal discussions with the Americans on BMD is an encouraging sign;
it is the first, clear indication that critics within the NACD community
have lost the battle to keep Canada out. Ironically, if they succeeded it
would have stripped from the Canadian government an excellent opportunity to live up to its most important obligation – to protect Canadians from emerging threats associated with WMD proliferation.[23]
Fortunately for Canada, the U.S. would have been more than happy to
fulfil that obligation by providing Canadian security on Ottawa's be-
half. Unfortunately for Canada, many officials in Ottawa continue to
view BMD deployment as just another dangerous manifestation of
American unilateralism. Not dangerous in the global security sense,
but dangerous because it threatens to dismantle the cornerstones of
arms control and irreparably damage the framework of agreements,
treaties, and conventions that encompass the nuclear non-proliferation,
arms control and disarmament regime. The NACD regime and the new
Global Partnership Program remain the focal points of every significant
Canadian foreign and defence policy initiative in the area of non-
proliferation, so any unilateral move by the U.S. to systematically
marginalize this framework cannot help having profound and long-
term implications for Canadian policies. But officials in Ottawa who
continue to defend these multilateral approaches have an important
obligation to inform the Canadian public that this activity has not
prevented WMD and ballistic missile technology from getting to states
who want it – presumably the benchmark for evaluating our ongoing
commitment to these policies. Rejecting any and all unilateral options
simply because they are unilateral, or because of some hope that, with
enough time, multilateralism can be made to work is not a credible
solution. Repeated efforts by Canadian officials to convince Washington that American security requires that the American public remain
vulnerable to nuclear annihilation are not only outdated but patroniz-
ing and insulting.

Lloyd Axworthy and Michael Byers's reaction to Prime Minister
Chrétien's decision to begin formal talks with the Americans on BMD is
noteworthy: '[C]halk up yet another potential casualty of the Iraq war:
Canada's impending loss of an independent voice on matters of peace

and security.'[24] As is typical with BMD critics, independence is always framed in terms of Canada's position *vis-à-vis* the U.S., and almost never in terms of the difference between the values we espouse and those of, for example, Russia, China, or Iran. The authors go on to claim that what is at stake 'is our freedom of action in international affairs, including not least our ability to play a constructive role in a global system based on co-operation and agreed rules rather than the threat and use of armed force.' But these same critics almost never provide any compelling evidence that nation states consistently agree on the rules, or offer any reasonable suggestions for what we should do when states disagree. They rarely provide proof that we live in a global system based on cooperation and not one that occasionally depends on armed force for stability and peace. And they cannot provide much evidence that Canadian officials are serious about paying for and playing (as opposed to fabricating) a constructive role in the international community by making meaningful contributions to defence, foreign aid, and diplomacy. Instead, what we tend to get from former DFAIT officials, such as Axworthy, is simply an assertion of multilateral benefits followed by a diatribe against the evils of unilateralism. But when the costs, risks, benefits, and unintended consequences of multilateral and unilateral approaches to terrorism and proliferation are juxtaposed, as the preceding seven chapters have attempted to do, the case in favour of multilateralism is considerably less obvious.

Canada-U.S. Relations: Revisiting First Principles after 9/11

Dispelling the 'Don't Worry, Be Happy' Myth

The prevailing opinion among many Europeans and Canadians is that the drift towards American unilateralism is an unfortunate mistake. Contemporary American security imperatives are temporary phenomena, simply the products of the ideological predisposition of a conservative president, surrounded by a particularly hawkish cabinet, committed to constructing a new American empire. Sooner or later, according to this prediction, officials in Washington will wake up, come to their collective senses, and begin once again to realize the value of multilateralism and collective action. 'In short,' as Louis Delvoie concludes, 'the unilateralist trend in U.S. foreign policy is neither inevitable nor irreversible,' and Canada has a responsibility to 'convince the US administration that in an interdependent world, the long term interests of the United States are best served by multilateralism rather than unilateralism.'[25] There are several problems with this view, many of which are apparent from the evidence compiled throughout this book.

The battle Canada is fighting today should not be against unilateralism, it should be against terrorism and proliferation. Officials in Ottawa do not have a responsibility to lecture the U.S. administration on the perils of unilateralism in an interdependent world; they have a responsibility to clearly demonstrate that the alternative approach they recommend can accomplish (or has accomplished) core security objectives in an interdependent world. To assume that there is only one way to address these threats, and that multilateralism is the answer, is not only dangerously superficial but ignores the overwhelming body of evidence that contemporary multilateralism has deficiencies that are compounded by the forces of globalization and interdependence. There is no logically compelling reason to expect that input from more states in the search for multilateral consensus will necessarily make a positive contribution to conflict resolution, counter-terrorism, or the fight against human rights abuses, WMD proliferation, genocide, and ethnic cleansing, and so on. The challenge for proponents of multilateralism is not to argue for its potential in an ideal world (that's easy), but to acknowledge its deficiencies and admit that, on occasion, alternatives will be needed, and threatening to apply alternatives will be crucial.

There is no empirical evidence to support the view that exclusive reliance on contemporary multilateralism, with all its deficiencies, is the best approach to addressing all aspects of new and emerging threats of terrorism and proliferation. Delvoie is certainly correct that 'in pursuing some of the objectives of the war [on terror – e.g., enhancing airline security and controlling money laundering] the U.S. administration might reasonably conclude that it could not rely solely on the assistance of a few select countries, but would be better served by marshalling efforts on a much broader basis.' Obviously a state's interests will be better served by marshalling efforts on a much broader basis, but is that always possible in a world in which perceptions of core security and economic threats are not shared? Are there instances in which an exclusive commitment to multilateral consensus could be catastrophic? The case in favour of multilateralism can not rest solely on its potential to facilitate collective action in some instances, or by selecting evidence that biases the case in favour of identifying relatively easy, cost-free examples of collective action (such as sharing intelligence or enhancing airline security). There are many complex and difficult challenges confronting policy makers faced with the obligation to stop terrorism and proliferation. Some of the challenges will require solutions that are not likely to generate multilateral consensus because they jeopardize the unilateral economic and security interests of other states. A refusal to acknowledge fundamental weaknesses with exclusive reliance on one or another approach (whether unilateral or multilateral) is dangerous. How responsible is it, for example, to defend a multilateral approach to arms control and disarmament that for decades has been so dysfunctional and, for many in the arms control movement, a failure?

Perhaps the most fundamental error in Delvoie's thesis is the underlying assumption that the drift towards American unilateralism is a temporary phenomenon. I would argue that there are insurmountable domestic and systemic pressures (described in detail in chapter 6 and outlined in table 6.1) that are driving the new security environment, and this *collection* of interdependent forces is not likely to evaporate when a different political party takes over in the White House or Congress. When it comes to America's new strategic imperatives no U.S. administration will survive unless it prioritizes security by demonstrating to the American public that it is willing to sacrifice a great deal, including traditional alliances, multilateral consensus, international institutions, and the economy (including Canada-U.S. economic trade relations) to protect American citizens. Any other approach will fail politically. In the end, it is the nexus between globalized terrorism and WMD proliferation that explains American foreign and security priorities, not the character, diplomatic skills, or ideological predisposition of one person and his/her immediate advisers. If only international politics was that simple, and that easy to fix.

Recalibrating Canada's Moral and Diplomatic Compass

Any real power and influence Canada has in the world today does not come from our contribution to multilateral institutions; it comes from our relationship with the U.S. Ironically, the health of that relationship accounts for most of our credibility in multilateral forums. If Ottawa remains committed to a set of priorities derived from illusions of our own importance, independence, and distinction in the world, or if Canadians begin to take seriously the prevailing assumption that we have a single mission in life, namely, to multilteralize the world, then Canada will continue to fade as a relevant ally to the United States and ultimately fade from relevance more generally. 'Canada has always struggled for attention in Washington,' Andrew Cohen reminds us, 'but it will now have to work harder to remain on the radar screen. The danger is that as Canada fades as a power in the world – in the reach of its military, the impact of its foreign aid, the influence of its diplomacy, the absence of foreign intelligence gathering – it risks becoming a fading presence in Washington, too.'[26]

Canada's declining status and influence in the world is not simply a product of what Canadian officials are doing wrong but what other emerging powers are doing right. To deal with this decline Ottawa has to reshape its perceptions of itself and apply its strengths in a post-9/11 environment more strategically. In that context, Canada has only one key asset that other emerging states do not – our relationship with the United States. As Michael Ignatieff says,

> In our relations with the Americans, we have to understand this. We have something they want – they need legitimacy. It is not the case that the Americans are comfortable, either domestically or internationally, with projecting force abroad unilaterally. They don't like it, they feel exposed, they want friends to come along. Our presence in Afghanistan may seem symbolic, but it is extremely important in producing legitimacy for the operation. So we have legitimacy to sell. And if we have legitimacy to sell then we shouldn't sell it cheap, we should be proud of what we bring to the table and we should tell the Americans: 'If you want our support, here are the conditions.' We have, it seems to me, a much too deep inferiority complex to operate effectively in an empire. We have to be tougher.[27]

Yet Ottawa does not appear to be managing the Canada-U.S. relationship as effectively, and strategically, as it should. While the default strategy of distinction over security may have been relatively cost-free to Canada before 9/11, Ottawa can no longer afford to miss valuable opportunities to support U.S. policies at the right time; bad timing and missed opportunities to gain important concessions from the Americans on a variety of economic and security issues make no sense, especially if we invariably support U.S. policies anyway, and for the right reasons.

Canada's support for U.S. policies at the right time will go a long way towards convincing American officials (and the American public) that Canada matters as a friend and ally. And Americans are looking for allies – France, Germany, and Belgium lost the status as allies when it counted most to the U.S., says John Ibbitson.[28] As a result of its decision to withhold its support Canada missed another valuable opportunity to recalibrate its diplomatic strategy. Chrétien failed to understand 'that our interests, our values, our future, lie in fighting terror and advancing democracy, in concert with the other English-speaking nations and their allies. The alternative is to join old Europe in standing aside, and receding with them into irrelevance.'[29] In essence, Ottawa failed to manage the U.S. relationship in a way that would have enhanced Canadian interests, values, and influence in the world – a thesis shared by individuals on the right and left, Liberals and Conservatives, nationalists and integrationist, Canadians and Americans.[30]

Perhaps, given their training, Canadian diplomats can't conceive of putting their demands on the table, but this is precisely what every other state did in the period leading up to the 2003 Iraq war, and they were brutally honest about their demands (see table 3.1). If security trumps economics south of the border, Canada could have achieved important concessions from the Americans in return for the legitimacy that comes from Canada's support. In response to a similar recommendation, Canada's foreign affairs minister, Bill Graham, commented, 'As far as Iraq is concerned, I'm not persuaded by those who say we must absolutely go with the United States and we will be punished if we don't. Because I say to them, what happens if we do? Will they settle the softwood dispute? Or wheat? The border and access question? No, that's not their system. Their system is their system.'[31] Graham and his predecessors can now add SARS and mad cow disease, along with virtually hundreds of other trade and security issues that will require polite requests from Ottawa for Washington to consider our 'special' relationship. The problem is, we don't have one. Moreover, it isn't entirely clear what Graham meant by 'their system,' since the U.S. political system is the one that produced billions of dollars in military and aid packages for those states that supported the U.S.-U.K. coalition (consider all the evidence on horse trading over Iraq outlined in chapter 3). The U.S. system is one in which the president, his cabinet, and every member of Congress use their bargaining leverage repeatedly to get the concessions they need for the objectives they seek. Everyone, except for Canada, appears to be in the right bargaining game. Not only did Canadian officials miss an opportunity to establish linkages between domestic trade priorities and Iraq, but the prime minister and president never once communicated throughout the entire crisis.

If security trumps economics, for reasons outlined in chapter 6, the legitimacy Canada can provide to American initiatives (initiatives that invariably serve Canada's values and interests anyway, such as in Iraq) will become more important to Washington in its never-ending search for security. Perhaps more important, the legitimacy Canada has to offer is becoming increasingly crucial to ensure that America's search for security doesn't trump Canada-U.S. economic relations every time.

Canada will obviously reserve the right as a sovereign country to take an independent position on any issue, even if it conflicts with the U.S. But Ottawa needs to improve its capacity (and inclination) to engage Americans in the right debates, with compelling arguments derived from sound logic and evidence, and based on a set of principles and values Canadian officials should be (and claim to be) defending. From the point of view of our American allies, rejecting unilateralism without explaining precisely how multilateral solutions will address these very real security threats is not particularly helpful. As Robert Fulford observes, 'Canadians love to lecture Americans on their short-comings in world affairs, not because the Americans listen but because it makes us feel we are part of great events and bring to them a superior wisdom. The idea of dealing even-handedly with both sides holds a particular appeal for Canadians ... Unfortunately, it may also leave us incapable of the one act that has always been essential to survival, distinguishing friends from enemies.'[32] If Ottawa remains committed to a default strategy of privileging distinction over security, officials in Washington will continue to ignore the mounting costs to Canada produced by America's ongoing imperative to unilaterally respond to emerging threats. Unless Canadian officials reconsider the utility of what is obviously a failing strategy, Canada and Canadian interests will continue to fade.

Appendix A

American Multilateralism in Iraq, 2003

(Jay Nathwani, 2003)[1]

Afghanistan
Explicit
- Political support for the war by joining the 'Coalition for the Immediate Disarmament of Iraq,' better known as the 'Coalition of the Willing' (hereafter 'Coalition')[2]

Albania
Explicit
- Political support: joined Coalition, signatory to letter of support for U.S. policy on Iraq[3]
- Permitting overflight rights to U.S. military aircraft[4]
- Offered troops in symbolic gesture[5]

Angola
Explicit
- Political support: joined Coalition[6]

Australia
Explicit
- Political support: joined Coalition[7]
- Sent 2000 troops to Iraq, 14 F-18 Hornet fighter jets, transport ships;[8] troop contribution included elite SAS troops[9]

Azerbaijan
Explicit
- Political support: joined Coalition[10]
Tactical
- Offered overflight rights[11]

Bahrain
Implicit
- Supported U.A.E. call for Saddam Hussein to step down[12]
- Headquarters of U.S. Fifth Fleet, as well as a special operations command. Approximately 1200 aircraft including fighters, surveillance, transport, and support aircraft based at the Muharraq Airfield;[13] as of 18 March 2003, there were more than 4000 U.S. military personnel in Bahrain[14]
- Provided naval and land forces to support Kuwaiti defences against an Iraqi attack[15]

Belgium
Tactical
- Allowed movement of troops and *matériel* from bases in Germany through port at Antwerp en route to Persian Gulf[16]
- Granted overflight rights[17]

Bulgaria
Explicit
- Political support: joined Coalition, signatory to letter in support of U.S. policy on Iraq[18]
- Military: sent 150 biological, chemical, nuclear warfare specialists to Kuwait[19]

Tactical
- Granted overflight rights, use of bases and refuelling for U.S. planes[20]

Canada
Implicit/Tactical
- Canada's troops were directly involved in planning the war against Iraq. For months before the war, Canadian military planners worked with U.S. Central Command, devising strategy for a war against Iraq, a privilege afforded few members of the Coalition. Canada transferred twenty-five senior officers to camp As Sayliyah in Qatar when CentCom moved there.[21]
- Some Canadian troops were actually placed in units that saw action in the Iraq campaign. Thirty-one troops were on exchange with British and American units. At least six of them were in battle zones, and one was with the British 7th Armoured Division, which saw heavy fighting around Basra.[22]
- Canadian Forces personnel operate as part of U.S. AWACS crews, managing the skies over Iraq and other Middle Eastern countries, and guiding fighters and bombers to their targets.[23]
- Canadian ships provided a significant boost to the war effort. Canada de-

ployed the destroyer HMCS *Iroquois* to lead Task Force 151, made up of about twenty ships from six countries. Canada contributed four frigates to the task group, in addition to the *Iroquois*.[24] While the ships were operating as part of Operation Apollo,[25] Defence Minister John McCallum stated that the ships might be 'double-hatted' to offer support in a war against Iraq.[26] Irrespective of any 'double-hatting,' Canada's ships contributed to the coalition war effort. Significantly, the area of operations for the ships of Task Force 151 was redefined when Canada took control. The ships had been tasked to protect allied vessels and interdict terrorists in the Gulf of Oman and the Strait of Hormuz. They had not operated in the Persian Gulf. When Canada took over, the area of operations was expanded to cover virtually the whole Persian Gulf, up to the twenty-ninth parallel, the southernmost point in Kuwait.[27] The ships were charged with helping to protect U.S. aircraft carriers,[28] so the expansion of their area of operations meant that they were in a position to lend direct assistance to coalition forces in transit to or engaged in operations against Iraq. Indeed, Comdr. Roger Girouard, the Canadian who took control of Task Force 151, was charged with protecting all allied warships except carriers and their escorts operating in the Strait of Hormuz *and south of Kuwait in the Persian Gulf.* Such a commitment freed up American ships to prosecute the war.[29]

- In the run-up to the Iraq war, Canada announced that its troops would return to Afghanistan. Canada's military had been badly stretched during the previous six-month deployment of eight hundred soldiers – 3rd Battalion Princess Patricia's Canadian Light Infantry. On 12 February 2003, Defence Minister John McCallum announced that Canada would send two consecutive rotations of fifteen hundred troops as part of the UN-mandated International Security Assistance Force.[30] The final force commitment was eighteen hundred troops in the first rotation from the Royal Canadian Regiment, with a second rotation to follow from the Royal 22nd Regiment (Vandoos).[31] At the time of the announcement, no one knew how long the war in Iraq would last, and it was anticipated that Canada's contribution would free up key U.S. logistical and combat troops for combat in Iraq.[32] Even without ongoing combat in Iraq, the contribution to Afghanistan would lighten the load on the U.S. military there, allowing it to redeploy assets to Iraq.
- Canada provided overflight rights and refuelling to U.S. planes bound for Iraq and the gulf region. Roughly two to three flights a day and about a thousand troops came through Gander airport in Newfoundland.[33]
- Canada would have supported a war against Iraq if a second UNSC resolution was blocked by a veto of only one of the permanent members.[34]
- Canada did not oppose a war against Iraq. What we opposed was a war

against Iraq that was not sanctioned by the UN Security Council. Prime Minister Jean Chrétien said that he would be willing to lend military support to a war against Iraq if the UNSC authorized it with a follow-up resolution to 1441, which demanded full and voluntary Iraqi disarmament.[35] Canada's position was not that Iraq had lived up to the terms of 1441, merely, in Chrétien's words, that '[o]ver the last few weeks the Security Council has been unable to agree on a new resolution authorizing military action.'[36] Thus, Canada's position was not that the U.S. concern over Iraqi WMD was illegitimate, nor that the use of military force was inappropriate or immoral; it was that the French and Russians were not onside, and so we could not be. Canada did not oppose a war against Iraq – rather, it abrogated its freedom of choice to the members of the Security Council and made our policy dependent on the Security Council. Canada did not make a policy determination; it took on a position of indeterminacy, and its opposition would have vanished, and the means and ends of a war legitimized in its mind, by no more than a change of heart by a few countries.

Colombia
Explicit
• Political support: joined Coalition[37]

Costa Rica
Explicit
• Political support: joined Coalition[38]

Croatia
Explicit
• Political support: joined Coalition, signatory to letter in support of U.S. policy on Iraq[39]
Tactical
• Granted overflight rights and refuelling stops at airports for Coalition planes[40]

Czech Republic
Explicit
• Political support: joined Coalition, signatory to letter in support of U.S. policy on Iraq[41]
• Sent about two hundred biological, chemical, and nuclear decontamination specialists to gulf to protect troops and civilians from attack[42]
Tactical
• Granted overflight rights[43]

Denmark
Explicit
- Political support: joined Coalition, signatory to letter in support of U.S. policy on Iraq[44]
- Military: offered corvette, submarine, and medical team[45]

Dominican Republic
Explicit
- Political support: joined Coalition[46]

Egypt
Implicit
- Rather than condemn the United States, as many leaders in the Arab world did, Egyptian President Hosni Mubarak said, 'I welcome the door the United States opened for the United Nations, especially the Security Council, to play a pivotal role ...' Placing the responsibility for a resolution of the conflict on Iraq, he said, 'I call upon the Iraqi leadership to seize this opportunity ... and avoid serious repercussions.'[47]

Tactical
- Quietly granted the U.S. overflight rights for its military aircraft[48]
- Allowed U.S. forces access to airbases and the Suez canal[49]

El Salvador
Explicit
- Political: joined Coalition[50]
- Military: before the war had begun, offered troops for peacekeeping duties at war's conclusion[51]

Eritrea
Explicit
- Political: joined Coalition[52]

Tactical
- Granted overflight rights[53]

Estonia
Explicit
- Political: joined Coalition, signatory to letter in support of U.S. policy on Iraq[54]

Ethiopia
Explicit
- Political: joined Coalition[55]
Tactical
- Granted overflight rights and use of air bases[56]

France
Implicit/Tactical
- France, which long denied that action was needed to deal with Iraq's weapons of mass destruction, and which argued that there was insufficient evidence of a WMD program to act on, nevertheless tacitly accepted the stated U.S. motivation for acting and signalled their recognition of the danger posed by Iraq's WMD, when, on 18 March, the French ambassador to the United States, Jean-David Levitte, said any use of chemical or biological weapons by Iraq would 'change completely the situation for the French president and for the French government.' French Foreign Minister Dominique de Villepin also said, 'If the U.S. and our allies face a new and unforeseen situation in a new crisis, France would obviously be on their side to show solidarity in the face of an exceptional crisis.'[57]
- Ships participated in Task Force 151, which protected U.S. carriers in the Arabian Sea, Strait of Hormuz, and Persian Gulf.[58]

Georgia
Explicit
- Political support: joined coalition;[59] sent letter to U.S. supporting campaign against Iraq[60]
Tactical
- Offered possible use of Georgian airfields[61]

Germany
Implicit/Tactical
- Even as German Chancellor Gerhard Schröeder condemned war on 18 March 2003, he boosted the number of troops guarding U.S. bases in Germany.[62] This freed up U.S. resources to take part in military action against Iraq.
- Germany made no move to prevent the U.S. from using its German bases as a staging area.[63]
- Germany kept its specialists aboard NATO AWACS planes operating over Turkey.[64] This attitude stands in contrast to France, which attempted to prevent the stationing of NATO Patriot missile batteries in Turkey before any war had begun.

- Germany had chemical decontamination experts in Kuwait and increased the number of those troops to between 200 and 250.[65]

Greece
Tactical
- Ships participated in Task Force 151, which protected U.S. carriers in the Arabian Sea, Strait of Hormuz and Persian Gulf[66]

Honduras
Explicit
- Political support: joined Coalition[67]

Hungary
Explicit
- Political support: joined Coalition; signatory to letter supporting U.S. policy on Iraq[68]

Implicit
- Hosts U.S. base where Iraqi exiles trained for roles in a post-war administration[69]

Tactical
- Granted overflight rights[70]
- Parliament authorized movement of U.S. troops and freight across the country to defend Turkey. Vote was 335–3 in a 368-seat parliament.[71]

Iceland
Explicit
- Political support: joined Coalition[72]

Iran
Implicit
- Declared its neutrality – not its opposition – with respect to the war[73]

Israel
Explicit
- Political: Israel offered staunch support for a policy of regime change. Prime Minister Ariel Sharon repeatedly urged the U.S. not to delay military action.[74] Israel was left off the list of coalition members so as not to inflame Arab sentiment.
- Military: Israeli special forces operated inside Iraq, pinpointing Scud locations in the western desert as early as September 2002.[75] In October, a U.S.

official confirmed that the U.S. was considering an Israeli offer to send commandos into Iraq to knock out missile sites.[76]

Italy
Explicit
- Political support: joined Coalition, signatory to letter supporting U.S. policy on Iraq.[77] Along with Britain and Spain, offered the most politically important explicit support for the war. Prime Minister Silvio Berlusconi was a prominent political ally of the United States in Europe.

Tactical
- Ships participated in Task Force 151, which protected U.S. carriers in the Arabian Sea, Strait of Hormuz, and Persian Gulf[78]
- Granted overflight rights; allowed Coalition aircraft use of military bases, though not for direct attacks on Iraq[79]
- Offered logistical help[80]

Japan
Explicit
- Political support: joined Coalition[81]
- Before war started, offered to help rebuild Iraq[82]

Jordan
Implicit/Tactical
- In January 2003, Jordanian officials and diplomats confirmed that Jordan had agreed to host U.S. troops to man air defences, man a military hospital, and conduct search-and-rescue operations using special forces. Jordan had also agreed to grant the U.S. use of its airspace during a war with Iraq. This was confirmed by an American official but publicly denied by Jordanian Information Minister Mohammed Affash Adwan.[83]
- American officials also privately confirmed in January that Jordan had allowed the stationing of special forces for operations against Iraq, not merely in a search-and-rescue role.[84]
- In February, Jordan admitted that it was allowing U.S. troops into the country. Prime Minister Ali Abu al-Rahgeb said that several hundred soldiers were in the country to protect Jordan. Their duties would include manning AWACS planes.[85]
- In March, the government declared an exclusion zone. There were reports from construction workers of large extensions to military runways near the Iraqi border.[86]
- While continuing to publicly deny the presence of large numbers of U.S. troops, a Jordanian government official privately acknowledged that be-

tween two and three thousand were stationed in the country.[87] That number was much lower than estimates by military analysts.[88]

- As early as August 2002, some news outlets were reporting that eight thousand U.S. troops were being moved into Jordan, in preparation for a war against Iraq.[89]
- Army, air force and marine troops were stationed at the Jordanian air base at Safawi. The rumble of B-1 bombers was audible.[90]
- U.S. troops in Jordan carried out search-and-rescue operations in western Iraq.[91]
- U.S. troops also manned three Patriot missile batteries in Jordan.[92] Hundreds of troops were used to operate these alone.[93]
- Jordan granted the Coalition overflight rights.[94]

Kuwait
Explicit
- Political support: joined Coalition.[95] Supported U.A.E. call for Saddam Hussein to step down.[96]
- Main staging area for invasion.[97] Kuwait cordoned off large areas in the north and west of the country for massing troops, closing about one-third of its territory by November 2002.[98] In February 2003, the government announced that the northern half of the country would be off limits to all but military personnel.[99] By March, 130,000 troops were stationed in northern Kuwait.[100] As of 18 March 2003, the U.S. military presence in Kuwait consisted of at least the following:

  - approximately sixteen thousand troops of the U.S. 101st Airborne Division alone
  - at least four armoured brigades, with about nine hundred M1A1 Abrams tanks and Bradley fighting vehicles; the brigades also include 120 mm mortars and 155 mm howitzers
  - about eighty-five aircraft, including twenty-four Apache helicopter gunships; F-15, F-16, and A-10 aircraft, operating from two Kuwaiti air bases, Al-Jaber and Ali Salem
  - at least two Patriot anti-missile batteries
  - a new $200 million logistics base, Camp Arifjan, south of Kuwait City
  - Navy Seabees in Kuwait for construction duties at two bases[101]

Latvia
Explicit
- Political support: joined Coalition[102]
- Military: Latvian government sought authorization from parliament to deploy a small number of troops[103]

Lithuania

Explicit

- Political support: joined Coalition, signatory to letter in support of U.S. policy on Iraq[104]

Tactical

- Granted overflight rights[105]

Macedonia

Explicit

- Political support: joined Coalition, signatory to letter in support of U.S. policy on Iraq[106]

Tactical

- Granted overflight rights[107]

Marshall Islands

Explicit

- Political support: joined Coalition[108]

Micronesia

Explicit

- Political support: joined Coalition[109]

Mongolia

Explicit

- Political support: joined Coalition[110]

Netherlands

Explicit

- Political support: joined Coalition[111]
- Military: sent three Patriot missile batteries and 360 soldiers to defend Turkey in case of counterattack[112]

Nicaragua

Explicit

- Political support: joined Coalition[113]

Oman

Implicit

- As of 18 March permitted the stationing of more than two thousand U.S. military personnel and about forty aircraft.[114]
- New base under construction with runway of fourteen thousand feet.[115]

- In a sign of the good relationship that exists between the U.S. and Oman and the assistance the Oman continues to provide, President Bush, in a March phone call to Omani Sultan Qaboos bin Said, thanked Oman for 'years of reliable and steady friendship and support for the United States.'[116]

Tactical
- Omani ports and airfields served as important points for the transfer of *matériel* into the Gulf region.[117]

Palau
Explicit
- Political support: signatory to letter in support of U.S. policy on Iraq[118]

Panama
Explicit
- Political support: joined Coalition[119]

Pakistan
Implicit
- The military establishment favoured the U.S. course on Iraq, but President Musharraf was unable to offer explicit support because Prime Minister Zafarullah Khan Jamali's government urged Musharraf not to vote with the U.S. at the UN and because public opinion was overwhelmingly against the war. The prime minister said, 'We have supported America against terrorism and we have no differences with it.' Pakistan was preparing to abstain from the UN vote, according to the government.[120]

Philippines
Explicit
- Political support: joined Coalition[121]
- Before conflict, offered post-war peacekeeping assistance[122]

Poland
Explicit
- Political support: joined Coalition, signatory to letter in support of U.S. policy on Iraq[123]
- Military: 2,350 troops

Portugal
Explicit
- Political support: joined Coalition, signatory to letter supporting U.S. policy on Iraq[124]

- Hosted pre-war summit of leaders from United States, Britain, and Spain on 15 and 16 March 2003[125]

Tactical

- Granted overflight rights and permission to use Lajes Field base in the Azores Islands, a transatlantic refuelling stop[126]

Qatar
Implicit and tactical

- Political: backed U.A.E. call for Saddam Hussein to step down.[127]
- Qatar was home to the U.S. Central Command's mobile headquarters at Camp As Sayliyah.[128] This CentCom was staffed by sixteen hundred U.S. and U.K. personnel.[129] The U.S. constructed a new, high-tech command headquarters at Al Udeid Air Base near the capital, Doha. The Pentagon shifted fully a third of its key military planners to the command base in Qatar as early as September 2002.[130] As of 18 March 2003, there were approximately thirty-five hundred U.S. military personnel in the country, according to some estimates.[131] Others put the number as high as seven thousand.[132]
- In December, Qatar hosted a simulation for a war against Iraq at the mobile command centre, involving a thousand military personnel.[133]
- At Al Udeid, the United States constructed a fifteen-thousand-foot runway, the longest in the gulf region and long enough for the heaviest bombers to take off fully loaded. Hangars for one hundred aircraft were built, as were a high-tech air-operations centre and ammunition dumps.[134]
- Al-Udeid airbase was opened for an in-flight refuelling squadron, an F-15 fighter wing, and aircraft maintenance.[135]
- Dozens of combat aircraft and tankers flew out of Al Budded airbase.[136]
- Two dozen F-117 Nighthawk stealth fighters were stationed in Qatar, including the aircraft used in the opening attack of the war.[137]

Romania
Explicit

- Political support: joined Coalition[138]
- Military: offered 278 non-combat biological, chemical, nuclear decontamination specialists, military police, and de-miners[139]

Rwanda
Explicit

- Political support: joined Coalition[140]

Saudi Arabia
Implicit

- Political: backed U.A.E. call for Saddam Hussein to step down.[141]

- About four thousand U.S. troops were stationed at Prince Sultan airbase, running an operations centre that was key to coordinating the campaign.[142] Saudi Arabia gave private assurance in December 2002 that the centre could be used to coordinate the air-campaign. They also gave assurance that the U.S. could launch air support missions from Saudi territory.[143] According to reports, the Saudis agreed to allow AWACS and JSTAR radar aircraft to fly from the base, with the possibility of jet fighters and bombers flying missions as well, in exchange for a phased U.S. withdrawal from Saudi Arabia once the war ended.[144] According to some reports, combat patrols from Prince Sultan airbase continued during the war.[145]
- Also gave the U.S. permission to use the huge and state-of-the-art Prince Bandar airbase to attack Iraq.[146]
- Saudi Arabia shut down its civilian airport at Araar (or Arar) on the Iraq border. The airport was reported to house U.S. surveillance and search-and-rescue teams.[147] There were reports of U.S. troops pouring into the base and into the garrison town of Tabuk. Saudi officials did not dismiss the possibility of combat missions being launched from the base at Araar.[148]
- Granted overflight rights to coalition aircraft.[149]
- Military presence in Saudi Arabia as of 18 March 2003 included:

  - between five and seven thousand U.S. personnel
  - about seventy-five aircraft
  - two Patriot anti-missile batteries[150]

Singapore
Explicit
- Political support: joined Coalition[151]

Slovakia
Explicit
- Political support: joined Coalition[152]
- Military: committed non-combat troops specializing in chemical decontamination[153]

Slovenia
Explicit
- Joined 'Vilnius Group' of countries – Albania, Bulgaria, Croatia, Estonia, Latvia, Lithuania, Macedonia, Romania, Slovakia – in stating its readiness to help disarm Iraq.[154]

Soloman Islands
Explicit
- Political support: joined Coalition[155]

South Korea
Explicit
- Political support: joined Coalition[156]
- Military: sent seven hundred non-combat troops, including doctors and engineers, to assist[157]

Spain
Explicit
- Political support: joined Coalition[158]
- Spain was a key ally at the United Nations and participated, along with the United States and Britain, in a summit in Portugal on 15 and 16 March to assess the status of diplomatic efforts and discuss strategy. Spain was arguably the most important European ally, next to Britain, in political terms.[159]
- Military: committed a vessel equipped with biological, chemical, and nuclear decontamination facilities, as well as a backup vessel and nine hundred accompanying troops; opened NATO bases to coalition aircraft.[160]

Syria
Implicit
- Syria voted in favour of UN Security Council Resolution 1441, which authorized the return of arms inspectors to Iraq, deplored Iraqi obstruction of weapons inspections, gave Iraq a 'final opportunity to comply with disarmament obligations' and warned Iraq of 'serious consequences as a result of its continued violations of its obligations.'[161]

Taiwan
Explicit
- Political: Taiwanese Foreign Affairs Minister Eugene Chien said on 10 March 2003, 'We support war on terrorism and UN Resolution No 1441 that demands Iraq dismantle all weapons of mass destruction. It is our hope that Iraq will destroy all weapons of mass destruction. But if a war is inevitable, the ROC will support any action taken by the United States to disarm Iraq.'[162]
- On 21 March the caucus of the ruling Democratic Progressive Party stated that Iraq was in violation of UNSC Resolution 1441.[163]

Tonga
Explicit
- Political support: joined Coalition[164]

Turkey
Explicit
- Political support: joined Coalition.[165] The government (cabinet) wanted parliament to authorize the deployment of U.S. troops on Turkish soil.[166] The country's top civilian and military leaders had agreed to a deployment of thirty-eight thousand U.S. troops in the country to open a northern front.[167]
- When the Turkish parliament voted on the motion to approve the deployment of U.S. troops, the vote was 264–250 in favour. The motion failed because 264 votes fell 4 short of a majority of those present in the chamber, owing to abstentions. The government had pushed hard for authorizing U.S. military action through Turkey.[168] A few days after the vote, a senior member of the governing party, Vahit Erdem, indicated that many members had 'changed their opinion and will vote in favour if a second vote is held,' and 'if a second vote is held it will be positive this time.'[169] No second vote was held.
- On 6 March the powerful Turkish military signalled its support for a war against Iraq. General Hilmi Ozkok, the chief of the Turkish general staff, backed efforts to reconsider the 1 March vote.[170] His comments made explicit that he thought it was in Turkey's interest to join the war.[171]

Implicit/Tactical
- Political support: In December of 2002, the Turkish foreign minister, Yasar Yakis, rather than reject military action, said only that his country sought UN approval before it would open its bases and airspace to the United States.[172]
- On 6 February 2003, the Turkish parliament voted to provide initial support to the U.S., permitting specialists into the country to begin upgrades to ports and bases. The vote was 308 to 193 in the 550-seat parliament.[173]
- On 20 March 2003, Turkey's parliament granted overflight rights to coalition aircraft.[174] The vote passed 332 to 202.[175]

Uganda
Explicit
- Political support: joined Coalition[176]

Ukraine
Explicit
- Political support: joined Coalition.[177]

- Military: Ukraine sent up to five hundred chemical and nuclear decontamination experts.[178]

United Arab Emirates

Implicit

- Political: The U.A.E. proposed on 1 March 2003 at the Gulf Cooperation Council that Saddam Hussein step down to avoid war. Kuwait, Saudi Arabia, Bahrain, and Qatar supported the suggestion,[179] as did Turkey subsequently.[180]
- The United States uses the airport and naval facilities at the U.A.E. army base at Dhafrah. The U.S. No. 763 and No. 4413 air-refuelling squadrons are based there. Under a bilateral agreement, the U.S. may also station military equipment and supplies and an army brigade in the country.[181] The U.S. also has surveillance aircraft stationed in the country.[182] Approximately three thousand Western troops are stationed in U.A.E.[183]
- The U.A.E. sent tanks, amphibious armoured vehicles, attack helicopters, gunships, a missile boat, a frigate, and about four thousand troops to defend Kuwait from Iraqi attacks.[184]

United Kingdom

Explicit

- Political support: joined Coalition, offered crucial political support for the United States.[185] Britain was obviously the most significant coalition partner next to the United States. Tony Blair was a leading advocate of action to disarm Iraq and endured a backbench rebellion in the course of gaining parliamentary approval for action. Tony Blair's speech to the House on 18 March 2003 was the most eloquent call for action from any statesman.[186]
- Military: Britain committed forty-five thousand troops to action in Iraq, or about a third of its armed forces.[187] That included:

  - a Naval Task Group, including an aircraft carrier, helicopter carrier, four destroyers, two frigates, and a submarine
  - an amphibious assault force including the whole of three Commando Brigade Royal Marines, and elements of forty Commando and forty-two Commando
  - a twenty-six-thousand strong land force, including 1 (U.K.) Armoured Division, 7th Armoured Brigade, 16 Air Assault Brigade, 102 Logistics Brigade
  - an air component of about 100 aircraft, 27 support helicopters, and 7000 men[188]

- Britain owns the island of Diego Garcia in the Indian Ocean, on which the United States maintains an airbase. The base had about nineteen hundred personnel from the 40th Air Expeditionary Wing as of 18 March 2003. Planes that operated from Diego Garcia against Iraq included B-2 Spirit stealth bombers and B-52s.[189]
- U.S. B-52 bombers flew from an RAF base in Fairford, England, to bomb targets in Iraq.[190]

Uzbekistan
Explicit
- Political support: joined Coalition.[191] Addressing the Uzbek parliament, President Islam Karimov said on 24 April 2003, 'We unequivocally supported, and are continuing to support, the USA's position on disarming Iraq.'[192]

Yemen
Implicit
- Intensified security around foreign embassies and locations frequented by Westerners just ahead of the war's start. Some American and British citizens were transferred to more secure locations, and the Ministry of Information cautioned that correspondents should be cautious because 'irresponsible threatening statements against the war will be issued,' reports said.[193]

Appendix B

U.S. Legislation Related to 11 September 2001

(http://thomas.loc.gov/home/terrorleg.htm)

Bills and Joint Resolutions Signed into Law

HR2882 Public Safety Officer Benefits bill

HR2883 Intelligence Authorization Act for Fiscal Year 2002

HR2884 Victims of Terrorism Relief Act of 2001

HR2888 2001 Emergency Supplemental Appropriations Act for Recovery from and Response to Terrorist Attacks on the United States

HR2926 Air Transportation Safety and System Stabilization Act

HR3162 Uniting and Strengthening America by Providing Appropriate Tools Required to Intercept and Obstruct Terrorism (USA PATRIOT ACT) Act of 2001

HR3448 Bioterrorism Response Act of 2001

HR3525 Enhanced Border Security and Visa Entry Reform Act of 2002

HR3986 To extend the period of availability of unemployment assistance under the Robert T. Stafford Disaster Relief and Emergency Assistance Act in the case of victims of the terrorist attacks of 11 September 2001

H.J.Res. 71 Designating 11 September as Patriot Day

S1424 A bill to amend the Immigration and Nationality Act to provide permanent authority for the admission of 'S' visa non-immigrants

S1438 National Defense Authorization Act for Fiscal Year 2002

S1447 Aviation and Transportation Security Act

S1465 A bill to authorize the President to exercise waivers of foreign assistance restrictions with respect to Pakistan through 30 September 2003 and for other purposes

S1573 Afghan Women and Children Relief Act of 2001

S1793 Higher Education Relief Opportunities for Students Act of 2001

S2431 Mychal Judge Police and Fire Chaplains Public Safety Officers' Benefit Act of 2002

S.J.Res. 22 A joint resolution expressing the sense of the Senate and House of Representatives regarding the terrorist attacks launched against the United States on 11 September 2001

S.J.Res. 23 Authorization for Use of Military Force

Legislation with Floor Action

HR2899 Freedom Bonds Act of 2001

HR2975 Provide Appropriate Tools Required to Intercept and Obstruct Terrorism (PATRIOT) Act of 2001

HR2976 Healing Opportunities in Parks and the Environment Pass Act

HR3004 Financial Anti-Terrorism Act of 2001

HR3008 To reauthorize the trade adjustment assistance program under the Trade Act of 1974, and for other purposes

HR3054 True American Heroes Act

HR3086 Higher Education Relief Opportunities for Students Act of 2001

HR3090 Temporary Extended Unemployment Compensation Act of 2002

HR3150 Airport Security Federalization Act of 2001

HR3160 To amend the Antiterrorism and Effective Death Penalty Act of 1996 with respect to the responsibilities of the Secretary of Health and Human Services regarding biological agents and toxins, and to amend title 18, United States Code, with respect to such agents and toxins

HR3210 Terrorism Risk Protection Act

HR3240 Reservists Education Protection Act of 2001

HR3253 Department of Veterans Affairs Emergency Preparedness Research, Education, and Bio-Terrorism Prevention Act of 2002

HR3275 Terrorist Bombings Convention Implementation Act of 2001

HR3375 Embassy Employee Compensation Act

HR3423 Arlington National Cemetery bill

HR3479 National Aviation Capacity Expansion Act

HR3529 Economic Security and Worker Assistance Act of 2001

HR3994 Afghanistan Freedom Support Act of 2002

HR4598 Homeland Security Information Sharing Act

HR4775 2002 Supplemental Appropriations Act for Further Recovery from and Response to Terrorist Attacks on the United States (American Servicemembers' Protection Act of 2002)

HR5018 Capitol Police Retention, Recruitment, and Authorization Act of 2002

H.Con.Res. 228 Put Our Children First Resolution of 2001

H.J.Res. 64 Authorization for Use of Military Force

S1214 Port and Maritime Security Act of 2001

S1372 Export-Import Bank Reauthorization Act of 2002

S1426 2001 Emergency Supplemental Appropriations Act for Recovery from and Response to Terrorist Attacks on the United States

S1450 Air Transportation Safety and System Stabilization Act

S1499 American Small Business Emergency Relief and Recovery Act of 2001

S1510 Uniting and Strengthening America Act (USA Act of 2001)

S1519 A bill to amend the Consolidated Farm and Rural Development Act to provide farm credit assistance for activated reservists

S1622 Extended Unemployment Compensation bill

S1637 A bill to waive certain limitations in the case of use of the emergency fund authorized by section 125 of title 23, United States Code, to pay the costs of projects in response to the attack on the World Trade Center in New York City that occurred on 11 September 2001

S1729 Post-Terrorism Mental Health Improvement Act

S1770 Terrorist Bombings Convention Implementation Act of 2001

S1858 Terrorist Victims' Courtroom Access Act

S1880 Afghanistan Freedom and Reconstruction Act of 2001

S1981 Enhanced Penalties for Enabling Terrorists Act of 2002

S2621 A bill to provide a definition of vehicle for purposes of criminal penalties relating to terrorist attacks and other acts of violence against mass transportation systems

S3009 Andean Trade Preference Expansion Act

H.AMDT. 361 to HR2883 Amendment revises Sec. 306 and establishes a 'Commission on National Security Readiness' to identify structural readiness of the United States to identify structural impediments to the effective collection, analysis, and sharing of information on national security threats, particularly terrorism

H.AMDT. 362 to HR2883 Amendment sought to reduce the Presidential appointments to 2; to remove the Commission's ability to hold hearings and to administer oaths or affirmations to witnesses appearing before it; to remove subpoena power; and to strike the requirement for termination of the Commission

H.AMDT. 383 TO HR3150 Amendment that makes various technical changes, authorizes airlines to use technologies to create a secure and expedited passenger screening process, and extends $1.5 billion authorized for airport security to 2003

S.AMDT. 1560 to HR2500 To express the sense of the Senate regarding discrimination against Arab Americans

S.AMDT. 1562 to HR2500 Combating Terrorism Act of 2001

S.AMDT.1573 to HR2590 To authorize the Secretary of the Treasury to issue War Bonds in support of recovery and response efforts relating to the 11 September 2001 hijackings and attacks on the Pentagon and the World Trade Center

S.AMDT.1574 to HR2590 To authorize the issuance of Unity Bonds in response to the acts of terrorism perpetrated against the United States on 11 September 2001

S.AMDT.1583 to HR2590 To provide that the Postal Service may issue a special commemorative postage stamp in order to provide financial assistance to the families of emergency relief personnel killed or permanently disabled in the line of duty in connection with the terrorist attacks against the United States on 11 September 2001

S.AMDT.1661 to S1438 To authorize emergency supplemental appropriations for fiscal year 2001 for recovery from and response to terrorist attacks on the United States

S.AMDT.1807 to S1438 To authorize the acceptance of contributions for the repair of the damage to the Pentagon Reservation caused by the terrorist attack on 11 September 2001 or establishment of a memorial of the attack at the Pentagon

S.AMDT.1815 to S1438 To express the sense of the Senate that the Secretary of the Treasury should immediately issue savings bonds, to be designated as 'Unity Bonds,' in response to the terrorist attacks against the United States on 11 September 2001

S.AMDT.2067 to HR3061 To express the sense of the Senate concerning the provision of assistance for airport career centers to enable such centers to serve workers in the airline and related industries who have been dislocated as a result of the 11 September 2001 attack on the World Trade Center

S.AMDT.2068 to HR3061 To express the sense of the Senate concerning assistance for individuals with disabilities who require vocational rehabilitation services as a result of the 11 September 2001 attack on the World Trade Center

S.AMDT.2082 to HR3061 To make funding available under title V of the Public Health Service Act for mental health providers serving public safety workers affected by the terrorist attacks of 11 September 2001

S.AMDT.2163 to HR2884 To amend the Internal Revenue Code of 1986 to provide tax relief for victims of the terrorist attacks against the United States, and for other purposes

S.AMDT. 2376 to HR3338 To authorize the burial in Arlington National Cemetery of any former Reservist who died in the 11 September 2001 terrorist attacks and would have been eligible for burial in Arlington National Cemetery but for age at time of death

S.AMDT. 2439 to HR3338 To establish a program to name national and community service projects in honor of victims killed as a result of the terrorist attacks on 11 September 2001

S.AMDT. 2452 to HR3338 To direct the Secretary of the Smithsonian Institution to collect and preserve in the National Museum of American History artifacts relating to the 11 September attacks on the World Trade Center and the Pentagon

Other Resolutions Approved

H.Con.Res. 225 Expressing the sense of the Congress that, as a symbol of solidarity following the terrorist attacks on the United States on 11 September 2001, every United States citizen is encouraged to display the flag of the United States

H.Con.Res. 231 Providing for a joint session of Congress to receive a message from the President

H.Con.Res. 243 Expressing the sense of the Congress that the Public Safety Officer Medal of Valor should be presented to the public safety officers who have perished and select other public safety officers who deserve special recognition for outstanding valor above and beyond the call of duty in the aftermath of the terrorist attacks in the United States on 11 September 2001

H.Res. 238 Condemning any price gouging with respect to motor fuels during the hours and days after the terrorist acts of 11 September 2001

H.Res. 239 Terrorist Victims Flag Memorial Resolution of 2001

H.Res. 309 Honoring the United States Capitol Police for their commitment to security at the Capitol

H.Res. 384 Honoring the men and women of the United States Secret Service New York field office for their extraordinary performance and commitment to service during and immediately following the terrorist attacks on the World Trade Center on 11 September 2001

H.Res. 385 Honoring the men and women of the United States Customs Service, 6 World Trade Center offices, for their hard work, commitment, and compassion during and immediately following the terrorist attacks on the World Trade Center on 11 September 2001

H.Res. 424 Paying tribute to the workers in New York City for their rescue, recovery, and clean-up efforts at the site of the World Trade Center

S.Res. 170 A resolution honoring the United States Capitol Police for their commitment to security at the United States Capitol, particularly on and since 11 September 2001

S.Res. 173 A resolution condemning violence and discrimination against Iranian Americans in the wake of the 11 September 2001 terrorist attacks

S.Res. 177 A resolution expressing the sense of the Senate that United States Postal Service employees should be commended for their outstanding service and dedication since the terrorist attacks of 11 September 2001

S.Res. 182 A resolution expressing the sense of the Senate that the United States must allocate significantly more resources to combat global poverty and that the President's decision to establish the Millennium Challenge Account is a step in the right direction

S.Res. 187 A resolution commending the staffs of Members of Congress, the Capitol Police, the Office of the Attending Physician and his health care staff, and other members of the Capitol Hill community for their courage and professionalism during the days and weeks following the release of anthrax in Senator Daschle's office

H.Con.Res. 232 Expressing the sense of the Congress in honoring the crew and passengers of United Airlines Flight 93

H.Con.Res. 233 Expressing the profound sorrow of the Congress for the death and injuries suffered by first responders as they endeavored to save innocent people in the aftermath of the terrorist attacks on the World Trade Center and the Pentagon on 11 September 2001

H.Con.Res. 259 Expressing the sense of Congress regarding the relief efforts undertaken by charitable organizations and the people of the United States in the aftermath of the terrorist attacks against the United States that occurred on 11 September 2001

H.Con.Res. 273 Reaffirming the special relationship between the United States and the Republic of the Philippines

H.Con.Res. 378 Commending the District of Columbia National Guard, the National Guard Bureau, and the entire Department of Defense for the assistance provided to the United States Capitol Police and the entire Congressional community in response to the terrorist and anthrax attacks of September and October 2001

H.J.Res. 61 Expressing the sense of the Senate and House of Representatives regarding the terrorist attacks launched against the United States on 11 September 2001

S.Con.Res. 66 A concurrent resolution to express the sense of the Congress that the Public Safety Officer Medal of Valor should be awarded to public safety officers killed in the line of duty in the aftermath of the terrorist attacks of 11 September 2001

S.Con.Res. 73 A concurrent resolution expressing the profound sorrow of Congress for the deaths and injuries suffered by first responders as they endeavored to save innocent people in the aftermath of the terrorist attacks on the World Trade Center and the Pentagon on 11 September 2001

S.Con.Res. 74 A concurrent resolution condemning bigotry and violence against Sikh-Americans in the wake of terrorist attacks in New York City and Washington, D.C., on 11 September 2001

S.Con.Res. 75 A concurrent resolution to express the sense of the Congress that the Public Safety Officer Medal of Valor should be presented to public safety officers killed or seriously injured as a result of the terrorist attacks perpetrated against the United States on 11 September 2001 and to those who participated in the search, rescue, and recovery efforts in the aftermath of those attacks

S.Con.Res. 76 A concurrent resolution honoring the law enforcement officers, firefighters, emergency rescue personnel, and health care professionals who have worked tirelessly to search for and rescue the victims of the horrific attacks on the United States on 11 September 2001

S.Con.Res. 87 A concurrent resolution expressing the sense of Congress regarding the crash of American Airlines Flight 587

S.Con.Res. 91 A concurrent resolution expressing deep gratitude to the government and the people of the Philippines for their sympathy and support since 11 September 2001 and for other purposes

S.Con.Res. 93 A concurrent resolution recognizing and honoring the National Guard on the occasion of the 365th anniversary of its historic beginning with the founding of the militia of the Massachusetts Bay Colony

S.Con.Res. 112 A concurrent resolution expressing the sense of Congress regarding the designation of the week beginning 19 May 2002 as 'National Medical Services Week'

S.J.Res. 25 National Day of Remembrance Act of 2001

S.Res. 264 A resolution expressing the sense of the Senate that small business participation is vital to the defense of our Nation, and that Federal, State, and local governments should aggressively seek out and purchase innovative technologies and services from American small businesses to help in homeland defense and the fight against terrorism

Notes

Introduction

1 For an excellent review of competing post–cold war predictions offered by 'euphoric liberals' and 'doom mongers,' see John Peterson 'Europe, America and 11 September,' *Irish Studies in International Affairs* 13 (2002): 1–20. Peterson observed that 9/11 tested these rival perspectives, and the 'liberal-optimistic' paradigm appeared to have failed – 'instead of the predicted spread of peaceful, liberal internationalism, the "end of history" yielded intense hatred of America and Western commercialism amongst the world's disaffected. In its more pessimistic guises the paradigm assumed that the transatlantic alliance had become brittle and was likely to crack, even before it became subject to new strains arising from the war on terrorism.' See Francis Fukuyama, *The Great Disruption: Human Nature and the Reconstitution of Social Order* (London: Oxford University Press, 1999); Ian Clark, *The Post–Cold War Order: The Spoils of Peace* (London: Oxford University Press, 2001); G. John Ikenberry, *After Victory: Institutions, Strategic Restraint and the Rebuilding of Order after Major Wars* (Princeton: Princeton University Press, 2001); Charles A. Kupchan, *The End of the American Era: U.S. Foreign Policy and the Geopolitics of the Twenty-first Century* (New York: Alfred A. Knopf, 2003).

2 See Robert Kaplan, *The Coming Anarchy: Shattering the Dreams of the Post Cold War* (New York: Vintage, 2000); John J. Mearsheimer 'The Future of the American Pacifier,' *Foreign Affairs* 80, no. 5 (2001); Samuel R. Huntington, *The Clash of Civilizations and the Remaking of World Order* (London: Oxford University Press, 1997).

3 Ivo Daalder, 'Are the United States and Europe Heading for Divorce?' *International Affairs* 77, no. 3 (2001): 553–67; David Pryce-Jones, 'Bananas

Are the Beginning: The Looming War between America and Europe,'
National Review, April 1999; Stephen M. Walt, 'The Ties That Fray,' *National Interest* 54 (Winter 1998): 3–11.

4 John Ikenberry, 'Institutions, Strategic Restraint, and the Persistence of American Postwar Order,' *International Security* 23, no. 3 (1998): 45–78; Ikenberry, *After Victory*; Ikenberry, ed. 'American Unipolarity: The Sources of Persistence and Decline,' in *America Unrivaled: The Future of the Balance of Power* (Ithaca: Cornell University Press, 2003), 284–310.

5 Ikenberry, *America Unrivaled*, 214.

6 William C. Wohlforth, 'U.S. Strategy in a Unipolar World' in Ikenberry, *America Unrivaled*, 98.

7 Stephen Walt, 'Keeping the World "Off Balance": Self Restraint and U.S. Foreign Policy,' in Ikenberry, *America Unrivaled*, 139.

8 Josef Joffe, 'Defying History and Theory: The United States as the "Last Remaining Superpower," in Ikenberry, *America Unrivaled*, 156.

9 Kupchan, *The End of the Amercian Era*, 29.

10 Ibid., 160.

1. Linking Globalism, Terrorism, and Proliferation

1 Throughout this chapter the terms *globalism* and *globalization* will be used synonymously and interchangeably. However, the former typically encompasses a combination of forces that go beyond the trade, investment, macro-economic, and global finance focus of the globalization / anti-globalization literature. As such, globalization represents only a subset of the globalism puzzle. For a more detailed discussion of the distinction between globalism and globalization, see Robert Keohane and Joseph S. Nye Jr., 'Globalization: What's New? What's Not? (And So What?),' *Foreign Policy* 118 (Spring 2000); available at www.foreignpolicy.com. According to the authors, the term *globalism* refers to the system-wide pressures that are currently transforming international and domestic politics, whereas *globalization* refers to the degree (i.e., 'thickness' or 'thinness') of globalism at particular points in time, or with respect to specific issue areas (e.g., trade, finance).

2 For an excellent review of contemporary work on globalization, see Joseph S. Nye Jr., *The Paradox of American Power: Why the World's Only Superpower Can't Go It Alone* (London: Oxford University Press, 2002). See also Patrick O'Meara, Howard D. Mehlinger, and Mathew Krain, eds., *Globalization and the Challenge of a New Century: A Reader* (Bloomington: Indiana University Press, 2000); Thomas L. Friedman, *The Lexus and the Olive Tree: Understand-*

ing Globalization (New York: Doubleday, 2000); George Soros, *On Globalization* (Washington: Public Affairs Press, 2002); Tomatsu Aoki, Peter L. Berger, and Samuel Huntington, eds., *Many Globalizations: Cultural Diversity in the Contemporary World* (London: Oxford University Press, 2002); Anthony Giddens, *Runaway World: How Globalisation Is Re-shaping Our World* (New York: Routledge, 2000).

For research on globalism and international conflict, see Benjamin Barber, *Jihad vs. McWorld: How Globalization and Tribalism Are Reshaping the World* (New York: Ballantine Books Inc., 1996); Benjamin Barber, 'Jihad vs. McWorld,' *Atlantic Monthly*, March 1992; Samuel Huntington, 'The Clash of Civilizations?' *Foreign Affairs* 72, no. 3 (1993); Robert Kaplan 'The Coming Anarchy.' *Atlantic Monthly*, February 1994.

On the positive and negative effects of globalism, see John Micklethwait and Adrian Wooldridge, *A Future Perfect: The Hidden Promise and Peril of Globalization* (New York: Crown Publishers, 1999); James H. Mittleman, *The Globalization Syndrome: Transformation and Resistance* (Princeton: Princeton University Press, 2000); Ann Kelleher and Laura Klein, *Global Perspectives: A Handbook for Understanding Global Issues* (New York: Prentice Hall, 1999); Zdenek Drabek, ed., *Globalisation under Threat: The Stability of Trade Policy and Multilateral Agreements* (Cheltenham, U.K.: Edward Elgar Publishers, 2001); Robert Gilpin, *The Challenge of Global Capitalism* (Princeton: Princeton University Press, 1999); Lael Brainard and Robert E. Litan, *Globalization: What Now?* (Washington: Brookings Institution Press, 2002); Kevin Danaher and Roger Burbach, eds., *Globalize This! The Battle against the World Trade Organization* (Monroe, ME: Common Courage Press, 2000).

Several works focus on domestic and international governance issues – e.g., Joseph S. Nye and John D. Donahue, eds., *Governance in a Globalizing World* (Washington: Brookings Institute Press, 2000); Jeremy Brecher, Tim Costello, and Brendan Smith, eds., *Globalization from Below* (Cambridge, MA: South End Press, 2000).

For a relatively controversial discussion of globalism and its impact on immigration, see Patrick J. Buchanan, *The Death of the West: How Mass Immigration, Depopulation, and a Dying Faith Are Killing Our Culture and Country* (New York/London: St. Martin's Press, 2001). For slightly more balanced assessments of the relationship, see Brian Nichiporuk, *The Security Dynamics of Demographic Factors* (Santa Monica: Rand, 2000); Stephen Castles and Alastair Davidson, *Citizenship and Migration: Globalization and the Politics of Belonging* (New York: Routledge, 2002).

3 Lael Brainard, 'Globalization in the Aftermath: Target, Casualty, Callous Bystander?' Analysis Paper #12 (Washington: Brookings Institution Press,

28 November 2001), available at http://brookings.org/views/papers/brainard/20011128.htm.

4 Keohane and Nye, 'Globalization: What's New? What's Not? (And So What?).'

5 Steve Smith and John Baylis, Introduction in *The Globalization of World Politics: An Introduction to International Relations* (London: Oxford University Press, 2001), 11.

6 Ibid.

7 One of the primary texts on the subject devotes only one chapter to security – see Patrick O'Meara, Howard D. Mehlinger, Mathew Krain, and Roxanna Ma Newman, eds., *Globalization and the Challenges of a New Century: A Reader* (Bloomington: Indiana University Press, 2000).

8 Brainard, 'Globalization in the Aftermath.'

9 Joseph Nye, 'The New Rome Meets the New Barbarians: How America Should Wield Its Power,' *Economist*, 23 March 2002.

10 Brainard, 'Globalization in the Aftermath.'

11 For an excellent analysis of the complex linkages between globalization and terrorism, and the implications for U.S. foreign policy, see Audrey Kurth Cronin, 'Behind the Curve: Globalization and International Terrorism,' *International Security* 27, no. 3 (2002–3): 30–58.

12 Ashton B. Carter and William J. Perry, *Preventive Defense: A New Security Strategy for America* (Washington: Brookings Institution Press, 1999), 151.

13 Thomas Homer-Dixon, 'The Rise of Complex Terrorism,' *Foreign Policy* (January–February 2002), available at http://www.foreignpolicy.com/issue_janfeb_2002/homer-dixon.html. For another excellent study of the security challenges tied to computer encryption, see *Strategic Assessment 1999*, Institute for National and Strategic Studies, National Defence University – especially chapter 2 'Economic Globalization: Stability or Conflict,' available at www.ndu.edu/inss/sa99/sa99cont.html.

14 Homer-Dixon, 'The Rise of Complex Terrorism,' identifies three technological advances that account for the greater destructive capacities of terrorist groups and individuals: 'more powerful weapons, the dramatic progress in communications and information processing, and more abundant opportunities to divert non-weapon technologies to destructive ends.'

15 With respect to interconnectedness, Homer-Dixon notes that globalization increases 'the number of nodes, the density of links among the nodes and the speed at which materials, energy, and information are pushed along these links. Moreover, the nodes themselves become more complex as the people who create, operate, and manage them strive for better perfor-

mance ... Complex and interconnected networks sometimes have features that make their behavior unstable and unpredictable. In particular, they can have feedback loops that produce vicious cycles ... Networks can also be tightly coupled, which means that links among the nodes are short, therefore making it more likely that problems with one node will spread to others ... [which means] a small shock or perturbation to the network produces a disproportionately large disruption.' This, in turn, increases the aggregate level of vulnerability, sensitivity, and ripple effects associated with any terrorist attack: 'Growing complexity and interdependence, especially in the energy and communications infrastructures, create an increased possibility that a rather minor and routine disturbance can cascade into a regional outage.' The 14 August 2003 blackout in the United States and Canada demonstrates the serious consequences of interconnectedness and mutual vulnerability. For information on the impact of this and other power failures, see http://mt.sopris.net/mpc/industrial/power.system .failures.html.

16 On the issue of geographic concentration, 'the range of possible terrorist attacks has expanded due to ... the rising concentration of high-value assets in geographically small locations. Advanced societies concentrate valuable things and people in order to achieve economies of scale. Companies in capital-intensive industries can usually reduce the per-unit cost of their goods by building larger production facilities. Moreover, placing expensive equipment and highly skilled people in a single location provides easier access, more efficiencies, and synergies that constitute an important source of wealth. That is why we build places like the World Trade Center' (Homer-Dixon, 'The Rise of Complex Terrorism').

17 Paul Pillar, *Terrorism and U.S. Foreign Policy* (Washington: Brookings Institution Press, 2001).

18 Simon Reeve, *The New Jackals* (London: Oxford University Press, 1999).

19 The ancient *hawala* system is cited to illustrate the Internet's contribution to efficiency: 'it is widely used in Middle Eastern and Asian societies. The system, which relies on brokers linked together by clan-based networks of trust, has become faster and more effective through the use of the Internet.' With respect to information processing, state-of-the-art encryption technology can be accessed from a number of sources on the Internet: 'Sometimes less advanced computer technologies are just as effective. For instance, individuals can use a method called steganography ("hidden writing") to embed messages into digital photographs or music clips. Posted on publicly available Web sites, the photos or clips are downloaded by collaborators as necessary. (This technique was reportedly used by recently arrested

terrorists when they planned to blow up the U.S. Embassy in Paris.) At latest count, 140 easy-to-use steganography tools were available on the Internet. Many other off-the-shelf technologies – such as "spread-spectrum" radios that randomly switch their broadcasting and receiving signals – allow terrorists to obscure their messages and make themselves invisible' (Homer-Dixon, 'The Rise of Complex Terrorism').

20 For a discussion of how computer-literate recruits to extremist groups like Hamas and Hezbollah now throw 'virtual electronic stones,' see Arnaud de Borchgrave, Frank J. Cilluffo, Sharon L. Cardash, and Michèle M. Ledgerwood, *Cyber Threats and Information Security: Meeting the 21st Century Challenge* (Washington: Center for Strategic and International Studies, December 2000). For an excellent series of articles on the subject of cyber-terrorism, see Jerrold Post, Kevin Ruby, and Eric Shaw, 'From Car Bombs to Logic Bombs: The Growing Threat of Information Terrorism,' *Terrorism and Political Violence* 12, no. 2 (2000): 97–122.

21 Frank Cilluffo, Joseph J. Collins, Arnaud de Borchgrave, Daniel Gouré, and Michael Horowitz, 'Executive Summary,' in *Defending America in the 21st Century: New Challenges, New Organizations, and New Policies* (Washington: Center for Strategic and International Studies, 2001). See also Frank Cilluffo and Bruce Berkowitz, eds., *Cybercrime, Cyberterrorism, and Cyberwarfare: Averting an Electronic Waterloo* (Washington: CSIS, 1998), 1–72.

22 Jim Wolf, 'Hacking of Pentagon Computers Persists; Pace Undiminished This Year, Complicating Security Efforts,' *Washington Post*, 23 August 2000, A23; and comments by Richard Clarke at a lecture for the MIT Alumni Association, 10 October 2000.

23 Yahya Sadowski, *The Myth of Global Chaos* (Washington: Brookings Institution Press, 1998: 'Some thought that globalization was triggering a wave of internal conflicts, civil wars and ethnic conflicts. Others thought it was fueling clashes between civilization, mammoth struggles involving terrorism and Weapons of Mass Destruction, with the future of the world hanging in the balance,' vii.

24 Michelle Locke, 'Computers Would Track Terror Threat,' Associated Press, 23 May 2002. See also Saul B. Wilen, 'Counter-terrorism, Bit by Bit,' *Washington Times*, 3 April 2002. According to Wilen, 'The prevention weapon consists of computer-based systems that utilize the multiple existing databases – many housed in different government agencies – domestically and globally in real-time. Information will continue to be gathered, incorporated and shared on multiple levels. The new and unique horizontal integration technology allows the data to be subdivided immediately into essential elements for evaluation to support effective decision-

making and action. The information presently stored in the multiple agency and entity databases, such as that relating to terrorist attacks, government buildings, and biological agents, would be merged into a functional "one-source-reservoir" used by these agencies for strategic prevention analysis. This technology exists now and can be readily implemented. But the job, logical as it seems, has yet to be done, while future terrorist attacks are being planned.'

25 Paul R. Pillar, *Terrorism and U.S. Foreign Policy* (Washington: Brookings Institution Press, 2001), 47.

26 Ann Scott Tyson, 'Is More Terror in US Inevitable? "Matrix" behind Warnings: How the President Gauges Threats,' *Christian Science Monitor*, 23 May 2002, available at http://www.csmonitor.com/2002/0523/p01s03–uspo.html.

27 Ibid. Many of the messages are hidden in digital photographs and sent via email.

28 Reuel Marc Gerecht, 'The Counterterrorist Myth,' *Atlantic Monthly*, July–August 2001, available at http://www.theatlantic.com/issues/2001/07/gerecht.htm.

29 Richard E. Hayes and Gary Wheatley, 'Information Warfare and Deterrence.' NDU Paper number 87, Institute for National Strategic Studies, National Defence University, October 1996, available at www.ndu.edu/inss/strforum/forum87.html. For another excellent study on the application of deterrence strategies to prevent cyber-crime, see Walter Tirenen, *White Paper for a Strategic Cyber Defense Concept: Deterrence through Attacker Identification* (University of Southern California: Information Science Institute, 2002), available at http://www.isi.edu/gost/cctws/tirenen.html. According to the author, '[W]hile the methods for identifying perpetrators of crimes in the law enforcement context, and attackers in the military context, are well-developed, similar capabilities do not currently exist for the networked cyber realm. Thus, while deterrence is recognized as a highly effective defensive strategy, its applicability to defense against attacks on our nation's information infrastructures has not been clear, mainly due to the inability to link attackers with attacks.' As Tirenen points out, '[F]orensics could help to provide the bridge between the defensive and offensive elements of an overall cyber defense strategy. Accurate and timely forensic techniques would also enable the effective use of the three elements of deterrence. Otherwise, attackers can act with impunity, feeling confident that they need not fear the consequences of their actions ... [S]uch a capability implies the need for laws specifying authorization to conduct cyberspace pursuits, and cooperative agreements

with foreign governments and organizations. A second suggestion is for the development of a tamper-proof, aircraft-like "black box" recording device to ensure that when an incident occurs and is not detected in real time, the trail back to the perpetrator does not become lost.'

30 According to Brainard, in 'Globalization in the Aftermath,' '[E]ven before the terrorist attacks, a number of globalization's key components showed signs of setting a slower pace ... In the aftermath of the attacks, however, even these dire predictions appear optimistic, as heightened security concerns compel nations to tighten their borders. Rising public anxiety and new travel restrictions, for example, appear likely to curtail travel between countries, perhaps leading to a decline in global tourism for the first time in the last 50 years ... These trends are likely to affect dispropor- tionately those countries that are closely integrated into the world econ- omy ... Singapore, with its small domestic economy and heavy reliance on trade, has already begun to feel the pain of globalization's downturn, with exports and imports alike plummeting more rapidly than at any time since the country's independence in 1965. Other countries in which trade ac- counts for the lion's share of integration with global markets – including Malaysia, Slovakia, Panama, Thailand, and the Philippines – may also see globalization levels affected over the coming year ... International Mon- etary Fund projections showed global economic growth slowing from 4.7 percent in 2000 to a mere 2.4 percent in 2001, just below levels consid- ered to be recessionary. Similarly, global trade growth in 2001 was ex- pected to remain nearly flat, while predictions showed FDI flows dropping more than 40 percent from the record highs in 2000.'

31 See also 'Globalization's Last Hurrah?' http://www.foreignpolicy.com/ issue_janfeb_2002/global_index.html.

32 Brainard, 'Globalization in the Aftermath.'

33 Colin S. Gray, 'The Definitions and Assumptions of Deterrence: Questions of Theory and Practice,' *Journal of Strategic Studies* 13 (December 1990): 6– 13; Christopher H. Achen and Duncan Snidal, 'Rational Deterrence Theory and Comparative Case Studies,' *World Politics* 41(January 1989); Edward Rhodes, 'Nuclear Weapons and Credibility: Deterrence Theory Beyond Rationality,' *Review of International Studies* 14 (January 1988).

34 See Colin S. Gray, *The Second Nuclear Age* (New York: Lynne Rienner, 1999); Keith B. Payne, *Deterrence in the Second Nuclear Age* (Lexington: University Press of Kentucky 1996).

35 *The National Security Strategy of the United States of America* (Washington: The White House, September 2002), 15.

36 Scott Peterson, 'US Grand Plan Proves Hard Sell Yesterday, Russia Re-
    buffed a Bush Proposal That Would Lift Barrier against US National
    Missile Shield,' *Christian Science Monitor*, 14 August 2001.

37 Leon Fuerth, 'Tampering with Strategic Stability,' *Washington Post*, 20 Feb-
    ruary 2001, A23. Leon Fuerth was the former national security adviser to
    former vice president Al Gore.

38 See John Rhinelander, 'ABM Treaty – Anachronism or Cornerstone?'
    Presentation to U.S. Senate Staff, 5 November 1999, available at
    www.nyu.edu/globalbeat/nuclear/Isaacs121799.html.

39 For academics who emphasize numbers, see www.mi.infn.it/~landnet/
    NMD/kapralov.pdf; for editors of major newspapers, www.clw.org/
    coalition/briefv5n14.htm; for Russian generals, www.chinadaily.net/
    highlights/docs/2001–04–29/2711.html; for Senate Democratic leader
    John Kerry, www.politicsol.com/guest-commentaries/2001-05-14.html.
    Carl Kaysen, Robert McNamara, and George W. Rathjens, 'Nuclear Weap-
    ons after the Cold War,' *Foreign Affairs* 70, no. 4 (Fall 1991): 106–7, con-
    cludes that 'although we have no quarrel with estimates on how much is
    enough, nor with most of the other discussions of minimum deterrence in
    the last three decades, the size of the force required for the purpose de-
    pends on the size of others' nuclear.'

40 Fuerth, 'Tampering with Strategic Stability,' A23.

41 Senator Jesse Helms, written statement issued by Office of Senator Jesse
    Helms, 1 May 2001, available at http://www.acronym.org.uk/bush1.htm.

42 McGeorge Bundy, 'To Cap the Volcano,' *Foreign Affairs* 48 (October 1969):
    9–10.

43 Colin S. Gray, 'Nuclear Strategy: A Case for a Theory of Victory,' *Interna-
    tional Security* 4 (Summer 1979): 54–87.

44 Rajesh Rajagopalan, 'Nuclear Strategy and Small Nuclear Forces: The
    Conceptual,' *Strategic Analysis* 23 (October 1999), available at www.idsa-
    india.org/an-oct9–5.html.

45 See 'A Post–Cold War Nuclear Strategy Model – Introduction: Framing the
    Question,' available at www.usafa.af.mil/inss/ocp20.htm. Of course, this
    is not to suggest that numbers are irrelevant – domestic imperatives will
    force Russian officials to continue to push the numbers game, but that has
    more to do with politics than strategic stability.

46 Alexei G. Arbitov, 'The Future of Strategic Deterrence and Nuclear Pos-
    tures of Great Powers,' Institute for National Strategic Studies, National
    Defense University, Washington, DC, available at http://isuisse.ifrance
    .com/emmaf2/USRUS/usrp13.html.

47 A more reliable indicator of the relationship than stockpiles can be found in Russian procurement and deployment plans. See Wolfgang K.H. Panofsky, 'The Remaining Unique Role of Nuclear Weapons in Post–Cold War Deterrence,' in *Post–Cold War Conflict Deterrence* (Washington: Naval Studies Board Commission on Physical Sciences, Mathematics, and Applications, National Research Council, National Academy Press 1997). According to Panofsky, 'Today such figures are overestimates for required deterrent forces against a possible reemergence of a Russian nuclear threat: Russia contains only parts of the former economic assets of the former Soviet Union and its military basing structure. Thus the core purpose against the reemergence of an aggressive Russia requires forces only a small fraction of those contemplated for START II. Under the core deterrent role of nuclear weapons, the "hedge" provided by the NPR is unnecessary and large reductions below START II levels are feasible.'

48 On the propensity to overplay and exaggerate Russia's concerns about a U.S. first strike, see www.russiajournal.com/index.htm?obj=4510: 'Of all the Russian politicians, military officials and experts who say the American plan represents a military threat to Russia, it's unlikely any of them seriously believe this. There's no reason to believe that Bush needs NMD in order to be able to launch a sudden first strike against Russia without fear of retaliation.' On transformations in American-Russian relations, see www.mi.infn.it/~landnet/NMD/fisher.pdf.

49 *The National Security Strategy of the United States of America* (Washington: The White House, September 2002), 13.

50 Quoted in Susan B. Glasser, 'Russia Has Warning, and Overture, on Missile Plan,' *Washington Post*, 19 December 2002, A29.

51 James M. Lindsay and Michael E. O'Hanlon, *Defending America: The Case for Limited National Missile Defense* (Washington: Brookings Institution Press, 2001).

52 On the many reasons China will modernize its nuclear program regardless of U.S. BMD deployment, see, for example, Frank P. Harvey, 'National Missile Defence Revisited, Again: A Response to David Mutimer,' *International Journal* 56 (Spring 2001): 347–60; and Harvey, 'Proliferation, Rogue-State Threats and National Missile Defence: Assessing European Concerns and Interests,' *Canadian Military Journal* 1 (Winter 2000–1): 69–79.

53 Robert O. Keohane and Joseph S. Nye Jr., *Power and Interdependence: World Politics in Transition*, 2d ed (Boston: Little, Brown, 1989).

54 John Mueller, *Retreat from Doomsday: The Obsolescence of Major War* (New York: Basic Books, 1989).

55 Benjamin Rinkle, 'The ABM Treaty and Arms Race Myths,' *Washington Times*, 21 December 2001.

56 Jim Hoagland, 'An Ally's Terrorism,' *Washington Post*, 3 October 2001, B7.

57 For two excellent studies on the unintended consequences of short-term, unilateral alliance shifts towards Pakistan and Iran (and other coalitions of convenience), see Robert Kagan and William Kristol, 'The Coalition Trap,' *Weekly Standard 7*, no. 15 (15 October 2001). See also Jim Hoagland, 'An Ally's Terrorism,' *Washington Post*, 3 October 2001, A31.

58 Thomas L. Friedman, *The Lexus and the Olive Tree: Understanding Globalization* (New York: Doubleday, 2000).

59 For a discussion of emerging terrorist network in U.S. and Canada, see John K. Cooley, *Unholy Wars: Afghanistan, America, and International Terrorism* (London: Pluto Press, 2001). Several books focus on the origins and evolution of terrorism: Walter Reich, *Origins of Terrorism: Psychologies, Ideologies, Theologies, States of Mind* (Washington: Woodrow Wilson Center Press, 1998); Bruce Hoffman, *Inside Terrorism* (New York: Columbia University Press, 1998); Jessica Stern, *The Ultimate Terrorists* (Boston: Harvard University Press, 2000); Glenn E. Schweitzer, *A Faceless Enemy: The Origins of Modern Terrorism* (Oxford, U.K.: Perseus Books Group, 2002); Walter Laqueur, *The New Terrorism: Fanatacism and the Arms of Mass Destruction* (London: Oxford University Press, 2001); Ian O. Lesser, Bruce Hoffman, John Arquilla, David F. Ronfeldt, Michele Zanini, and Brian Michael Jenkins, eds., *Countering the New Terrorism* (Santa Monica: Rand, 1999), available at http://www.rand.org/publications/MR/MR989/; John Arquilla and David Ronfeldt, eds., *Networks and Netwars: The Future of Terror, Crime, and Militancy* (especially chapter 2, 'The Networking of Terror in the Information Age') (Santa Monica: Rand, 2001), available at http://www.rand.org/ publications/ MR/MR1382/; Meghan O'Sullivan, *Shrewd Sanctions: Economic Statecraft in an Age of Global Terrorism* (Washington: Brookings Institution Press, 2002); Ivo H. Daalder, I.M. Destler, Robert E. Litan, Michael E. O'Hanlon, and Peter R. Orszag, eds., *Achieving Homeland Security* (Washington: Brookings Institution Press, 2002); Seymon Brown, *The Illusion of Control: Force and Foreign Policy in the 21st Century* (Washington: Brookings Institution Press, 2001).

60 Ihekwoaga Onwudiwe, *The Globalization of Terrorism* (New York: Ashgate Publishing, 2000); Robert Mandel, *Deadly Transfers and the Global Playground: Transnational Security Threats in a Disorderly World* (Westport, CT: Greenwood Publishing Group, 1999). The latter book highlights the

failures of international organizations to address the problem of transnational terrorism and associated problems with current approaches and strategies. The international environment (or 'global playground') within which these threats are spreading points to the many challenges the international community faces; Harry Henderson, *Global Terrorism: The Complete Reference Guide* (New York: Checkmark Books, 2002).

61 Quoted in Arnaud de Borchgrave, Frank J. Cilluffo, Sharon L. Cardash, and Michèle M. Ledgerwood, *Cyber Threats and Information Security: Meeting the 21st Century Challenge* (Washington: Center for Strategic and International Studies, 2000).

62 Warren E. Walker, *Uncertainty: The Challenge for Policy Analysis in the 21st Century* (Santa Monica: Rand, 2001).

2. Linking Globalism, Unilateralism, and Multilateralism

1 Frank Cilluffo, Joseph J. Collins, Arnaud de Borchgrave, Daniel Gouré, and Michael Horowitz, *Defending America in the 21st Century: New Challenges, New Organizations, and New Policies* (Washington: Center for Strategic and International Studies, 2001). See also Frank Cilluffo and Bruce Berkowitz, eds., *Cybercrime, Cyberterrorism, and Cyberwarfare: Averting an Electronic Waterloo* (Washington: CSIS, 1998), 1–72.

2 Benjamin Barber, 'Beyond Jihad Vs. McWorld,' *Nation*, 21 January 2002, available at http://www.thenation.com/doc.mhtml?i=20020121&s=barber.

3 Kishore Mahbubani, 'The United Nations and the United States: An Indispensible Partnership' (139–52); Ramesh Thakur, 'UN Peace Operations and U.S. Unilateralism and Mulitlateralism' (153–80); Ekaterina Stepanova, 'The Unilateral and Multilateral Use of Force' (181–200); Qingguo Jia, 'In Search of Absolute Security: U.S. Nuclear Policy' (201–16); and Kanto Bajbai, 'U.S. Nonproliferation Policy After the Cold War' (217–50) – all in David M. Malone and Yuen Foong Khong, *Unilateralism and U.S. Foreign Policy: International Perspectives* (London: Lynne Rienner Publishers, 2003).

4 Lael Brainard, 'Globalization in the Aftermath: Target, Casualty, Callous Bystander?' Analysis Paper #12 (Washington: Brookings Institution Press, 2001). 'The aftermath of September 11 confronts America with countervailing pressures. When a sense of safety previously taken for granted is profoundly undermined, there is a natural tendency to pull up the drawbridges and pull back from the world. And when jobs and economic security are put at risk, there is a tendency to look towards protectionist solutions.'

5 Campaign for UN Reform, 'U.S. Funding to the U.N.,' available at http://www.cunr.org/priorities/Arrears.htm.

6 *The National Security Strategy of the United States of America* (Washington: The White House, September 2002), 31.

7 *A National Security Strategy for a New Century* (Washington: The White House, December 1999), 1, 14, 19, 20.

8 Charles Krauthammer, 'The Real New World Order: The American and the Islamic Challenge,' *Weekly Standard* 7, no. 9 (12 November 2001). For an equally compelling contribution to the debate over an emerging U.S. empire, see Robert Kaplan, *Warrior Politics: Why Leadership Demands a Pagan Ethos* (New York: Random House, 2001). Similarly, Emily Eakin, 'It Takes an Empire,' *New York Times*, 2 April 2002, argues that the prevailing opinion in the U.S. today is that 9/11 was a product of 'insufficient American involvement and ambition; the solution is to be more expansive in our goals and more assertive in their implementation.'

9 With respect to evidence on the failures of unilateralism, see Serge Sur, ed., 'Disarmament and Limitation of Armaments: Unilateral Measures and Policies' (New York: United Nations Publications, 1993); William Rose, 'U.S. Unilateral Arms Control Initiatives: When Do They Work?' Contributions in *Military Studies*, no. 82 (Westport, CT: Greenwood Press, 1988); Ernest H. Preeg, 'Feeling Good or Doing Good with Sanctions: Unilateral Economic Sanctions and the U.S. National Interest,' *Significant Issues Series* 21, no. 3 (1999). One in a series on economic sanctions sponsored by the Center for Strategic and International Studies is Joseph J. Collins and Gabrielle D. Bowdoin, 'Beyond Unilateral Economic Sanctions: Better Alternatives for U.S. Foreign Policy,' (Washington: Center for Strategic and International Studies, April 1999). According to these CSIS reports, without exception unilateral sanctions have failed to achieve their primary political objectives. As a strategy they are likely to become even less successful as a result of globalization; see Douglas Johnston and Sidney Weintraub, 'Altering U.S. Sanctions Policy: Final Report of the CSIS Project on Unilateral Economic Sanctions' (Washington: Center for Strategic and International Studies, April 1999); Elisabeth Zoeller, 'Peacetime Unilateral Remedies: An Analysis of Countermeasures' (Ardsley, NY: Transnational Publishers, April 1984); Christian M. Scholz and Frank Stähler, 'Unilateral Environmental Policy and International Competitiveness' (Kiel, Germany: Kiel Institute of World Economics, January 2000).

10 Robert Keohane, 'The Globalization of Informal Violence: Theories of World Politics and the Liberalism of Fear,' paper prepared for delivery at the annual meeting of the American Political Science Association, Boston,

29 August–September 2002, available at http://www.iyoco.org/911/
911keohane.htm.

11 Thakur, 'UN Peace Operations and U.S. Unilateralism and Multilateralism'
in Malone and Yuen Foong Khong, *Unilateralism and U.S. Foreign Policy*,
176.

12 Ibid., 172.

13 Samuel Berger, 'A Foreign Policy for the Global Age,' *Foreign Affairs* 79,
no. 6 (2000).

14 Allan Gotlieb, 'The Chrétien Doctrine: By Blindly Following the UN, the
Prime Minister Is Hurting Canada,' *Maclean's*, 31 March 2003.

15 Peter Worthington, 'Moral Causes and the UN Are Odd Companions,'
Toronto Sun, 23 March 2003.

16 Adrian Karatnycky, 'Libya Unfit to Head UN Human Rights Group,'
Newsday, 19 September 2002. According to Freedom House reports cited in
the article, typical human rights abuses include 'imprisonment of hundreds of political and civic activists, death sentences for nonviolent activists, torture, disappearances and suspected assassinations of numerous
political opponents around the world.'

17 Marian L. Tupy, 'South Africa Helps Libya Gain U.N. Human Rights Seat,'
2 February 2003, available at http://www.cato.org/dailys/02-02-03.html.

18 Michael Mandelbaum, 'The Inadequacy of American Power,' *Foreign
Affairs* 81, no. 5 (2002). Also in Michael Mandelbaum, *The Ideas That Con-
quered the World: Peace, Democracy and Free Markets in the Twenty-first
Century* (Washington: Public Affairs Press, 2002).

19 Mandelbaum, 'The Inadequacy of American Power.'

20 Tim Hames, 'Arrogance, Ignorance and the Real New World Order,' *Times*
(London), 15 February 2002, available at www.thetimes.co.uk. As Hames
correctly points out, 'Genuine multilateralism requires a multipolar order.
That can only be achieved when authority is distributed evenly across a
number of players (a transient event in human history so far) or if the
largest power chooses, for some reason, to shrink itself to meet the occasion. That was the essence of American foreign policy in the decade
between the Gulf War and September 11.' It is important to distinguish
debates over the structure of (and power distribution within) the contemporary international system (unipolarity vs. multipolarity) from debates
over the presence/absence of unilateral and multilateral approaches to
security. The former debate is typically framed with reference to arguments put forward by Charles Krauthammer, 'The Unipolar Moment:
America and the World,' *Foreign Affairs* 70, no. 1 (1990–1991); and Samuel
Huntington 'The Lonely Superpower,' *Foreign Affairs* 78, no. 2 (1999). For

an excellent treatment of the subject, see Stewart Patrick and Shepard Forman, eds., *Multilateralism and U.S. Foreign Policy: Ambivalent Engagement* (Washington: Center on International Cooperation Studies in Multilateralism, 2001). Several excellent contributions to the series highlight many of the arguments on each side of the unilateral-multilateral divide through several case studies covering a variety of relevant issue areas. See, for example, G.J. Ikenberry, 'Multilateralism and U.S. Grand Strategy'; R. Wedgwood, 'Unilateral Action in a Multilateral World'; M. Mastanduno, 'Extraterritorial Sanctions: Managing "Hyper-Unilateralism" in U.S. Foreign Policy'; B.S. Brown, 'Unilateralism, Multilateralism, and the International Criminal Court'; Andrew Moravcsik, 'Why Is U.S. Human Rights Policy So Unilateralist?'

21 For a good example of the evidence that could be cited to question the claims of unilateralism in U.S. foreign policy, see David Limbaugh, 'The Phony Charge of "Unilateralism,"' 16 July 2002, available at http:// www.worldnetdaily.com. A slightly different approach to measuring U.S. unilateralism focuses on the extent to which, prior to 9/11, U.S. officials were decidedly disengaged internationally. As Andrew Cohen points out, 'Until Sept. 11, less foreign news was presented in America in recent years, as measured in minutes on television news and space in newspapers and magazines, than was the case in the past. By one measure *Time* magazine ran fewer cover stories on foreign subjects in the 1990s than it did in the 1960s.' See Andrew Cohen, 'Canadian-American Relations: Does Canada Matter in Washington? Does It Matter If Canada Doesn't Matter?' in Norman Hilmer and Maureen Appel Molot, eds., *Canada among Nations: A Fading Power* (London: Oxford University Press, 2002), 44.

22 Peter Spiro uses the term *internationalism à la carte* to describe the school of thought that assumes the U.S., given its power and influence, 'can pick and choose the international conventions and laws that serve its purpose and reject those that do not.' See Peter J. Spiro, 'The New Sovereigntists: American Exceptionalism and Its False Prophets,' *Foreign Affairs* 79, no. 6 (2000).

23 John Gerard Ruggie, ed., *Multilateralism Matters: The Theory and Praxis of an Institutional Form* (New York: Columbia University Press, 1993).

24 David M. Malone and Yeun Foong Khong, 'Unilateralism and U.S. Foreign Policy: International Perspectives,' in Malone and Yuen Foong Khong, *Unilateralism and U.S. Foreign Policy*, 3.

25 Ekaterina Stepanova, 'The Unilateral and Multilateral Use of Force by the United States,' in Malone and Yuen Foong Khong, *Unilateralism and U.S. Foreign Policy*.

26 Robert Kagan, 'The Benevolent Empire,' *Foreign Policy* 12, no. 111 (1998), 24.

27 Michael Kelly, 'Immorality on the March,' *Washington Post*, 19 February 2003, A29.

28 Andrew Mack and Oliver Rohls, 'The Great Canadian Divide,' *Globe and Mail* (Toronto), 28 March 2003, A19.

29 Stepanova, 'The Unilateral and Multilateral Use of Force by the United States,' in Malone and Yuen Foong Khong, *Unilateralism and U.S. Foreign Policy*, 190–1.

30 Brainard, 'Globalization in the Aftermath.'

31 John Peterson, 'Europe, America and the War on Terrorism,' *Irish Studies in International Affairs* 13 (2002): 23–42.

32 Jose E. Alvarez, 'Multilateralism and its Discontents,' *European Journal of International Law* 14, no. 1 (2002). See also Alvarez, 'The United States Financial Veto,' in *Proceedings of the 90th Annual Meeting, 27–30 March 1996* (Washington: American Society of International Law, 1997), 319. For a critique of the ICC's concept of 'complementarity,' see Alvarez, 'Crimes of States/Crimes of Hate: Lessons from Rwanda,' *Yale Journal of International Law* 24, no. 365 (1999): 476–79.

33 UN Security Council Resolution 955, UN SCOR, 49th year, 3453d mtg, at 1, UN Doc. S/RES/955 (1994).

34 Alvarez, 'Crimes of States/Crimes of Hate.' As Alvarez explains, 'The ICTR's [International Criminal Tribunal for Rwanda] version of justice is, at least from a Rwandan perspective, seriously compromised since the message that tribunal conveys is that the more culpable governmental elites – defendants like the former Prime Minister responsible for the deaths of thousands – are entitled to all the benefits of international due process and serve their terms in comparatively comfortable prisons made available by UN member states like the Netherlands, while those responsible for lesser crimes, beneath the notice of the ICTR, get (at best) expedited local justice and perhaps the death penalty ... The tendency to stress the virtues of multilateral solutions, narrowly understood to mean liberal institutions on the model of the UN, artificially restricts the range of available prescriptions for modern human rights dilemmas ... The current obsession with international criminal accountability, shown by the inclination to replicate international tribunals on the model of the ICTR and ICTY [International Criminal Tribunal for the Former Yugoslavia] shows a regrettable tendency to cast the issue in an either/or fashion: either an international criminal trial with the full panoplies of international justice or nothing at all.'

35 Alvarez, 'Multilateralism and its Discontents'; Alvarez, 'The United States Financial Veto,' 319.

36 Elizabeth Nickson, 'Don't Turn Iraq over to the UN,' *National Post* (Toronto), 24 April 2003.

37 Mark Steyn, 'The United Nations: Unfit to Govern,' *National Post* (Toronto), 28 April 2003.

38 George Jonas, 'The UN Is Its Own Worst Enemy,' *National Post* (Toronto), 19 March 2003.

3. Gulf War II: Unilateralism and Multilateralism in Practice

1 John Humphrys, 'The Gigantic Bluff of UN Power Has Been Called,' *Sunday Times* (London), 9 March 2003, available at www.timesonline.co.uk.

2 James A. Paul, 'Oil in Iraq: The Heart of the Crisis,' Global Policy Forum, New York, NY, December 2002, available at www.globalpolicy.org/security/oil/2002/12heart.htm.

3 James A. Paul, 'Oil in Iraq.' See also Ed Vulliamy, Paul Webster, and Nick Paton, 'Scramble to Carve Up Iraqi Oil Reserves,' *Observer* (London) 6 October 2002, available at http://www.observer.co.uk/; Diane Francis, 'Iraqi Liberation Will Unleash Untapped Wealth,' *Financial Post* (Toronto), 1 April 2003; and Helle Dale, 'No Blood for French Oil,' *Washington Times*, 5 March 2003.

4 Francis, 'Iraqi Liberation Will Unleash Untapped Wealth.'

5 Dale, 'No Blood for French Oil.'

6 Vulliamy, Webster, and Paton, 'Scramble to Carve Up Iraqi Oil Reserves.'

7 Dale, 'No Blood for French Oil.' The author goes on to point out that China 'also controls 5.8 percent of Iraq's annual imports. Under the food-for-oil program, China's Aero-Technology Import-Export Co. (CATIC) has contracted to sell "metereological satellite" and "surface observation" equipment to Iraq. CATIC has also received U.N. approval to sell optic cables to Iraq, which can be used for secure data and communications links for military installations. From 1981–2001, China was second only to Russia in sales of arms to Iraq.'

8 Alan Freeman, 'Old, New Europe Spar Bitterly over Iraq,' *Globe and Mail* (Toronto), 19 February 2003, A24.

9 Quoted in Mark Steyn, 'M. le Président's Imperiousness,' *National Post* (Toronto), 20 February 2003. See also Freeman, 'Old, New Europe Spar Bitterly Over Iraq.'

10 Freeman, 'Old, New Europe Spar Bitterly Over Iraq, ' A24.

11 'Last Chance for NATO,' editorial in *Telegraph* (London), 25 April 2003, available at www.telegraph.co.uk.

12 Quoted in Alexandra Richie, 'New Europeans Know Who Their Allies Are,' *National Post* (Toronto), 6 March 2003.

13 Emily Wax, 'France's Tentative Role in a Civil War: Troops in Ivory Coast Clash with Some Rebels While Staying Clear of Others,' *Washington Post*, 10 January 2003, A12.

14 Charles Krauthammer, 'Lift the Sanctions Now,' *Washington Post*, 21 April 2003, A23; William Safire, 'Follow the Money,' *New York Times*, 21 April 2003; Sol W. Sanders, 'Cleaning the U.N. Trough,' *Washington Times*, 22 April 2003.

15 Jan Cienski, 'Washington Trades Arms for Friends Military Aid: War on Terrorism Has Redefined America's Interests,' *National Post* (Toronto), 27 January 2003; Marguerite Michaels and Karen Tumulty, 'Horse Trading on Iraq,' *Time*, 10 March 2003, available at http://www.time.com/time/archive/preview/from_redirect/0,10987,1101030310-428053,00.html; Roland Watson and Richard Beeston, 'America Pulls the Rug from under Turkey,' *Times Online* (Beaver and Alegueny Counties, Penn.), 7 March 2003, available at www.timesonline.com.

16 Glenn Kessler and Philip P. Pan, 'Missteps with Turkey Prove Costly: Diplomatic Debacle Denied U.S. a Strong Northern Thrust in Iraq,' *Washington Post*, 28 March 2003, A01.

17 Quoted in ibid.

18 *The National Security Strategy of the United States of America* (Washington: The White House, September 2002), 7.

19 Watson and Beeston, 'America Pulls the Rug from under Turkey.'

20 Claudia Rosett, 'Oil, Food and a Whole Lot of Questions,' *New York Times*, 18 April 2003. According to another report on the UN's profits, 'Since 1996, it has helped sell US$55-billion in Iraqi oil, but authorized just US$34-billion in aid shipments back into Iraq. The US$21-billion surplus is thought to be sitting in UN accounts, mostly in French banks. In 2002 alone, the UN is believed to have sold US$16-billion in oil, turned a blind eye to a further US$5.3-billion in illegal sales, ... and pocketed US$483-million in commissions in the process.' 'The Iraqi Files,' editorial in *National Post* (Toronto), 29 April 2003.

21 Christopher A. Hitchens, 'Multilateralism and Unilateralism: A Self-Cancelling Complaint,' 18 December 2002, available at www.slate.com.

22 Ibid.

23 Robert Kagan, 'Multilateralism, American Style,' *Washington Post*, 13 September 2002.

24 Sean Loughlin, 'White House Touts International Support for Military Campaign,' CNN Washington Bureau, 20 March 2003, available at www.cnn.com.

25 I would like to thank Jay Nathwani for his excellent research assistance in compiling the information in this report.

26 Paul Wells, 'Germany, France: The Sum of All Fears,' *National Post* (Toronto), 30 January 2003, A13.

27 Karen DeYoung, 'U.S. and Saudis Agree on Cooperation: Gulf Nation to Permit Expanded Use of Military Facilities in Event of War,' *Washington Post*, 26 February 2003, A01.

28 Robert Kagan, 'Power and Weakness,' *Policy Review*, no. 113 (June and July 2002).

29 Robert Kagan, 'The Benevolent Empire,' *Foreign Policy* 12, no. 111 (1998), 24.

30 Walter B. Slocombe, 'Preemptive Military Action and the Legitimate Use of Force: An American Perspective,' prepared for meeting at the CEPS/IISS (Centre for European Policy Studies/ International Institute for Strategic Studies) European Security Forum, Brussels, 13 January 2003.

31 See Frank Harvey and Ben Mor, *Conflict in World Politics: Advances in the Study of Crisis War and Peace* (New York: Macmillan/St. Martin's Press, 1998). For a more complete discussion of the methodological and evidenciary requirements for evaluating success and failure, see Frank Harvey and David Carment, *Using Force to Prevent Ethnic Violence* (Westport, CT: Praeger, 2001) for assessments of the relative utility of state vs. international organization applications of those strategies.

32 Steven Everts, *Unilateral America, Lightweight Europe? Managing Divergence in Transatlantic Foreign Policy* (London: Centre for European Reform, 2001), 10.

33 Ibid.

34 Robert Kagan, 'Multilateralism, American Style,' *Washington Post*, 13 September 2002.

35 Brendan O'Neill, 'Don't Mention the U-word,' 21 February 2003, available at http://www.spiked-online.com/Articles/00000006DC77.htm.

36 Kagan, 'Multilateralism, American Style.' See also Kagan, 'Power and Weakness'; Robert Kagan, *Of Paradise and Power: America and Europe in the New World Order* (New York: Random House, 2003).

37 Kagan, 'Multilateralism, American Style.'

4. WMD Proliferation: The Case for Unilateral Ballistic Missile Defence

1 The following recent studies on ballistic missile defence cover virtually all the criticisms addressed in this chapter: Michael O'Hanlon, James M. Lindsay, Michael H. Armacost, *Defending America:The Case for Limited National Missile Defense* (Washington: Brookings Institution Press, 2001). The book is among the strongest arguments available for a limited missile defence system, with an excellent assessment of the costs and benefits of alterna-

tive platforms. See also Richard Butler, *Fatal Choice: Nuclear Weapons and the Illusion of Missile Defense* (Boulder, CO: Westview Press, 2001). Butler's critique of BMD also includes a comprehensive survey of the nature of the nuclear weapons threat on the planet and the implications for global security. Butler focuses on the destabilizing potential of BMD but also acknowledges limitation of the potential for abolition – two competing facts that create enormous problems for policy officials. The author goes on to develop a strong argument for abolition based on the need to create additional barriers to terrorist access to WMD weapons. Roger Handberg, *Ballistic Missile Defense and the Future of American Security: Agendas, Perceptions, Technology, and Policy* (New York: Praeger Publishers, 2002) is an excellent overview of who in the U.S. government advocates BMD and some of the economic/constituency reasons why BMD makes sense. The book goes well beyond the typical assessment of costs and risks to explore the long-term implications for U.S. foreign and security policy. See also Iain Smith, *The European Case for Missile Defense* (Washington: The Heritage Foundation, March 2002). Melvin A. Goodman, Gerald E. Marsh, and Craig R. Eisendrath, *The Phantom Defense: America's Pursuit of the Star Wars Illusion* (New York: Praeger Publisher, 2001) is a strong critique of BMD deployment plans under George W. Bush and a good overview of the arguments assessed in this book. Among the key points raised are the automatic arms race and that the technology is bound to fail, especially as it relates to identifying countermeasures. Also covered are arguments explaining BMD as the result of economic power and control of large defence contractors.

Other excellent books on missile defence include Robert L. Pfaltzgraff, ed., *Security Strategy and Missile Defense*, a special report produced for the Institute for Foreign Policy Analysis (Washington, April 1998); Erin V. Causewell, ed., *National Missile Defense: Issues and Developments* (Nova Science Publishers, February 2002); Ashton Carter and David Schwartz, *Ballistic Missile Defense* (Washington: Brookings Institution Press, 1984); Michael J. Mazarr, *Missile Defenses and Asian-Pacific Security* (London: Palgrave, 1989); Michael Swaine, Rachael Swanger, and Takashi Kawakami, *Japan and Ballistic Missile Defense* (Washington: Rand Corporation, 2001) addresses the underlying economic and security motivations, in 1998–99, for Japan to begin exploring ways to develop and deploy BMD through joint U.S.-Japan programs; Gordon Mitchell, *Strategic Deception: Rhetoric, Science, and Politics in Missile Defense Advocacy* (East Lansing: Michigan State University Press, 2000) offers another attempt to uncover the underlying economic motivations for renewed interest in BMD, focusing on the pressure from defence contractors to push for hawkish policies on rogue states. BMD represents the latest and most successful effort to create new threats in a post–cold war environment. They refer to this as strategic deception. See also James Anderson, *America at Risk: The Citizen's Guide to Missile Defense* (Washington: Heritage Foundation, April 1999), which makes the case that the threat to the United States from rogue states is anything but fabricated; Ernest Yanarella, *Missile Defense Controversy: Technology in Search of a Mission* (Lexington: University Press of Kentucky, October 2002); The Heritage Foundation's Commission on Missile Defense, 'Defending America: A Plan to Meet the Urgent Missile Threat' (Washington: Heritage Foundation, March 1999); Scott Sagan and Kenneth Waltz, *The Spread of Nuclear Weapons: A Debate Renewed* (New York: W.W. Norton and Co., 2001) is perhaps one of the best overviews of the divide between pessimists and optimists regarding the proliferation of nuclear weapons, focusing on the logic and evidence presented by both sides of this debate; Keith Payne, *Missile Defense in the 21st Century: Protection against Limited Threats* (Washington: National Institute for Public Policy, July 1991); David Deboon, *Ballistic Missile Defense in the Post–Cold War Era* (Westport, CT: Westview Press, July 1995); K. Scott McMahon, *Pursuit of the Shield: The U.S. Quest for Limited Ballistic Missile Defense* (New York: Rowman & Littlefield Publishing, 1997); E. Larson, *A New Methodology for Assessing Multi-Layer Missile Defense Options* (Washington: Rand Corporation, 1994); Helen Caldicott, *The Coming Nuclear War: Manhattan Project II, The National Missile Defense, and How to Avert Catastrophe* (New York: Free Press, 2001).

2 Representative Richard A. Gephardt, Letter from Office of Representative Richard A. Gephardt, 1 May 2001, available at http://www.acronym.org .uk/bush1.htm.

3 'Misrepresenting the ABM Treaty,' *New York Times*, 15 June 2001.

4 Relevant charts can be obtained from the Natural Resources Defense Council Web page at http://www.nrdc.org/nuclear/nudb/datainx.asp.

5 Andrew Coyne, 'Taking Potshots at the Missile Shield,' *National Post* (Toronto), 26 February 2001.

6 The following collection of articles and books summarize these WMD proliferation patterns: Zachary S. Davis and Benjamin Frankel, eds., *The Proliferation Puzzle: Why Nuclear Weapons Spread and What Results* (London: Frank Cass, 1993); William C. Potter and Harlan W. Jencks, eds., *The International Missile Bazaar: The New Suppliers' Network* (London: Westview Press, 1994); Rensselaer W. Lee, *Smuggling Armageddon: The Nuclear Black Market in the Former Soviet Union and Europe* (New York: St. Martin's Press,

1998); Carl Ungerer and Marianne Hanson, *The Politics of Nuclear Non-Proliferation* (London: Allen & Unwin, 2002); Martin Van Crevald, *Nuclear Proliferation and the Future of Conflict* (New York: Free Press, 1993); Leonard Spector and Mark G. McDonough, *Tracking Nuclear Proliferation: A Guide in Maps and Charts* (Washington: Carnegie Endowment for International Peace, 1995); Ronald J. Bee, *Nuclear Proliferation: The Post–Cold War Challenge* (New York: Foreign Policy Association, 1995); Robert Blackwill and Albert Carnesale, eds., *New Nuclear Nations: Consequences for U.S. Policy* (New York: Council on Foreign Relations Press, 1993); Chritoph Bluth, *Arms Control and Proliferation: Russia and International Security After the Cold War* (London: Brassey's, 1996); Peter Clausen, *Nonproliferation and the National Interest: America's Response to the Spread of Nuclear Weapons* (New York: HarperCollins College Publishers, 1993).

7 For additional details about Russia's SS-X-27 missile program, see http://www.softwar.net/ss27.html. According to the report, 'The first deployment was reported to be in a SS-19 silo complex located at Tatishchevo in January of 1998 ... The mobile Russian SS-27 also raises serious proliferation questions since the Moscow Institute of Thermal Technology is providing the SS-27 design to China. China intends to produce the TOPOL-M missile under the designation "Dong Feng" (East Wind) DF-41. The DF-41 is expected to be deployed with Chinese manufactured nuclear warheads also designed with the aid of U.S. super computers.'

8 For a more detailed description of the Russian position on these issues, see http://www.armscontrol.ru/start/publications/kapralov020601.htm and http://www.acronym.org.uk/44abm.htm.

9 Frank J. Gaffney Jr., 'Self-Deterred from Defending the U.S.?' *Washington Times*, 7 March 2001.

10 Keith B. Payne, Yuri Chkanikov, and Andrei Shoumikhin, 'A "Grand Compromise" with Russia on National Missile Defense?' *Defense News* (Fairfax, VA: National Institute for Public Policy, 8 May 2000). The authors go on to argue that 'the ABM Treaty is considered valuable not because Moscow actually views the treaty itself as strategically significant, but because it now provides significant leverage over Washington.'

11 Http://www.gsinstitute.org/laws.pdf.

12 For a detailed analysis of China's evolving nuclear policy, see Brad Roberts, Robert A. Manning, and Ronald N. Montaperto, 'China: The Forgotten Nuclear Power,' *Foreign Affairs* 79, no. 4 (2000). For an interesting discussion of China's position on bombing or invading Taiwan, see http://www.washingtonpost.com/wp-dyn/articles/A7467–2000Jul8.html.

13 Dingli Shen, 'China's Concern over National Missile Defence,' Waging

Peace.org, 2001, available at http://www.wagingpeace.org/articles/
00.06/0006ShenChinasconcern.html.

14 Robert Kagan, 'A Real Case for Missile Defense,' *Washington Post*, 21 May
2001, B07.

15 Testimony of Dr John Steinbruner, director of the Center for International
Security Studies at the University of Maryland, before the Standing Committee on National Defence and Veterans Affairs / Comité Permanent de
la Défense Nationale et des Anciens Combattants, Canada. Evidence,
recorded by electronic apparatus, 29 February 2000; available at http://
www.parl.gc.ca/InfoComDoc/36/2/NDVA/Meetings/Evidence/
ndvaev19-e.htm.

16 Senator Thomas Daschle, comments to the press, released by his office,
1 May 2001.

17 Joseph A. Bosco, 'Has China Turned into a Frankenstein?' *Los Angeles
Times*, 5 March 2001.

18 Ibid.

19 David Mutimer, 'Good Grief!: The Politics of Debating National Missile
Defence,' *International Journal* 56, no. 2 (2001): 330–46.

20 Fareed Zakaria, 'Misapprehensions about Missile Defense,' *Washington
Post*, 7 May 2001, A19.

21 The overview is from William Broad, 'A Missile Defense with Limits: The
ABC's of the Clinton Plan,' *New York Times*, 30 June 2000, available at
www10.nytimes.com/ library/world/Americas /063000missile-plan
.html. For additional details of BMD architecture, refer to http://
www.acq.osd.mil/bmdo/bmdolink/html.

22 The detailed description of BMD technology is taken from Peter Pae, 'Kill
Vehicle Scores a Hit with Proponents of Missile Defense Weapons: The
Pentagon Says the Successful Tests May Restore Credibility to the Pro-
gram,' *Los Angeles Times*, 26 March 2002.

23 Senator John Kerry, 'New Era Not Here Yet, Say Critics,' *Washington Post*,
2 May 2001.

24 David Warren, 'Shooting Down the Criticisms of Bush's Shield: Common
Arguments against the U.S. Missile Defence Plan, and Why They'll Never
Work,' *Ottawa Citizen*, 3 May 2001.

25 Burton Richter, 'It Doesn't Take Rocket Science,' *Washington Post*, 23 July
2000, B02, available at http://www.washingtonpost.com/wp-dyn/
articles/A25940–2000Jul22.html.

26 James Hackett, 'Missile Defense: Deploy When?' *Washington Times*,
6 March 2003.

27 Statements by Lt. Gen. Ronald T. Kadish, USAF, Director, Ballistic Missile

Defense Organization. Testimony before House Subcommittee on National Security, Veterans Affairs, and International Relations Committee on Government Reform, 8 September 2000, available at http://www.acq.osd.mil/bmdo/bmdolink/html/ kadish8sep00.html.

28 Ibid. Provides several additional illustrations from the U.S. space program: 'A series of launch mishaps occurred in the 1980s and 1990s involving several of America's *operational* space launch vehicles. Between 1984 and 1987, catastrophic failures and mission-ending glitches in our Atlas, Titan, Delta, and space shuttle launchers destroyed or rendered useless critical satellite payloads for enhancing national communications, intelligence-gathering, and weather-monitoring missions. The tragic loss of the Challenger and its crew in 1986 caused the entire shuttle fleet to be grounded for many months thereafter. Indeed, for much of 1986, as a result of these failures, the United States lost its ability to place heavy objects in orbit.'

29 Ibid.

30 Defense Secretary Donald Rumsfeld, transcript, remarks on missile defense (Washington: U.S. Department of State, 2 May 2001).

31 A brief history of BMD successes and failures follows. 'The first intercept test of the GMD system research and development program, on October 3, 1999, resulted in the successful intercept of a ballistic missile target. The second test, on January 19, 2000, missed an intercept because of a clogged cooling pipe on the "kill vehicle," but officials said the launch did provide a good test of the integrated system of elements. The third test, on July 8, 2000, missed an intercept because of unsuccessful separation of the "kill vehicle" from the booster rocket. Both the fourth test, on July 14, 2001, and the December test successfully intercepted ballistic missile targets.' From http://europe.cnn.com/2002/US/03/15/missile.defense.test/.

Out of six tests since 1999, 15 March 2002 was the fourth to be successful in U.S. efforts to develop a shield. 'The Missile Defense Agency (MDA) announced today it has successfully completed a test involving a planned intercept of an intercontinental ballistic missile target. The test took place over the central Pacific Ocean. A modified Minuteman intercontinental ballistic missile (ICBM) target vehicle was launched from Vandenberg Air Force Base, Calif., at 9:11 p.m. EST, and a prototype interceptor was launched approximately 20 minutes later and 4,800 miles away from the Ronald Reagan Missile Site, Kwajalein Atoll, in the Republic of the Marshall Islands. The intercept took place approximately 10 minutes after the interceptor was launched, at an altitude in excess of 140 miles above the earth and during the midcourse phase of the target warhead's flight. This was the fourth successful intercept for the Ground-based Midcourse

Defense (GMD) Segment, formerly known as National Missile Defense ...
The test successfully demonstrated exoatmospheric kill vehicle (EKV)
flight performance and "hit to kill" technology to intercept and destroy a
long-range ballistic missile target. In addition to the EKV locating, tracking, and intercepting the target resulting in its destruction using only the
body-to-body impact, this test also demonstrated the ability of system
elements to work together as an integrated system.' From http://www
.cnn.com/ 2002/US/03/16/ missile. defense.test/. See also http://
www.dod.gov/news/Mar2002/ b03152002_ bt127–02.html.

On 14 June 2002, a ship-based missile test succeeded, bringing the count
to five successes out of seven attempts. 'The United States has successfully
destroyed a missile in space with a rocket fired from a Navy ship, hours
after a treaty with Russia ending a ban on missile defence systems came
into effect. Pentagon officials said the exercise showed an incoming missile
could be intercepted by a rocket guided by a warship's radar. The test
gave an important boost to President George W. Bush's plans to build a
protective shield against a foreign missile attack. But with five successful
missile tests in a row, the Pentagon is determined to push ahead with its
plans. Work will begin next weekend on construction of six underground
silos for missile interceptors, prohibited while the ABM Treaty was in
force. Military officials say a rudimentary missile defence system should
be in place over Alaska by the year 2004.' From http://news.bbc.co.uk/1/
hi/world/americas/2044289.stm.
32 The calculation is relatively straightforward – simply add the differences
after allowing for a 71 per cent success rate for each of the four interceptors:
Interceptor 1 = 71% of 100% (100 – 71 = 29% failure rate remaining) +
Interceptor 2 = 71% of 29% (29 – 20.6 = 8.4% failure rate remaining) +
Interceptor 3 = 71% of 8.4% (8.4 – 5.96 = 2.44% failure rate remaining) +
Interceptor 4 = 71% of 2.44% (2.44 – 1.73 = 0.71% failure rate remaining)
   If the remaining failure rate after four interceptors is 0.71 per cent, then
the corresponding success rate is 99.29 per cent.
33 The overview is from Broad, 'A Missile Defense with Limits.'
34 Quoted in ibid.
35 Consider, for example, the hundreds of separate elements in a typical
BMD testing and evaluation program, only a few of which are listed here:
boost-phase missile defense, airborne lasers, kinetic energy kills, ground-
based testing, sea-based midcourse, terminal-phase defense, theatre high-
altitude area defense (THAAD), PATRIOT advanced capability-3, medium
extended air defense systems, technology segment sensors, and so on.

As Warren correctly points out in 'Shooting Down the Criticisms of Bush's Shield,' '[T]he problem is making what works in theory work in practice. It is the standard technological problem, and all invention starts this way. Anything that truly works in theory can be made to work in practice, given sufficient time, money and desire. In the drudgery-laden real world you have to plug away. Take the Patriot system, for example, which enjoyed a good press during the Gulf War, then a bad press slightly after. It did half-work, even 10 years ago. It kind of intercepted Iraqi missiles, but a little late; and then it didn't really do what it was supposed to do at the big moment. Today, I gather, it kind of works a lot better, getting there almost every time and very often doing its thing on arrival. Patience, patience: we are making progress. And remember, we have a little time. The Americans' best estimates are five or more years before there's a serious threat of intercontinental ballistics from little countries. It makes sense to use this time.'

36 James Hackett, 'Missile Defense Safely on Track,' *Washington Times*, 19 December 2002.

37 Ibid. The Pentagon's integrated system test 'included the use of satellite-based missile warning, ground-based early warning radar, the new X-band radar at Kwajalein Atoll in the Pacific, and the battle management facilities at Kwajalein and Colorado Springs ... The SPY-1 radar of an Aegis cruiser, the USS Lake Erie, also successfully tracked the target missile. And in this test there was a new element – the Boeing 747 being outfitted as the first airborne-laser system succeeded in locating and tracking the target missile in the boost phase of its flight with what a senior defense official described as spectacular results ... The test used radars and other sensors at multiple locations on the ground, in space, in the air and on ships at sea. Much of this was banned by the ABM treaty, which had made such a test impossible. The treaty prohibited using either the ship-based Aegis radar or the airborne laser's sensors to support a defense of the nation. Only now, in the wake of U.S. withdrawal from the treaty, can all these capabilities be brought together in a rational and cost-effective way.'

38 For more on this point, see James T. Hackett, 'The Countermeasures Debate,' *Washington Post*, 9 August 2000. 'Missile defense technology is not static. It is constantly being improved with faster computers and new components. The Pentagon is well aware of the possible use of countermeasures and is working on some two dozen defenses against them. But that work of necessity is highly classified, making it difficult to defend against irresponsible charges. It is ridiculous to suggest a developing country could design missiles more advanced than this country can design de-

fenses. Besides, the primary goal of missile defense is deterrence – to discourage adversaries from acquiring missiles and using them to threaten or blackmail. It is the absence of defenses that encourages missile proliferation and blackmail. The solution is not to dream up phony reasons why defenses may not work, but to do the hard engineering needed to make sure they do.'

39 Caleb Carr, 'The Myth of a Perfect Defense,' *New York Times*, 7 August 2001.

40 For a comprehensive review of the costs and benefits of alternative BMD platforms, see Michael E. O'Hanlon, 'Beyond Missile Defense: Countering Terrorism and Weapons of Mass Destruction,' Policy Brief #86 (Washington: Brookings Institution Press, August 2001), available at www.brookings/comm/policybriefs/pb86.pdf.

41 Jack Spencer, 'Debating the Dollars for Missile Defense,' *Washington Times*, 20 August 2001.

42 Http://www.cdi.org/nuclear/nukecost.html.

43 As Bruce Herschensohn notes, the fact that more Americans die each year of heart disease does not imply we should ignore, for example, AIDS research – 'History Hasn't Stood Still for ABM Treaty,' *Los Angeles Times*, 18 June 2001.

44 Warren, 'Shooting down the Criticisms of Bush's Shield.'

45 Michael Wallace, 'Ballistic Missile Defense: The View from the Cheap Seats,' 2001, available at http://www.wagingpeace.org/articles/bmd/Wallace_BMD_View_from_cheap_seats.htm.

46 Center for Arms Control and Non-Proliferation, Washington, available at http://www.armscontrolcenter.org/nmd/md101.html.

47 Http://www.genevabriefingbook.com/chapters/uscontrib.html.

48 Roger Boyes, 'German Cities in "Firing Line" of Rogue States,' *Times* (London), 16 February 2001.

49 Warren, 'Shooting Down the Criticisms of Bush's Shield': 'It is argued more knowingly that the Pentagon can be counted on to gum up almost anything it is instructed to touch, for it is a large, wayward public bureaucracy. I'm sure my reader is with me in the desire to privatize as much of the missile shield as possible – money for results, and all that. But in the meanwhile, one must use what one has, and breathe as much fire down Pentagon necks as possible. The Bush administration includes a pretty fair selection of tried-and-true old action-tested fire-breathers; people who don't want to be made to look stupid, and will do what it takes to succeed. Trust them to get the bureaucrats hopping.'

50 David Mutimer, 'The Politics of Debating NMD,' *International Journal 56*, no. 2 (2001).

51 Richard Gwyn, 'Bush Fear of "Rogue" States Is Laughable,' *Toronto Star*, 18 February 2001.
52 *The National Security Strategy of the United States of America* (Washington: The White House, September 2002), 15.
53 Scott Petersen, 'Iran's Nuclear Challenge: Deter, Not Antagonize,' *Christian Science Monitor*, 21 February 2002, available at http://www.csmonitor.com/.

5. WMD Proliferation: The Case against Multilateral Arms Control and Disarmament

1 David Mutimer, 'The Politics of Debating NMD,' *International Journal* 56, no. 2 (2001).
2 In fact, the only example of real progress on nuclear disarmament in the past half century was a product of unilateral moves by the Bush administration to exchange the U.S. withdrawal from the ABM Treaty for deep cuts in nuclear forces to between 1700 and 2200.
3 On the 'utter bankruptcy' of multilateral arms control, see Charles Krauthammer, 'The Real New World Order: The American and the Islamic Challenge,' *Weekly Standard* no. 7, 12 November 2001.
4 Many of the most important arms control treaties are bilateral, not multilateral, agreements. For example, the Strategic Offensive Reductions Treaty; ABM Treaty; Strategic Arms Limitation Talks (SALT I and II); START I, II, III, and IV; Intermediate-Range Nuclear Forces Treaty.
5 For a summary of evidence compiled through public access Web pages on the nature and extent of WMD proliferation threats, see the WMD proliferation table compiled by Centre for Foreign Policy Studies, Halifax, Nova Scotia, available at http://www.dal.ca/~centre/nmdchart.pdf.
6 The extent and nature of threats associated with WMD proliferation are covered in Zachary S. Davis and Benjamin Frankel, eds., *The Proliferation Puzzle: Why Nuclear Weapons Spread and What Results* (London: Frank Cass, 1993). See also William C. Potter and Harlan W. Jencks, eds., *The International Missile Bazaar: The New Suppliers' Network* (London: Westview Press, 1994); Rensselaer W. Lee, *Smuggling Armageddon: The Nuclear Black Market in the Former Soviet Union and Europe* (New York: St. Martin's Press, 1998); Carl Ungerer and Marianne Hanson, *The Politics of Nuclear Non-Proliferation* (London: Allen & Unwin, 2002); Martin Van Crevald, *Nuclear Proliferation and the Future of Conflict* (New York: Free Press, 1993); Leonard Spector and Mark G. McDonough, *Tracking Nuclear Proliferation: A Guide in Maps and Charts* (Washington: Carnegie Endowment for International Peace,

1995); Ronald J. Bee, *Nuclear Proliferation: The Post-Cold War Challenge* (New York: Foreign Policy Association, 1995); Robert Blackwill and Albert Carnesale, eds., *New Nuclear Nations: Consequences for U.S. Policy* (New York: Council on Foreign Relations Press, 1993); Chritoph Bluth, *Arms Control and Proliferation: Russia and International Security after the Cold War* (London: Brassey's, 1996); Peter Clausen, *Nonproliferation and the National Interest: America's Response to the Spread of Nuclear Weapons* (New York: HarperCollins College Publishers, 1993).

7 Gordon Giffin, 'Washington: When It Comes to Missile Defence, You Can Trust Us,' *Globe and Mail* (Toronto), 13 March 2001.

8 David Warren, 'Shooting Down the Criticisms of Bush's Shield: Common Arguments against the U.S. Missile Defence Plan, and Why They'll Never Work,' *Ottawa Citizen*, 3 May 2001.

9 Ian Traynor, 'West Scours Georgia for Nuclear Trash,' *Guardian* (Manchester), 27 March 2002.

10 Ann Scott Tyson, 'US, China Cautiously Rekindle Military Ties,' *Christian Science Monitor*, 20 February 2002.

11 On nuclear theft and related dangers, see www.stimson.org/policy/ nucleardangers.htm; on proliferation by rogue states, see Debra Mohanty, 'Defence Industries in a Changing World: Trends and Options,' *Strategic Analysis*, January 2000; on emerging regional threats, see the Strategic Assessment Reports, Briefings, and Books, available at www.csis.org/ stratassessment. For an arms control optimist's view, see Joseph Cirincione and Frank von Hippel, eds., 'The Last 15 Minutes: Ballistic Missile Defence in Perspective' (Washington: Coalition to Reduce Nuclear Danger, 1996).

12 'The Paradoxes of Post–Cold War US Defense Policy: An Agenda for 2001,' Quadrennial Defense Review, Project on Defense Alternatives, 'Briefing Memo #18,' 5 February 2001, available at www.comw.org/pda/ 0102bmemo18.html.

13 Glenn Kessler, 'No Support for Strikes against N. Korea,' *Washington Post*, 2 January 2003, A10. According to the report, 'Clinton and his top advisers were discussing options for deploying troops and military equipment to South Korea when (former president Jimmy) Carter called to say that North Korean leader Kim Il Sung had agreed to the freeze ... The resulting deal, known as the Agreed Framework, committed North Korea to shuttering its nuclear reactor in exchange for regular fuel oil shipments and the construction of two light-water nuclear reactors.'

14 Daniel Schorr, 'Preempting Preemptive Action,' *Christian Science Monitor*, 3 January 2003, available at http://www.csmonitor.com/2003/0103/ p09s02–cods.html.

15 John C. Hopkins and Steven A. Maaranen, 'Nuclear Weapons in Post–Cold War Deterrence,' appendix E in *Post Cold War Conflict Deterrence* (Washington: National Academy Press, 1997).

16 'A Pain in the Pyongyang,' *Washington Times*, 19 June 2001.

17 John Steinbruner, Testimony to Parliamentary committee on security, Ottawa, Canada, 2001, quoted in www.Ploughshares.com.

18 Walter B. Slocombe, 'Preemptive Military Action and the Legitimate Use of Force: An American Perspective,' prepared for meeting at the CEPS/IISS European Security Forum, Brussels, 13 January 2003.

19 Http://www.nautilus.org/papers/energy/ModernizingAF.pdf and http://www.fes.or.kr/K_Unification/U-paper7.htm.

20 David E. Sanger, 'Administration Divided Over North Korea,' *New York Times*, 20 April 2003.

21 'North Korea's Nukes,' *National Post* (Toronto), 28 April 2003.

22 Sanger, 'Administration Divided over North Korea.'

23 James E. Goodby, 'Try to Engage with Pyongyang,' *International Herald Tribune*, 6 January 2003. The author is affiliated with the Center for Northeast Asian Policy Studies at the Brookings Institution and was chief negotiator of Nunn-Lugar cooperative threat reduction agreements during the Clinton administration.

24 'Fidel Castro's Friends in Ottawa,' *National Post* (Toronto), 19 April 2003.

25 Andrew Mack, 'North Korea's Matches-Our Powder Keg,' *Globe and Mail* (Toronto), 15 January 2003, A15.

26 Keith B. Payne, 'Deterrence and US Strategic Force Requirements after the Cold War,' *Comparative Strategy* 9 (July–September 1990): 269–82; and Keith B. Payne and Lawrence Falk, 'Deterrence: Gambling on Perfection,' *Strategic Review*, Winter 1989: 25–40.

27 Keith B. Payne, 'Rational Requirements for U.S. Nuclear Forces and Arms Control: Executive Report,' *Comparative Strategy* 20 (May 2001): 105–28. See also the discussion of the evolution of nuclear deterrence theory after the cold war in James Dougherty and Robert Pfaltzgraff, eds., *Contending Theories of International Relations*, 5th ed. (New York: Longman, 2001), 374–86.

28 *The National Security Strategy of the United States of America* (Washington: The White House, September 2002), 14.

29 Http://www.istc.ru/istc/website.nsf/fm/z10ThreatRedProgr.

30 Http://www.cdi.org/nuclear/nukecost.html.

31 Http://www.mbe.doe.gov/budget/03budget/content/defnn/nuclnonp.pdf.

32 Http://www.stimson.org/fopo/pdf/FollowingtheMoney.pdf.

33 Shirley Williams, 'An Offensive Defence,' *Guardian* (Manchester), 24 February 2001. See also Frank J. Gaffney Jr., 'With Friends like These ...,' *Washington Times*, 5 June 2001. Gaffney interprets European views as being driven almost entirely by ideology. 'The leaders of these and most other governments in Western Europe (with the notable exception of the newly elected Berlusconi administration in Italy, which supports missile defenses) are individuals who cut their political teeth demonstrating their opposition to U.S. military power, the NATO alliance and America more generally. Germany´s Prime Minister Gerhard Schroeder and his Green Party foreign minister, Joschka Fischer, are pedigreed leftists who were active in the pro-Soviet European left´s campaign in the early 1980s aimed at preventing deployment of U.S. intermediate-range nuclear missiles in five allied countries. Ditto France´s Socialist premier, Lionel Jospin, and, for that matter Britain´s Tony Blair and his foreign minister, Robin Cook. Even the present and immediate past secretaries general of NATO, Britain´s George Robertson and Spain´s Javier Solana respectively, were determined opponents of the U.S. leadership of the Atlantic Alliance in the face of manifest Soviet threats.'

34 David Warren, 'Up with Your Missile Shield,' *Ottawa Citizen*, 22 February 2001.

35 Ben Barber, 'NATO Drops its Support of ABM Pact,' *Washington Times*, 30 May 2001.

36 David R. Sands, 'Europe Warms to Missile Defense,' *Washington Times*, 6 February 2001.

37 Helle Bering, 'Serious about Missile Defense,' *Washington Post*, 7 February 2001.

38 Keith B. Payne, 'Proliferation, Deterrence, Stability and Missile Defence,' *Comparative Strategy* 13 (January 1994); Lewis A. Dunn, 'Deterring the New Nuclear Powers,' *Washington Quarterly* 17 (Winter 1994); Richard K. Betts, 'The Concept of Deterrence in the Postwar Era,' *Security Studies*, Autumn 1991.

39 Sam Nunn and Bruce Blair, 'From Nuclear Deterrence to Mutual Safety,' *Washington Post*, 30 June 1997, 22.

40 For an excellent assessment of the risks considered by U.S. policy makers, see Keith Payne and Lawrence Falk, 'Deterrence without Defence: Gambling on Perfection,' *Strategic Review*, Winter 1989: 25–40.

41 Payne, 'Rational Requirements.'

42 Giffin, 'Washington: When It Comes to Missile Defence ...'

43 Bering, 'Serious about Missile Defense.'

44 Warren, 'Up with Your Missile Shield.'

6. The Inevitability of Terrorism, and American Unilateralism: Security Trumps Economics

1 Charles Krauthammer, 'Bush Doctrine,' *Washington Post*, 24 January 2003.
2 *The National Security Strategy of the United States of America* (Washington: The White House, September 2002), 13.
3 Joseph Nye, 'Divided We War,' *Globe and Mail* (Toronto), 24 March 2003, A15.
4 Frank Cilluffo, Joseph J. Collins, Arnaud de Borchgrave, Daniel Gouré, and Michael Horowitz, *Defending America in the 21st Century: New Challenges, New Organizations and New Policies* (Executive Summary). Washington: Center for Strategic and International Studies (http://www.911investigations.net/IMG/pdf/doc-340.pdf), 1–28. See also Frank Cilluffo and Bruce Berkowitz, eds., *Cybercrime, Cyberterrorism, and Cyberwarfare: Averting an Electronic Waterloo* (Washington: CSIS, 1998), 1–72.
5 Ibid.
6 Ibid.
7 Ibid.
8 Ibid, 7.
9 Dennis Pluchinsky, 'Deadly Puzzle of Terrorism,' *Washington Times*, 11 September 2002.
10 Ibid.
11 Louis Freeh quoted in 'Cyberspace Next Target for Terrorists: Experts,' CBC News Online, 30 April 2002.
12 Ron Fournier, 'Ridge Warns of Al Qaeda Threat,' *Boston Globe*, 30 April 2002, A15.
13 David A. Lake, 'Rational Extremism: Understanding Terrorism in the Twenty-First Century,' *International Organization*, Spring 2002: 15–19.
14 For a detailed treatment of the implications of this thesis, see Jack Levy, 'An Introduction to Prospect Theory in International Relations,' *Political Psychology* 13, no. 2 (1992): 171–86.
15 'US Citizens Likely to Give Government More Power to Act Unilaterally,' *Professionals for Cooperation* 5 (2002). Moscow: Russian-American Academic Exchanges Alumni Association.
16 *The National Security Strategy of the United States of America*, 15.
17 Wolfgang K.H. Panofsky, 'The Remaining Unique Role,' appendix D in *Post–Cold War Conflict Deterrence* (Washington: National Academy Press, 1997).
18 Paul H. Nitze and J.H. McCall, 'Contemporary Strategic Deterrence and

Precision-Guided Munitions,' chapter 5 in *Post–Cold War Conflict Deterrence*.

19 Andrew J. Goodpaster, C. Richard Nelson, and Seymour J. Deitchman, 'Deterrence: An Overview,' in *Post–Cold War Conflict Deterrence*. 'Devising relationships with many of these power centers remains a dynamic and changeable process. The problem of deterrence is thus more complex than it was, and the approaches to situations requiring deterrent actions must be even more measured and flexible.'

20 Ibid. With respect to developing appropriate deterrence capabilities, 'policy makers must carefully determine just what combination of deterrence capabilities – the visible and demonstrable power to punish serious violations of the norms of international behaviour, deny success to aggression, impose heavy costs and losses on the aggressor – should be created and sustained to provide a high likelihood of deterrence against a wide variety of potential threats and risks.'

21 Fred Ikle, Albert Wohlstetter, et al., *Discriminate Deterrence: Report of the Commission on Integrated Long-Term Strategy* (Washington: U.S. Government Printing Office for the Department of Defense, 1988).

22 They include, but are not limited to: (1) characteristics of specific adversaries are essential for determining locations, types, and number of targets; (2) targeting strategies determine the extent to which counter-value or counter-force targeting will be used (counter-force strategies require more targets); (3) survivability and vulnerability vary from context to context and dictate whether multiple platforms or basing modes will be required to counter surprise attacks; (4) both active (NMD/BMD) and passive (mobility, dispersal, redundancy, deception, concealment, hardening, and so on) remain important, but will vary from case to case; (5) intelligence about targets also affects the size of an arsenal (less intelligence means more targets), but intelligence gathering becomes more difficult as the number of rivals proliferates; (6) pre-launch survivability, system reliability, penetration capability, delivery accuracy, and so on all have an impact on numbers (for example, the more reliable the systems, the fewer weapons required).

23 Dana Milbank and Sharon LaFraniere, 'U.S., Russia Agree to Arms Pact: Nuclear Arsenals on Both Sides to Be Reduced by Two-Thirds,' *Washington Post*, 14 May 2002, A01. 'Honoring the Bush administration's desire for future flexibility, it contains no requirement to destroy warheads that are taken out of service. It puts no prohibition on the U.S. plan to build a missile defense system. The pact's expiration in 10 years allows either side to return to any level it desires, and even before the 10-year expiration it

allows the ability to pull out with 90 days' notice. In exchange, Bush granted only one concession: the notion of even having a treaty. The administration saw no need for a written agreement, and preferred any agreement not to take the form of a treaty requiring Senate ratification. Although the administration met Russia's request, the president did not agree to anything he had not pledged to do unilaterally.'

24 'No Frills Arms Control,' *New York Times*, 14 May 2002. The editorial goes on to point out that '[t]he choice of weapons systems to be cut, and the pace at which they are removed, will be left to each government to decide on its own. Russia will apparently be released from its previous commitment to scrap its land-based multiple-warhead missiles, long considered its most threatening weapons.'

25 Walter B. Slocombe, 'Preemptive Military Action and the Legitimate Use of Force: An American Perspective,' prepared for a meeting at the CEPS/IISS European Security Forum, Brussels, 13 January 2003.

26 Ibid.

27 Kenneth Janda, 'Global Terrorism, Domestic Order, and the United States,' available at http://uspolitics.org/student/terrorism/terrorism4.htm, in Anatoly Kulik, ed., *Professionals for Cooperation*, vol. 5 (Moscow: Russian-American Academic Exchanges Alumni Association, forthcoming), available at http://www.prof. msu.ru/eng/index.html.

28 Ibid.

7. The Moral Foundations of Canadian Multilateralism: Distinction Trumps Security

1 Mark Proudman, 'Soft Power Meets Hard: Canadian and European Responses to US Power,' in Norman Hilmer and Maureen Appel Molot, eds., *Canada Among Nations* (London: Oxford University Press, 2003).

2 Richard Gwyn, 'Our Foreign Policy Making Us Invisible,' *Toronto Star*, 23 February 2003.

3 Robert Kagan, 'The Benevolent Empire,' *Foreign Policy* 12, no. 111 (1998): 24.

4 Alan Krueger and Jitka Maleckova, 'Education, Poverty, Political Violence and Terrorism: Is There a Causal Connection' (2002), unpublished manuscript.

5 Charles Russell and Bowman Miller, 'Profile of a Terrorists,' reprinted in *Perspectives on Terrorism* (Wilmington: Scholarly Resources Inc, 1983), 45–60. Quoted in Krueger and Maleckova, 'Education, Poverty, Political Violence, and Terrorism,' 29.

6 See Anatol Lieven, 'The Roots of Terrorism, and a Strategy against It,'
   Prospect Magazine 68 (October 2001); Daniel Pipes, 'God and Mammon:
   Does Poverty Cause Militant Islam?' *National Interest*, no. 66 (Winter 2001–
   2): 14–21; and Michael Mousseau, 'Market Civilization and Its Clash with
   Terror,' *International Security* 27, no. 3 (2003): 5–29.
7 Thomas Friedman, 'Invade Iraq to Impose Democracy, Not Disarmament,'
   Globe and Mail (Toronto), 21 September 2002.
8 See Jeffrey Simpson, 'If Only Moral Superiority Counted as Foreign Aid,'
   Globe and Mail (Toronto), 21 January 2003, A17. According to Simpson, 'of
   the 22 countries that give foreign aid, Canada ranked 19th in aid as a share
   of gross national product. The dollar volume of our aid was 11th – this for
   a country that brags about being a G8 member and a light unto the developing nations.'
9 Ibid.
10 Ibid.
11 Maureen Appel Molot and Norman Hilmer, 'Diplomacy in Decline' in
   Hilmer and Molot, *Canada among Nations*, 1–33.
12 David Warren, 'Up with Your Missile Shield,' *Ottawa Citizen*, 22 February
   2001.
13 Allan Gotlieb, 'The Chrétien Doctrine,' *Maclean's*, 31 March 2003.
14 Randall Palmer, 'Canada to Push for Special Saddam Criminal Court,'
   Reuters, 27 March 2003.
15 For an excellent overview of the confusion produced by Ottawa's constantly shifting, contradictory, and disjointed policies on Iraq, please see
   Andrew Coyne, 'Canada on Iraq: A Clarification,' *National Post* (Toronto),
   31 March 2003. See also Rex Murphy, 'Canada, Two-faced? It All Depends,'
   Globe and Mail (Toronto), 29 March 2003, A23: 'Reading the Prime Minister
   on this issue over the past five or six months required the skills of a seance
   master and the deductive powers of a Sherlock Holmes. Both John
   McCallum and Bill Graham have had to be corrected in public on the
   government's position. If these two can't keep up, what hope for the rest
   of us, never mind the Americans?'
16 John Ibbitson, 'So How Do We Get Bush to Like Canada?' *Globe and Mail*
   (Toronto), 14 April 2003, A13.
17 Sheldon Alberts, 'Chrétien to Endorse U.S. "Mission,"' *National Post*
   (Toronto), 8 April 2003.
18 Sheldon Alberts, '"Saddam Is Dangerous": Foreign Minister Wishes Allied
   Troops "Godspeed,"' *National Post* (Toronto), 25 March 2003.
19 Ibid.

20 Andrew Coyne, 'PM's Decision Means Moral Free Ride Is Over,' *National Post* (Toronto), 19 March 2003.

21 Kofi Annan quoted in 'DR of Congo: UN investigates reports of massacres in troubled Ituri region,' United Nations News Services, 7 April 2003.

22 Quoted in Sheldon Alberts and Steven Edwards, 'Canada Unable to Help Congo, Martin Says,' *National Post* (Toronto), 17 May 2003.

23 Gotlieb, 'The Chrétien Doctrine.' According to Gotlieb, 'The UN Charter, membership and voting structure have made humanitarian intervention – the prevention of genocide, ethnic cleansing and gross violations of human rights – difficult, if not impossible, to achieve through the UN.'

24 Quoted in Shawn McCarthy and Daniel Leblanc, 'PM Scolds McCallum on Canada's Role In Iraq,' *Globe and Mail* (Toronto), 16 January 2003, A1.

25 Drew Fagan, 'Canada in the World: Heading Back Up?' *Globe and Mail* (Toronto), 15 April 2003, A17.

26 Interview given by Prime Minister Jean Chrétien to ABC's *This Week with George Stephanopoulos*, 9 March 2003, quoted in Prime Minister Speeches, available at http://www.pm.gc.ca/default.asp?language=E&page=newsoom&sub=newsreleases+Doc=thisweek.20030309_e.htm.

27 Bill Curry, 'Canada Has Backed Wars without UN,' *National Post* (Toronto), 9 April 2003. With respect to some of the atrocities committed by Saddam Hussein the author quotes figures from Canadian Alliance MP Jason Kenney: 'In Iraq, 185,000 Kurds, over a quarter of a million Shia Arabs, and countless thousands of other civilians have died as a result of ethnic cleansing or state oppression, and Saddam is responsible both through domestic terror and wars of aggression for the deaths of over a million of his own citizens, leading to the emigration of nearly four million Iraqis.' In contrast, NATO intervened in 1999 after it was estimated that Serb troops executed close to 10,000 ethnic Albanians (the estimate before the war was actually closer to about 2,500) and forced approximately 900,000 Kosovar Albanians into neighbouring countries.

28 'Canada Will Only Go to War with UN Approval,' CTV News, 9 May 2003.

29 John Ward, 'First Gulf War Left Problems Unsolved,' Canadian Press, Ottawa, 9 March 2003, available at http://cnews.canoe.ca/CNEWS/Canada/2003/03/09/39880–cp.html.

30 Jean Chrétien quoted in 'Conflict Narrowly Averted in Iraq,' *Disarmament Diplomacy*, no. 23 (February 1998).

31 Jean Chrétien, 'The Prime Minister's Position on the Iraq Crisis: 1998,' quoted in *National Network News* 10, no. 1 2003: 6–7.

32 Kenneth Katzman, 'Iraq: International Support For U.S. Policy,' Foreign

Affairs and National Defense Division. Congressional Research Service, Library of Congress Number 98–114 F, 19 February 1998.

33 Gotlieb, 'The Chrétien Doctrine.'

34 Jeffrey Simpson, 'Choose Your Side: Puerile or Servile?' *Globe and Mail* (Toronto), 4 April 2003, A15.

35 Carol Goar, 'It's about War, Not Loyalty, Sir,' *Toronto Star*, 28 March 2003.

36 Bill Graham quoted in Paul Wells, 'Minister Lectured on Foreign Policy,' *National Post* (Toronto), 15 February 2003.

37 Walter B. Slocombe, 'Preemptive Military Action and the Legitimate Use of Force: An American Perspective,' prepared for meeting at the CEPS/IISS European Security Forum, Brussels, 13 January 2003.

38 George Jonas, 'The UN Is Its Own Worst Enemy,' *National Post* (Toronto), 19 March 2003.

39 Jean Chrétien, 'Notes for an Address by Prime Minister Jean Chrétien.' Speech to the Chicago Council on Foreign Relations, 13 February 2003.

40 'The War Canada Missed,' editorial in *National Post* (Toronto), 20 March 2003.

41 The results were covered in a *National Post* editorial summarizing comments by Michael Marzolini (CEO of Pollara Inc.) in a speech to the Economic Club of Toronto, 'Canada's Place Is with the U.S.,' *National Post* (Toronto), 27 March 2003. As Marzolini is quoted as saying, 'Canadians don't "want" war,' but 'Canadians do "accept" war, reluctantly and grudgingly.'

42 Andrew Coyne, 'Imagine Where We'd Be with Some Leadership,' *National Post* (Toronto) 9 April 2003. The COMPAS poll included five hundred respondents, April 4–6; accurate within 4.5 per cent nineteen times out of twenty.

43 For an excellent version of this particular argument, see Barry Cooper and Ted Morton, 'Chrétien Has Put Party ahead of Country,' *National Post* (Toronto), 28 March 2003; and Gordon Gibson, 'Why Chrétien Put Us on the Sidelines,' *National Post* (Toronto), 3 April 2003.

44 Quoted in 'Canadian Troops in Iraq Invisible to Chrétien,' editorial in *Vancouver Sun*, 29 March 2003.

45 Quoted in 'Hiding the Troops,' editorial in *Globe and Mail* (Toronto), 2 April 2003, A16.

46 Ibid.

47 Lewis MacKenzie, 'Admit It, We're Engaged in Combat,' *National Post* (Toronto), 4 April 2003.

48 Stephen Thorne, 'Our Ships Not in War, McCallum Says,' *Halifax Herald*, 3 April 2003.

49 Kelly Toughill, 'Canadians Help U.S. Hunt in Gulf,' *Toronto Star*, 2 April 2003.

50 Ibid.

51 Ibid.

52 Quoted in Daniel Leblanc, 'Frigates "Critical" to Guard against Terror, Former Admiral Says,' *Globe and Mail* (Toronto), 21 March 2003, A12.

53 Ibid.

54 MacKenzie, 'Admit It, We're Engaged in Combat.'

55 Coyne, 'Canada on Iraq: A Clarification.'

56 Hans Blix, Transcript of Blix's Remarks, United Nations, 7 March 2003.

57 Ibid.

58 Andrew Coyne, 'The Council Is Too Easily Pleased,' *National Post* (Toronto), 17 February 2003. See also George F. Will, 'Shrinking the U.N.,' *Washington Post*, 20 February 2003, A39.

59 Michael Ignatieff, 'Canada in the Age of Terror: Multilateralism Meets a Moment of Truth,' *Policy Options*, February 2003: 14–18.

8. Recalibrating Canada's Moral and Diplomatic Compass

1 Mark Kingwell, 'What Distinguishes Us from the Americans?' *National Post* (Toronto), 5 March 2003.

2 Jeffrey Simpson, 'If Only Moral Superiority Counted as Foreign Aid,' *Globe and Mail* (Toronto), 21 January 2003, A17.

3 *Canada in the World: Canadian Foreign Policy Review* (Ottawa: DFAIT, 1995). See also 'A Dialogue on Foreign Policy,' 2003, available at www.foreign-policy-dialogue.ca.

4 Christopher Sands, 'Fading Power or Rising Power: 11 September and Lessons from the Section 110 Experience,' in Norman Hilmer and Maureen Appel Molot, eds., *Canada among Nations: A Fading Power* (London: Oxford University Press, 2002), 63–73, 71–2.

5 Bill Dymond and Michael Hart, 'Canada and the Global Challenge: Finding a Place to Stand,' *Border Papers*, no. 180 (Toronto: C.D. Howe Institute, March 2003), 13–14.

6 For a comprehensive overview of terrorist funding and networking in Canada please see Stewart Bell, 'Blood Money: International Terrorists Fundraising in Canada' in Hilmer and Molot, *Canada among Nations*, 172–90.

7 Ward Elcock quoted in 'Never Said Canada a Terrorist Haven: CSIS Boss,' editorial in *Toronto Star*, 29 April 2003.

8 David Frum, 'Quit Hemming and Hawing Over Hezbollah,' *National Post*

(Toronto), 10 December 2003. Frum states that 'for the Chrétien government, the crassest domestic political considerations have regularly taken priority over national security ... And so Mr. Manley has for 11 months opposed a Hezbollah ban ... Even now, we can doubt whether the ban is truly coming – and how much it will mean if it does come.'

9 Mark Steyn, 'Join America? Dream On,' *National Post* (Toronto), 20 January 2003.

10 Joseph Nye, 'The New Rome Meets the New Barbarians: How America Should Wield Its Power,' *Economist*, 23 March 2002.

11 Wendy Dobson, 'Shaping the Future of the North American Economic Space: A Framework for Action,' *C.D. Howe Institute Commentary* 162 (Toronto: C.D. Howe Institute, April 2002): 18.

12 Jeffrey Simpson, 'Timing Is Everything for PM's New York Trip,' *Globe and Mail* (Toronto), 28 September 2001. As Simpson correctly notes, '[R]arely does our government take an initiative vis-à-vis the United States ... Governments have historically preferred to react to pressures, proposals and developments coming from Washington. That way they can pick and choose among responses, trying all the while to protect themselves from a public opinion wary of a government being seen as "too American" ... The continuing Canadian hang-up in bilateral relations, much on display (in the Post September 11 crisis), is a persistent reluctance to take the lead in dealing with the United States, with the result that the Americans tend to take initiatives.'

13 Malcolm Gladwell, 'Safety in the Skies,' *New Yorker*, 1 October 2001. 'The better we are at preventing and solving the crimes before us, the more audacious criminals become. Put alarms and improved locks on cars, and criminals turn to the more dangerous sport of carjacking. Put guards and bulletproof screens in banks, and bank robbery gets taken over by high-tech hackers. In the face of resistance, crime falls in frequency but rises in severity, and few events better illustrate this tradeoff than the hijackings of September 11th ... The contemporary hijacker, in other words, must either be capable of devising a weapon that can get past security or be willing to go down with the plane (or both). Most terrorists have neither the cleverness to meet the first criterion nor the audacity to meet the second, which is why the total number of hijackings has been falling for the past thirty years.'

14 For a discussion of the unintended consequences of fighting the last war (i.e., focusing exclusively on preventing attacks like those that have already occurred), see Barry Rubin, 'Don't Fight the Last War,' *Jerusalem Post*, 28 September 2001.

15 According to Lloyd Skaalen and Migs Turner 'More than six million foreign maritime cargo containers pass through North American ports annually. According to the U.S. Coast Guard (USCG), the contents of less than 3% of these containers are physically inspected. And when they are "inspected," with potentially fraudulent certification, only one end of the container is seen.' See Lloyd Skaalen and Migs Turner, 'Put-up or Shut-up Canada!' *Journal of Homeland Security*, 22 March 2002. For an excellent account of the maritime dimensions of homeland security, see http://ifpafletchercambridge.info/USCGFR.pdf.

16 Ed Struzik, 'Biological Weapons Pose Threat to Canada, U.S., Scientist Says,' *Edmonton Journal*, 11 March 2001.

17 Ibid.

18 See Frank P. Harvey, 'National Missile Defence Revisited, Again: A Response to David Mutimer,' *International Journal*, July 2001; 'Proliferation, Rogue-State Threats and National Missile Defence: Assessing European Concerns and Interests,' *Canadian Military Journal* 1, no. 4 (2001: 69–79; 'The International Politics of National Missile Defense: A Response to The Critics,' *International Journal* 55, no. 4 (2000: 545–66; 'ABCs of BMDs and ABMs,' *Chronicle-Herald* (Halifax), 10 May 2001; 'Politics, Not Technology Explains U.S. Missile Defence Decision,' *Chronicle-Herald* (Halifax), 9 September 2000; 'North Korea: A Rogue by Any Other Name,' *National Post* (Toronto), 29 July 2000; 'Price Is Right for U.S. Missile Program,' *National Post* (Toronto), 22 July 2000; 'Good Defences Make Good Neighbours,' *Globe and Mail* (Toronto), 11 April 2000.

19 David Warren, 'Up with Your Missile Shield,' *Ottawa Citizen*, 22 February 2001.

20 Don MacNamara, BGEN (Ret'd), 'September 11, 2001–September 11, 2002,' *On Track* 7, no. 3 (2002): 11–15.

21 This assumes, of course, that Canada has options and choices in this regard, a luxury Canada appears to have lost as a result of several years of cutbacks in defence spending. For a comprehensive account of the many problems facing the Canadian military, please refer to Fergusson, Harvey, and Huebert, *To Secure a Nation: The Case for a New Defence White Paper*, report prepared for the Council for Canadian Security in the 21st Century (Calgary: Centre for Military and Strategic Studies, 2001).

22 Please refer to examples of relevant publications listed in note 18.

23 John McCallum DGPA Transcripts, Defence Minister Statement in the House of Commons: Debating Opposition Motion on Support for NORAD (29 May 2003), Reference: 03052901: 'Up until today Canada had not expressed an interest in participating in ballistic missile defence and therefore the Americans were going along without us, not in NORAD because

NORAD is binational but rather in Northern Command. But now as of today when we are announcing our interest to enter into discussions with the possibility of participation, we will be suggesting that ballistic missile defence be lodged in NORAD. And I believe very firmly that this will be in Canada's interest, that this represents the continuity that I have described since 1940, the constancy of our joint defence of the continent, the constancy of our binational efforts with the Americans to defend our joint land space but as well the evolution of the nature of that threat, the evolution of technology and hence the evolution of the detailed appropriate response from the time of Nazi Germany through the Cold War through the post-September 11th world where terrorism and other threats have become our main preoccupation.'

24 Lloyd Axworthy and Michael Byers, 'Say No to Missile Defence,' *Globe and Mail* (Toronto), 29 April 2003, A13.
25 Louis Delvoie, 'Multilateralism or Unilateralism: Whither American Foreign Policy,' *Policy Options*, November 2002, 12–14.
26 Andrew Cohen, 'Canadian-American Relations: Does Canada Matter in Washington? Does It Matter If Canada Doesn't Matter?' in Hilmer and Molot, eds., *Canada among Nations*, 34–47, 37.
27 Michael Ignatieff, 'Canada in the Age of Terror: Multilateralism Meets a Moment of Truth,' *Policy Options*, February 2003, 14–18.
28 John Ibbitson, 'Canada, Be Warned: A New Alliance Is Taking Shape,' *Globe and Mail* (Toronto), 24 February 2003, A13.
29 Ibid.
30 Jack L. Granatstein, 'Why Go to War? Because We Have To,' *National Post* (Toronto), 20 February 2003; Richard Gwyn, 'Our Foreign Policy Making Us Invisible,' *Toronto Star*, 23 February 2003; John Ibbitson, 'Canada, Be Warned'; Allan Gotlieb, 'The Chrétien Doctrine: By Blindly Following the UN, the Prime Minister Is Hurting Canada,' *National Post* (Toronto), 31 March 2003. See also Allan Gotlieb, 'A Grand Bargain with the U.S.,' *National Post* (Toronto), 5 March 2003; Paul Wells, 'Minister Lectured on Foreign Policy,' *National Post* (Toronto), 15 February 2003.
31 Wells, 'Minister Lectured on Foreign Policy.'
32 Robert Fulford, 'From Delusions to Destruction: How Sept. 11 Has Called into Question the Attitudes by Which Our Society Lives,' *National Post* (Toronto), 6 October 2001.

Appendix A: American Multilateralism in Iraq, 2003

1 Report prepared by Jay Nathwani for Dr. Frank Harvey, Director, Centre for Foreign Policy Studies, Dalhousie University, Halifax, May 2003. The

author would like to thank Jay Nathwani for his excellent research assistance in compiling this overview of American coalition building and for his observations about the French and Canadian contributions to the 2003 war in Iraq.

2 Ian MacLeod, 'The "Coalition of the Willing,"' *Ottawa Citizen*, 26 March 2003, 4.
3 Ibid
4 Ibid.
5 Ibid.
6 Pamela Hess, 'DOD: 13 Nations Join Coalition of Willing,' United Press International, 20 March 2003.
7 MacLeod, 'The "Coalition of the Willing."'
8 Ibid.
9 Australian Broadcasting Corporation, 'Factfile: Countries Offering Support for Attack on Iraq,' ABC Online, 20 March 2003, available at www.abc.net.au (accessed 20 March 2003).
10 MacLeod, 'The "Coalition of the Willing."'
11 Ibid.
12 Xinhua News Agency, 'Gulf Countries Endorse UAE Call for Saddam's Exile,' 3 March 2003.
13 Mark Thompson and Laura Brafford, 'Ready to Move In: U.S. Forces Could Be Primed to Start Fighting Iraq Again in Short Order,' *Time*, 2 December 2002: 40.
14 CBC News Online, 'The Military Buildup,' available at www.cbc.ca (accessed 20 March 2003).
15 Xinhua News Agency, 'First Batch of U.A.E. Forces Arrive in Kuwait,' 18 February 2003, and Xinhua News Agency, 'Bahraini Forces Arrive in Kuwait Amid Looming Iraq War,' 25 February 2003.
16 MacLeod, 'The "Coalition of the Willing."'
17 Associated Press, 'Nations Joining Anti-Iraq Coalition,' *Washington Post*, 18 March 2003, available at www.washingtonpost.com (accessed 20 March 2003).
18 MacLeod, 'The "Coalition of the Willing."'
19 MacLeod, 'The "Coalition of the Willing."' See also Jessica Guynn and Tom Infield, 'Of Nations Supporting U.S. Attack, Few Offering Troops for Battle,' Knight Ridder Newspapers, 18 March 2003, available at www.krwashington.com (accessed 20 March 2003).
20 MacLeod, 'The "Coalition of the Willing."' See also Australian Broadcasting Corporation, 'Factfile.'
21 Richard Sanders, 'Who Says We're Not at War?' *Globe and Mail* (Toronto),

31 March 2003, A15, and CTV.ca News Staff, 'Canadian to Command Allied Warships in the Gulf,' 11 February 2003, available at www.ctv.ca (accessed 20 March 2003).

22 Sanders, 'Who Says We're Not at War?'

23 Ibid.

24 Rick Mofina, 'Canadian Ships Won't Be Drawn into War: PM,' *Ottawa Citizen*, available at www.canada.com (accessed 20 March 2003).

25 Sheldon Alberts, 'Ships May Enter War by Back Door,' *National Post* (Toronto), 13 March 2003, available at www.nationalpost.com (accessed 1 May 2003).

26 Sheldon Alberts and Richard Foot, '40-year-old Sea King Forces Flagship Back,' *National Post*, 28 February 2003, available at www.nationalpost.com (accessed 20 March 2003).

27 Alberts, 'Ships May Enter War by Back Door.'

28 Sanders, 'Who Says We're Not at War?'

29 CTV.ca News Staff, 'Canadian to Command Allied Warships in the Gulf,' February 2003.

30 Daniel LeBlanc, 'Canada Takes Afghan Mission,' *Globe and Mail* (Toronto), 13 February 2003, A1.

31 Daniel LeBlanc, 'Canadian Troops Headed to Afghanistan,' *Globe and Mail* (Toronto), 6 May 2003, A5.

32 Sanders, 'Who Says We're Not at War?'

33 Ibid.

34 Allan Thompson, 'Call to Arms,' *Toronto Star*, 15 February 2003.

35 CTV.ca News Staff, 'Canadian to command allied warships in the Gulf.'

36 Shawn McCarthy, 'PM rejects war without UN,' *Globe and Mail* (18 March 2003). Available at www.globeandmail.com (Accessed 20 March 2003).

37 MacLeod, 'The "Coalition of the Willing."'

38 Http://www.whitehouse.gov/infocus/iraq/news/20030327-10.html.

39 MacLeod, 'The "Coalition of the Willing."'

40 Ibid. See also Associated Press 'Nations Joining Anti-Iraq Coalition,' 2003.

41 MacLeod, 'The "Coalition of the Willing."'

42 Ibid. See also Guynn and Infield, of Nations Supporting U.S. Attack ...' Guynn and Infield say that the troops are specialists in dealing only with chemical weapons. It is they who provide the number of troops deployed.

43 MacLeod, 'The "Coalition of the Willing."'

44 Ibid.

45 Australian Broadcasting Corporation, 'Factfile.' See also MacLeod, 'The "Coalition of the Willing."'

46 MacLeod, 'The "Coalition of the Willing."'
47 Paul Garwood, Associated Press, 'Egypt Supports U.S. call to Disarm,' *The Record* (New Jersey), 14 September 2002, A6.
48 CBC, 'The Military Buildup.'
49 MacLeod, 'The "Coalition of the Willing."'
50 Ibid.
51 Ibid.
52 Ibid.
53 Ibid.
54 Ibid.
55 Ibid.
56 Ibid. See also Associated Press, 'Ethiopia Grants Airspace Rights to U.S.,' *New York Times*, 21 March 2003, available at www.nytimes.com (accessed 21 March 2003).
57 CNN, 'Paris: We May Help in Chemical War,' CNN Online, 18 March 2003, available at www.cnn.com (accessed 25 April 2003).
58 CTV.ca News Staff, 'Canadian to Command Allied Warships in the Gulf.'
59 MacLeod, 'The "Coalition of the Willing."'
60 Interfax News Agency, 'Georgia Voices Support for U.S. War against Iraq,' BBC Monitoring Former Soviet Union – Political, 3 February 2003.
61 MacLeod, 'The "Coalition of the Willing."'
62 John Rossant, 'Europe Can't Afford to Stay Mad for Long,' Business Week Online, 20 March 2003, available at www.businessweek.com (accessed 20 March 2003).
63 Rossant, 'Europe Can't Afford to Stay Mad for Long.'
64 Ibid.
65 Australian Broadcasting Corporation, 'Factfile.'
66 CTV.ca News Staff, 'Canadian to Command Allied Warships in the Gulf.'
67 MacLeod, 'The "Coalition of the Willing."'
68 Ibid.
69 Ibid.
70 Ibid.
71 'Hungary Open to U.S.,' *Vancouver Province*, 25 February 2003, A6.
72 MacLeod, 'The "Coalition of the Willing."'
73 IRNA news agency, Tehran, 'Senior Official Reaffirms Iran's Neutrality in US War against Iraq,' BBC Monitoring Middle East – Political, 21 February 2003.
74 BBC News Online, 'Where the World Stands on Iraq,' 3 December 2002, available at news.bbc.co.uk (accessed 10 May 2003).

75 'Israeli Commandos "Inside Iraq."' *Sunday Herald Sun* (Melbourne), 29 September 2002, 33.

76 'Israel Offers to Knock Out Iraqi Missile Sites,' *The Record* (New Jersey), 19 October 2002, A6.

77 MacLeod, 'The "Coalition of the Willing."'

78 CTV.ca News Staff, 'Canadian to Command Allied Warships in the Gulf.'

79 MacLeod, 'The "Coalition of the Willing."'

80 Ibid.

81 Ibid.

82 Ibid.

83 Anthony Shadid, 'Jordan to Allow Limited Stationing of U.S. Troops,' *Washington Post*, 30 January 2003, A10, and *Ottawa Citizen*, 'Jordan Opens Bases, Airspace,' 31 January 2003, A8.

84 Robert Collier, 'Arab Nations Falling into Line on Iraq,' *San Francisco Chronicle*, 27 January 2003, A1.

85 CBC News Online, 'Jordan Admits Hosting U.S. Troops,' 25 February 2003, available at www.cbc.ca (accessed 21 March 2003).

86 CBC News Online, 'Evidence Jordan Helping U.S. Military,' 10 March 2003, available at www.cbc.ca (accessed 21 March 2003).

87 Tyler Marshall and Kim Murphy, 'Arab Nations Work to Veil Cooperation with U.S.,' *Los Angeles Times*, 20 March 2003, available at www.latimes.com (accessed 20 March 2003).

88 Vivienne Walt, 'U.S. Troops Keep Quiet on Iraq's Western Front,' *USA Today*, 17 March 2003, 05A.

89 DEBKAfile, 'More U.S. Troops Poised in Jordan for Iraq Invasion,' *Israel Faxx*, 14 August 2002, available at www.DEBKA.com.

90 Walt, 'U.S. Troops Keep Quiet on Iraq's Western Front.'

91 'Jordan Opens Bases, Airspace.' 31 January 2003, A8. *Ottawa Citizen*. See also Australian Broadcasting Corporation, 'Factfile.'

92 Ibid. See also CBC, 'Jordan Admits Hosting U.S. troops.'

93 Shadid, 'Jordan to Allow Limited Stationing of U.S. Troops.'

94 Australian Broadcasting Corporation, 'Factfile.'

95 MacLeod, 'The "Coalition of the Willing."'

96 Xinhua News Agency, 'Gulf Countries Endorse UAE Call for Saddam's Exile.'

97 MacLeod, 'The "Coalition of the Willing."'

98 BBC, 'Kuwait Seals Off Iraq Border Area,' 2 November 2002, available at news.bbc.co.uk (accessed 21 March 2003).

99 CBC, 'The Military Buildup.'

100 Peter Godspeed, 'Kuwait: The Waiting Is Almost Over,' *Vancouver Sun*, 17 March 2003, A9.
101 CBC, 'The Military Buildup'
102 MacLeod, 'The "Coalition of the Willing."'
103 Guynn and Infield, 'Of Nations Supporting U.S. attack ...'
104 MacLeod, 'The "Coalition of the Willing."'
105 Ibid.
106 Ibid.
107 Ibid.
108 Ibid.
109 Ibid.
110 Ibid.
111 Ibid.
112 Ibid.
113 Ibid.
114 CBC, 'The Military Buildup.'
115 CBC, 'The Military Buildup.'
116 Agence France Presse English, 'Bush Calls Leaders of China, Japan, South Africa, Oman as Iraq Vote Looms,' 10 March 2003.
117 CBC, 'The Military Buildup.'
118 MacLeod, 'The "Coalition of the Willing."'
119 http://www.whitehouse.gov/infocus/iraq/news/20030327–10.html.
120 United Press International, 'Pakistan Finds U.N. Vote Difficult,' 12 March 2003. For further comment on the effect of public opinion on Musharraf's position, see Liz Sly, 'Pakistan President Again Weighs Aiding U.S., Upsetting Populace.' Knight Ridder/Tribune News Service, 12 March 2003.
121 MacLeod, 'The "Coalition of the Willing."'
122 Ibid.
123 Ibid.
124 Ibid.
125 Ibid.
126 Ibid.
127 Xinhua News Agency, 'Gulf Countries Endorse UAE Call for Saddam's Exile.'
128 Marshall and Murphy, 'Arab Nations Work to Veil Cooperation with U.S.,' and MacLeod, 'The "Coalition of the Willing."'
129 CBC, 'The Military Buildup.'
130 Richard Wallace, 'Qatar Base for War on Iraq,' *Daily Record* (Glasgow, Scotland), 13 September 2002, 2.
131 CBC, 'The Military Buildup.'

132 'Support Concealed,' *Gazette* (Montreal), 21 March 2003, A16.

133 CBC, 'The Military Buildup.'

134 Wallace, 'Qatar base for war on Iraq,' and Richard Wallace, 'War Build-up Revealed by Satellite,' *The Mirror* (London), 7 August 2002. See also: CBC, 'The Military Buildup.'

135 MacLeod, 'The "Coalition of the Willing."'

136 CBC, 'The Military Buildup.'

137 Eric Schmitt, 'Back From Iraq, High-Tech Fighter Recounts Exploits,' *New York Times*, 23 April 2003, available at www.nytimes.com accessed 25 April 2003).

138 MacLeod, 'The "Coalition of the Willing."'

139 Guynn and Infield, 'Of Nations Supporting U.S. ...'

140 MacLeod, 'The "Coalition of the Willing."'

141 Xinhua News Agency, 'Gulf Countries Endorse UAE Call for Saddam's Exile.'

142 Marshall and Murphy, 'Arab Nations Work to Veil Cooperation with U.S.'

143 Robert Burns, 'U.S. Gets Saudi Assurances on Use of Bases for War,' Associated Press, 29 December 2002.

144 Tony Harnden, 'Bush and Saudis trade air bases for withdrawal,' *Telegraph* (London), 27 February 2003, available at www.telegraph.co.uk (accessed 25 April 2003).

145 Associated Press, 'Saudi Arabia Quietly Helping U.S. Prepare,' *New York Times*, 20 March 2003, available at www.nytimes.com (accessed 21 March 2003).

146 Collier, 'Arab Nations Falling into Line on Iraq.'

147 Marshall and Murphy, 'Arab Nations Work to Veil Cooperation with U.S.'

148 'Saudi Arabia Quietly Helping U.S. Prepare.'

149 CBC, 'The Military Buildup.'

150 Ibid.

151 MacLeod, 'The "Coalition of the Willing."'

152 Ibid.

153 Ibid. See also Australian Broadcasting Corporation, 'Factfile.'

154 Richard Whittle, 'U.S. Allies vs. Iraq Exceed British and Australians,' *Record* (New Jersey), 9 February 2003, A16.

155 MacLeod, 'The "Coalition of the Willing."'

156 Ibid.

157 'Meanwhile, in Pyongyang ...'. *The Economist Global Agenda*, 10 April 2003, available at www.economist.com (accessed 12 April 2003).

158 MacLeod, 'The "Coalition of the Willing."'

159 MacLeod, 'The "Coalition of the Willing,"' and CNN, 'Leaders of U.S.,

Britain, Spain to Meet on Resolution,' 14 March 2003, available at www.cnn.com.

160 Australian Broadcasting Corporation, 'Factfile.' See also MacLeod, 'The "Coalition of the Willing."'

161 Collier, 'Arab Nations Falling into Line on Iraq' and United Nations, www.un.org/Docs/scres/2002/sc2002.htm.

162 Central News Agency web site in English, 'Taiwan's Foreign Minister Chien Reaffirms Support for US Action against Iraq,' BBC Monitoring Asia Pacific – Political, 10 March 2003.

163 Central News Agency Web site, Taipei, in English, 'Taiwanese Ruling Party Says Iraq Violates UN Resolution,' BBC Monitoring Asia Pacific – Political, 21 March 2003.

164 Http://www.whitehouse.gov/infocus/iraq/news/20030327-10.html.

165 Ibid.

166 Agence France Presse English, 'Turkish Lawmakers Vote to Back US Preparations for Iraq War,' 6 February 2003.

167 Whittle, 'U.S. Allies vs. Iraq Exceed British and Australians.'

168 'Turkey Upsets U.S. Military Plans,' BBC News Online, 1 March 2003, available at news.bbc.co.uk (accessed 1 March 2003).

169 Agence France Presse English, 'Turkey Swinging in Favor of US War Plans,' 6 March 2003.

170 Dexter Filkins, 'Turkish Military Backs Role in U.S. Drive on Iraq,' *New York Times*, 6 March 2003, A18.

171 Agence France Presse English, 'Turkey Swinging in Favor of US War Plans.'

172 Associated Press, 'Turkey Approves U.S. Use of Bases If UN Authorizes a War against Iraq,' *Cape Breton Post*, 4 December 2002, B10.

173 Agence France Presse English, 'Turkish Lawmakers Vote to Back US Preparations for Iraq War,' March 2003.

174 Frank Bruni, 'Turkey Opens Airspace for U.S. Warplanes,' *New York Times*, 20 March 2003, available at www.nytimes.com (accessed 21 March 2003).

175 'Turkey Grants U.S. Right to Use airspace for Iraq War,' *Telegram* (St. John's), 21 March 2003, A7.

176 MacLeod, 'The "Coalition of the Willing."'

177 http://www.whitehouse.gov/infocus/iraq/news/20030327-10.html.

178 MacLeod, 'The "Coalition of the Willing."'

179 Xinhua News Agency, 'Gulf Countries Endorse UAE Call for Saddam's Exile,' 3 March 2003.

180 WAM (Emirates News Agency) Web site in English, 'Turkey Voices

Support for UAE Initiative on Iraq,' BBC Monitoring Middle East –
Political, 6 March 2003.

181 Xinhua News Agency, 'Backgrounder: U.A.E.'s Dhafrah Air Base,'
19 March 2003.

182 Whittle, 'U.S. Allies vs. Iraq Exceed British and Australians.'

183 Australian Broadcasting Corporation, 'Factfile.'

184 Xinhua News Agency, 'First Batch of U.A.E. Forces Arrive in Kuwait' and
'Gulf states send troops to protect Kuwait,' *Ottawa Citizen*, 13 February
2003, A8.

185 MacLeod, 'The "Coalition of the Willing."'

186 http://politics.guardian.co.uk/foreignaffairs/story/0,11538,916789,00
.html.

187 Jack Fairweather, 'Britain Calls Home Bulk of Ground Forces,' http://
www.telegraph.co.uk/news/main.jtml?xml=news/2003/04/24/
nbrit24.xml.

188 Ministry of Defence of the United Kingdom, http: www.operations
.mod.uk/telic/forces.htm (accessed 1 May 2003).

189 CBC, 'The Military Buildup.'

190 Lee Hulteng, 'B-52s Making 16-hour Trips,' *Gazette* (Montreal), 24 March
2003, A20.

191 MacLeod, 'The "Coalition of the Willing."'

192 Uzbek Radio First Programme, 'Uzbekistan Backed USA's War in Iraq
Despite Anti-war Moods in World – President,' BBC Monitoring
Newsfile, 24 April 2003.

193 Al-Sahwah Web site, Sanaa, in Arabic, 'Yemen Takes Steps to Protect US,
UK Citizens from Anti-war Backlash,' BBC Monitoring Middle East –
Political, 21 March 2003.

Bibliography

Achen, Christopher H., and Duncan Snidal. 'Rational Deterrence Theory and Comparative Case Studies.' *World Politics* 42 (January 1989).

Agence France Presse English, 'Bush Calls Leaders of China, Japan, South Africa, Oman as Iraq Vote Looms,' 10 March 2003.

Alberts, Sheldon. 'Chrétien to Endorse U.S. Mission.' *National Post* (Toronto), 8 April 2003.

Alberts, Sheldon, and Steven Edwards, 'Canada Unable to Help Congo, Martin Says,' *National Post* (Toronto), 17 May 2003.

– '"Saddam Is Dangerous": Foreign Minister Wishes Allied Troops "Godspeed,"' *National Post* (Toronto), 25 March 2003.

Alberts, Sheldon, and Richard Foot. 'Ships May Enter War by Back Door.' *National Post* (Toronto), 13 March 2003.

Alvarez, Jose E. 'Crimes of States/Crimes of Hate: Lessons from Rwanda.' *Yale Journal of International Law* 24, no. 365 (1999): 476–9.

– 'Multilateralism and its Discontents.' *European Journal of International Law* 14, no. 1 (2002).

– 'The United States Financial Veto,' in Proceedings of the 90th Annual Meeting, 27–30 March 1996, American Society of International Law, 1997, 319.

Anderson, James. 'America at Risk: The Citizen's Guide to Missile Defense.' Washington: Heritage Foundation, 1999.

Annan, Kofi, quoted in 'DR of Congo: UN Investigates Reports of Massacres in Troubled Ituri Region.' United Nations News Services, 7 April 2003.

Aoki, Thomas, Peter L. Berger, and Samuel Huntington, eds. *Many Globalizations: Cultural Diversity in the Contemporary World*. London: Oxford University Press, 2002.

Arbitov, Alexei G. 'The Future of Strategic Deterrence and Nuclear Postures of

Great Powers.' In *U.S.–Russian Partnership: Meeting the New Millennium*. Washington: National Defense University Press, 1999.

Arquilla, John, and David Ronfeldt, eds. *Networks and Netwars: The Future of Terror, Crime, and Militancy*. Santa Monica: Rand, 2001.

Associated Press. 'Nations Joining Anti-Iraq Coalition,' *Washington Post*, 18 March 2003. Available at www.washingtonpost.com (accessed 20 March 2003).

Australian Broadcasting Corporation. 'Factfile: Countries Offering Support for Attack on Iraq.' ABC Online. Available at www.abc.net.au (accessed 20 March 2003).

Axworthy, Lloyd and Michael Byers. 'Say No to Missile Defence.' *Globe and Mail* (Toronto), 29 April 2003, A13.

Barber, Benjamin. *Jihad vs. McWorld: How Globalization and Tribalism Are Reshaping the World*. New York: Ballentine Books, Inc., 1996.

– 'NATO Drops Its Support of ABM Pact.' *Washington Times*, 30 May 2001.

– 'Beyond Jihad vs. McWorld.' *Nation*, 21 January 2002.

Bee, Ronald J. *Nuclear Proliferation: The Post-Cold War Challenge*. New York: Foreign Policy Association, 1995.

Berger, Samuel. 'A Foreign Policy for the Global Age.' *Foreign Affairs* 79, no. 6 (2000).

Bering, Helle. 'Serious about Missile Defense.' *Washington Post*, 7 February 2001.

Bernier, Justin. 'Take the Missile Defense Debate to a Higher Level.' *Los Angeles Times*, 2 February 2001.

Betts, Richard K. 'The Concept of Deterrence in the Postwar Era.' *Security Studies*, Autumn 1991.

Blackwill, Robert, and Albert Carnesale, eds. *New Nuclear Nations: Consequences for U.S. Policy*. New York: Council on Foreign Relations Press, 1993.

Blix, Hans. Transcript of Blix's Remarks. New York: United Nations, 7 March 2003.

Bluth, Chritoph. *Arms Control and Proliferation: Russia and International Security after the Cold War*. London: Brassey's, Inc., 1996.

Bockhorn, Lee. 'A New Day for Missile Defense.' *Weekly Standard* (Washington), 12 October 2001.

Bosco, Joseph A. 'Has China Turned into a Frankenstein?' *Los Angeles Times*, 5 March 2001.

Boyes, Roger. 'German Cities "in Firing Line" of Rogue States.' *Times* (London), 16 February 2001.

Brainard, Lael. 'Globalization in the Aftermath: Target, Casualty, Callous Bystander?' Analysis Paper #12, Washington: Brookings Institution Press, 2001.

Brainard, Lael, and Robert E. Litan. *Globalization: What Now?* Washington: Brookings Institution Press, 2002.

Brecher, Jeremy, Tim Costello, and Brendan Smith, eds. *Globalization from Below.* Cambridge, MA: South End Press, 2000.

Broad, William. 'A Missile Defense with Limits: The ABC's of the Clinton Plan.' *New York Times*, 30 June 2000.

Brown, Seymon. *The Illusion of Control: Force and Foreign Policy in the 21st Century.* Washington: Brookings Institution Press, 2001.

Bruni, Frank. 'Turkey Opens Airspace for U.S. Warplanes.' *New York Times*, 20 March 2003.

Buchanan, Patrick J. *The Death of the West: How Mass Immigration, Depopulation, and a Dying Faith Are Killing Our Culture and Country.* New York/London: St. Martin's Press, 2001.

Bundy, McGeorge. 'To Cap the Volcano.' *Foreign Affairs* 48 (October 1969): 9–10.

Burns, Robert. 'U.S. Gets Saudi Assurances on Use of Bases for War.' Associated Press, 29 December 2002.

Butler, Richard. *Fatal Choice: Nuclear Weapons and the Illusion of Missile Defense.* Boulder, CO: Westview Press, 2001.

Caldicott, Helen. *The Coming Nuclear War: Manhattan Project II, The National Missile Defense, and How to Avert Catastrophe.* New York: Free Press, 2001.

Cambone, Stephen, Ivo Daalder, Stephen Hadley, and Christopher Makins. *European Views of National Missile Defense.* Washington: The Atlantic Council of the United States, September 2000.

Campaign for U.N. Reform. 'U.S. Funding to the U.N.' Available at http://www.cunr.org/priorities/Arrears.htm.

'Canada and Ballistic Missile Defence: Dilemmas and Options.' Available at http://www.ploughshares.ca/content/MONITOR/monm99e.html.

Canada in the World: Canadian Foreign Policy Review. Ottawa: DFAIT, 1995.

Carment, David, Fen Osler Hampson, and Norman Hilmer, eds. *Canada among Nations.* London: Oxford University Press, 2003.

Carr, Caleb. 'The Myth of a Perfect Defense.' *New York Times*, 7 August 2001.

Carter, Ashton, and David Schwartz. *Ballistic Missile Defense.* Washington: Brookings Institution Press, 1984.

Carter, Ashton, and William J. Perry. *Preventive Defense: A New Security Strategy for America.* Washington: Brookings Institution Press, 1999.

Castles, Stephen, and Alastair Davidson, *Citizenship & Migration: Globalization and the Politics of Belonging.* New York: Routledge, 2002.

Causewell, Erin V., ed. *National Missile Defense: Issues and Developments.* New York: Novinka Books, 2002.

CBC News Online, 'The Military Buildup.' Available at www.cbc.ca (accessed 20 March 2003).

Center for Arms Control and Non-Proliferation, Washington, D.C. Available at http://www.armscontrolcenter.org/nmd/md101.html.

Cheon, Seongwhun. 'The KEDO Process at the Crossroads.' Available at http://www.fes.or.kr/K_Unification/U-paper7.htm.

Chollet, Derek H., and James M. Goldgeier. 'Missile Defense in Perspective.' *Washington Times*, 22 May 2001.

Chomsky, Noam, Baker Spring, Lisbeth Gronlund, Stephen Young, Jack Spencer, David Nyhan, 'National Missile Defense System.' Available at http://www.prospect.org/webfeatures/2000/07/chomsky-n-07-18.html.

Chrétien, Jean. Notes for an Address by Prime Minister Jean Chrétien. Speech to the Chicago Council on Foreign Relations, Chicago, 13 February 2003.

Cienski, Jan. 'Washington Trades Arms for Friends Military Aid: War on Terrorism Has Redefined America's Interests.' *National Post* (Toronto), 27 January 2003.

Cilluffo, Frank, Joseph J. Collins, Arnaud de Borchgrave, Daniel Gouré, and Michael Horowitz. *Defending America in the 21st Century: New Challenges, New Organizations and New Policies.* Washington: Center for Strategic and International Studies, 2001.

Cirincione, Joseph, and Frank von Hippel. 'The Last 15 Minutes: Ballistic Missile Defence in Perspective.' On-line edition, http://www.stimson.org/coalitio/last15.htm. Washington: Stimson Center, October 2000.

Clark, Ian. *The Post–Cold War Order: The Spoils of Peace.* London: Oxford University Press, 2001

Clarkson, Stephen. *Uncle Sam and Us: Globalization, Neoconservatism, and the Canadian State.* Toronto: University of Toronto Press, 2002.

Clausen, Peter. *Nonproliferation and the National Interest: America's Response to the Spread of Nuclear Weapons.* New York: HarperCollins College Publishers, 1993.

Collins, Joseph J., and Gabrielle D. Bowdoin. 'Beyond Unilateral Economic Sanctions: Better Alternatives for U.S. Foreign Policy.' Washington: Center for Strategic & International Studies, April 1999.

Collier, Robert. 'Arab Nations Falling into Line on Iraq.' *San Francisco Chronicle*, 27 January 2003, A1.

Colvin, Marie. 'Iraqi Missile Launch Raises Arms Fears.' *Sunday Times* (London), 2 July 2000.

'Conflict Narrowly Averted in Iraq.' *Disarmament Diplomacy*, no. 23 (February 1998).

Cooley, John K. *Unholy Wars: Afghanistan, America, and International Terrorism.* London: Pluto Press, 2001.

Cooper, Barry, and Ted Morton. 'Chrétien Has Put Party ahead of Country.' *National Post* (Toronto), 28 March 2003.

Cordesman, Anthony. 'The Risks and Effects of Indirect, Covert, Terrorist, and Extremist Attacks with Weapons of Mass Destruction.' Washington: Center for Strategic and International Studies, 14 February 2001.

Coyne, Andrew. 'Canada on Iraq: a Clarification.' *National Post* (Toronto), 31 March 2003.

– 'Imagine Where We'd Be with Some Leadership.' *National Post* (Toronto), 9 April 2003.

– 'PM's Decision Means Moral Free Ride Is Over.' *National Post* (Toronto), 19 March 2003.

– 'The Council Is Too Easily Pleased.' *National Post* (Toronto), 17 February 2003.

– 'Taking Potshots at the Missile Shield.' *National Post* (Toronto), 26 February 2001.

Cronin, Audrey Kurth. 'Behind the Curve: Globalization and International Terrorism.' *International Security* 27, no. 3 (2002–03): 30–58.

CTV News. 'Canada Will Only Go to War with UN Approval,' 9 May 2003.

Curry, Bill. 'Canada Has Backed Wars without UN.' *National Post* (Toronto), 9 April 2003.

Daalder, Ivo. 'Are the United States and Europe Heading for Divorce?' *International Affairs* 77, no. 3 (2001): 553–67.

Daalder, Ivo, I.M. Destler, Robert E. Litan, Michael E. O'Hanlon, and Peter R. Orszag, eds. *Achieving Homeland Security.* Washington: Brookings Institution Press, 2002.

Daalder, Ivo, and James M. Lindsay. 'How to Get Europe and Russia into a Consensus on Defense.' *International Herald Tribune*, 13 June 2001.

– 'How to Get Europe and Russia into a Consensus on Defense.' *International Herald Tribune*, 13 June 2001.

Daalder, Ivo, and Christopher Makins. 'A Consensus on Missile Defence?' *Survival* 43, no. 3 (2001): 61–66.

Dale, Helle. 'No Blood for French Oil.' *Washington Times*, 5 March 2003.

Danaher, Kevin, and Roger Burbach, eds., *Globalize This! The Battle against the World Trade Organization.* Monroe, ME: Common Courage Press, 2000.

Davis, Zachary S., and Benjamin Frankel, eds. *The Proliferation Puzzle: Why Nuclear Weapons Spread and What Results.* London: Frank Cass, 1993.

de Borchgrave, Arnaud, Frank J. Cilluffo, Sharon L. Cardash and Michèle M.

Ledgerwood. *Cyber Threats and Information Security: Meeting the 21st Century Challenge*. Washington: Center for Strategic and International Studies, December 2000.

Deboon, David. *Ballistic Missile Defense in the Post-Cold War Era*. Westport, CT: Westview Press, 1995.

Delvoie, Louis. 'Multilateralism or Unilateralism: Whither American Foreign Policy,' *Policy Options*, November 2002: 12–14.

DeYoung, Karen. 'U.S. and Saudis Agree on Cooperation: Gulf Nation to Permit Expanded Use of Military Facilities in Event of War.' *Washington Post*, 26 February 2003, A01.

Dobson, Wendy. 'Shaping the Future of the North American Economic Space: A Framework for Action,' *C.D. Howe Institute Commentary* 162. Toronto: C.D. Howe Institute, April 2002.

Donahue, John D., and Joseph S. Nye, eds. *Governance in a Globalizing World*. Washington: Brookings Institution Press, 2000.

Dougherty, James, and Robert Pfaltzgraff, eds. *Contending Theories of International Relations*. 5th ed. New York: Longman, 2001.

Drabek, Zdenek, ed. *Globalisation Under Threat: The Stability of Trade Policy and Multilateral Agreements*. Cheltenham, U.K.: Edward Elgar Publishers, 2001.

Dunn, Lewis A. 'Deterring the New Nuclear Powers.' *Washington Quarterly* 17 (Winter 1994).

Dymond, Bill, and Michael Hart. 'Canada and the Global Challenge: Finding a Place to Stand,' *Border Papers*, no. 180. (Toronto: C.D. Howe Institute, March 2003, 13–14.

Eakin, Emily. 'It Takes an Empire.' *New York Times*, 2 April 2002.

Economist. 'Meanwhile, in Pyongyang ...,' 10 April 2003.

Everts, Steven. *Unilateral America, Lightweight Europe? Managing Divergence in Transatlantic Foreign Policy*. London: Centre for European Reform, 2001.

Fagan, Drew. 'Canada in the World: Heading Back Up?' *Globe and Mail* (Toronto), 15 April 2003, A17.

Fergusson, Jim, Frank Harvey, and Robert Huebert. *To Secure a Nation: The Case for a New Defence White Paper*. Report Prepared for the Council for Canadian Security in the 21st Century. University of Calgary: Centre for Military and Strategic Studies, 2001.

Filkins, Dexter. 'Turkish Military Backs Role in U.S. Drive on Iraq.' *New York Times*, 6 March 2003, A18.

Feulner, Edwin. 'The Arms Race Myth.' Washington: Heritage Foundation, 2000.

'Flying Madly Off in All Directions.' *Globe and Mail* (Toronto), 21 July 2000.

Fournier, Ron. 'Ridge Warns of Al Qaeda Threat.' *Boston Globe*, 30 April 2002, A15.

Francis, Diane. 'Iraqi Liberation Will Unleash Untapped Wealth.' *Financial Post* (Toronto), 1 April 2003.

Freeh, Louis, quoted in 'Cyberspace Next Target for Terrorists: Experts.' CBC News Online, 30 April 2002.

Freeman, Alan. 'Old, New Europe Spar Bitterly over Iraq.' *Globe and Mail* (Toronto), 19 February 2003, A24.

Friedman, Thomas. 'Invade Iraq to Impose Democracy, Not Disarmament.' *Globe and Mail* (Toronto), 21 September 2002.

– *The Lexus and the Olive Tree: Understanding Globalization*. New York: Doubleday, 2000.

Frum, David. 'Quit Hemming and Hawing Over Hezbollah.' *National Post* (Toronto), 10 December 2003.

Fuerth, Leon. 'Tampering with Strategic Stability.' *Washington Post*, 20 February 2001, A23.

Fukuyama, Francis. *The Great Disruption: Human Nature and the Reconstitution of Social Order*. London: Oxford University Press, 1999.

Fulford, Robert. 'From Delusions to Destruction: How Sept. 11 Has Called into Question the Attitudes by Which Our Society Lives.' *National Post* (Toronto), 6 October 2001.

Gaffney, Frank J., Jr. 'Critical Mass.' *Washington Times*, 20 March 2001.

– 'The Real Debate about Missile Defense.' *Washington Times*, 8 May 2001.

– 'Self-Deterred from Defending the U.S.?' *Washington Times*, 7 March 2001.

– 'With Friends like These ...' *Washington Times*, 5 June 2001.

Garwood, Paul. 'Egypt Supports U.S. Call to Disarm,' Associated Press, *Record* (Washington) 14 September 2002, A6.

Gazette (Montreal). 'Nukes No Longer News,' 15 May 2002.

Geoghegan, Ian. 'Canadian Stance May Be Easing on U.S. Missile Shield.' *National Post* (Toronto), 9 May 2001.

'George W. Bush and the Real Missile Threat.' *Globe and Mail* (Toronto), 3 February 2001.

Gephardt, Richard A. Letter from Office of Representative Richard A. Gephardt, 1 May 2001. Available at http://www.acronym.org.uk/bush1.htm.

Gerecht, Reuel Marc. 'The Counterterrorist Myth.' *Atlantic Monthly*, July–August 2001. Available at http://www.theatlantic.com/issues/2001/07/gerecht.htm.

Gertz, Bill. 'Cohen Sees Iran Making Progress with Missiles.' *Washington Times*, 18 July 2000.

– 'Pakistan Gets More Chinese Weapons.' *Washington Times*, 9 August 2000.
– 'Russia Sells Missile Technology to North Korea.' *Washington Times*, 30 June 2000.
Gibson, Gordon. 'Why Chrétien Put Us on the Sidelines.' *National Post* (Toronto), 3 April 2003.
Giddens, Anthony. *Runaway World: How Globalisation Is Re-Shaping our World.* New York: Routledge, 2000.
Giffin, Gordon. 'Washington: When It Comes to Missile Defence, You Can Trust Us.' *Globe and Mail* (Toronto), 13 March 2001.
Gilpin, Robert. *The Challenge of Global Capitalism.* Princeton: Princeton University Press, 1999.
Gladwell, Malcolm. 'Safety in the Skies.' *New Yorker*, 1 October 2001.
'Globalization's Last Hurrah?' *Foreign Policy* no. 128 (January–February 2002).
Globe and Mail. 'Hiding the Troops.' Editorial, 2 April 2003, A16.
Goar, Carol. 'It's about War, Not Loyalty, Sir.' *Toronto Star*, 28 March 2003.
Godspeed, Peter. 'Kuwait: The Waiting Is Almost Over,' *Vancouver Sun*, 17 March 2003, A9.
Goodby, James E. 'Try to Engage with Pyongyang.' *International Herald Tribune*, 6 January 2003.
Goodman, Melvin A., Gerald E. Marsh, and Craig R. Eisendrath. *The Phantom Defense: America's Pursuit of the Star Wars Illusion.* New York: Praeger, 2001.
Goodpaster, Andrew J., C. Richard Nelson, and Seymour J. Deitchman. 'Deterrence: An Overview.' In *Post Cold-War Conflict Deterrence.* Washington: National Academy Press, 1997.
Gotlieb, Allan. 'A Grand Bargain with the U.S.,' *National Post* (Toronto), 5 March 2003.
– 'The Chrétien Doctrine: By Blindly Following the UN, the Prime Minister Is Hurting Canada.' *Maclean's*, 31 March 2003.
Glasser, Susan B. 'Russia Has Warning, and Overture, on Missile Plan.' *Washington Post*, 19 December 2002, A29.
Graham, William R. and Keith Payne. 'The Pace and Threat of Missile Proliferation.' *Washington Times*, 18 August 1998, A15.
Granatstein, Jack L. 'Why Go to War? Because We Have To.' *National Post* (Toronto), 20 February 2003.
Gray, Colin S. 'The Definitions and Assumptions of Deterrence: Questions of Theory and Practice.' *Journal of Strategic Studies* 13, December 1990: 6–13.
– *The Second Nuclear Age.* New York: Lynne Reimer Publishers, 1999.
Gwyn, Richard. 'Bush Fear of "Rogue" States Is Laughable.' *Toronto Star*, 18 February 2001.
– 'Our Foreign Policy Making Us Invisible.' *Toronto Star*, 23 February 2003.

Hackett, James T. 'The Countermeasures Debate.' *Washington Post*, 9 August 2000.
– 'Missile Defense: Deploy When?' *Washington Times*, 6 March 2003.
Hames, Tim. 'Arrogance, Ignorance and the Real New World Order.' *Times* (London), 15 February 2002.
Handberg, Roger. *Ballistic Missile Defense and the Future of American Security: Agendas, Perceptions, Technology, and Policy.* New York: Praeger, 2002.
Hartung, William D. 'Star Wars Revisited: Still Dangerous, Costly, and Unworkable.' *Foreign Policy in Focus* 4, no. 24 (1999).
Harvey, Frank P. 'ABCs of BMDs and ABMs.' *Chronicle-Herald* (Halifax), 10 May 2001.
– 'Addicted to Security: Globalized Terrorism and the Inevitability of American Unilateralism,' *International Journal* 59, no. 1 (Winter 2003–2004): 27–58.
– 'Dispelling the Myth of Multilateral Security After 11 September and the Implications for Canada.' In David Carment, Fen Osler Hampson, and Norman Hilmer, eds., *Canada among Nations* (2003). London: Oxford University Press, 2003, 200–218.
– 'The Future of Strategic Stability and Nuclear Deferrence,' *International Journal* 58, no. 2 (Spring 2003): 321–46.
– 'Good Defences Make Good Neighbours.' *Globe and Mail*, 11 April 2000.
– 'The International Politics of National Missile Defence: A Response to the Critics.' *International Journal* 55, no. 4 (Autumn 2000): 545–66.
– 'National Missile Defence Revisited, Again: A Response to David Mutimer.' *International Journal* 56, no. 3 (Spring 2001): 347–60.
– 'North Korea: A Rogue by Any Other Name.' *National Post* (Toronto), 29 July 2000.
– 'Politics, Not Technology Explains U.S. Missile Defence Decision.' *Chronicle-Herald* (Halifax), 9 September 2000.
– 'Price Is Right for U.S. Missile Program.' *National Post* (Toronto), 22 July 2000.
– 'Proliferation, Rogue-State Threats and National Missile Defence: Assessing European Concerns and Interests.' *Canadian Military Journal* 1, no. 4 (Winter 2000–2001): 69–78.
Harvey, Frank P., and David Carment. *Using Force to Prevent Ethnic Violence.* New York: Praeger, 2001.
Harvey, Frank P., and Ben Mor. *Conflict in World Politics: Advances in the Study of Crisis War and Peace.* New York: Macmillan/St. Martin's Press, 1998.
Hayes, Richard E., and Gary Wheatley. 'Information Warfare and Deterrence.' Paper no. 87, Institute for National Strategic Studies. Washington: National Defence University, October 1996.

Helms, Jesse. Written statement issued by Office of Senator Jesse Helms, 1 May 2001. Available at http://www.acronym.org.uk/bush1.htm.

Henderson, Harry. *Global Terrorism: The Complete Reference Guide*. New York: Facts on File, Inc., 2002.

Heritage Foundation's Commission on Missile Defense, 'Defending America: A Plan to Meet the Urgent Missile Threat.' Washington: The Heritage Foundation, March 1999.

Herschensohn, Bruce. 'History Hasn't Stood Still for ABM Treaty.' *Los Angeles Times*, 18 June 2001.

Hess, Pamela. 'DoD: 13 Nations Join Coalition of the Willing.' United Press International, 20 March 2003.

Hilmer, Norman, and Maureen Appel Molot, eds., *Canada among Nations: A Fading Power*. London: Oxford University Press, 2002.

Hitchens, Christopher A. 'Multilateralism and Unilateralism: A Self-Cancelling Complaint.' 18 December 2002. Available at www.slate.com.

Hoagland, Jim. 'An Ally's Terrorism.' *Washington Post*, 3 October 2001.

Hoffman, Bruce. *Inside Terrorism*. New York: Columbia University Press, 1998.

Holmes, Kim R., and Baker Spring. 'Clinton's ABM Treaty Muddle.' Committee Brief No. 16. Washington: The Heritage Foundation, 7 July 1995.

Homer-Dixon, Thomas. 'The Rise of Complex Terrorism.' *Foreign Policy* no. 128 January–February 2002.

Hopkins, John C., and Steven A. Maaranen. 'Nuclear Weapons in Post–Cold War Deterrence.' In *Post Cold-War Conflict Deterrence*. Washington: National Academy Press, 1997.

Hulteng, Lee. 'B-52s making 16-hour trips.' *Gazette* (Montreal), 24 March 2003, A20.

Humphrys, John. 'The Gigantic Bluff of UN Power Has Been Called.' *Sunday Times* (London), 9 March 2003.

Huntington, Samuel R. 'The Clash of Civilizations?' *Foreign Affairs* 72, no. 3 (1993).

– *The Clash of Civilizations and the Remaking of World Order*. London: Oxford University Press, 1997.

– 'The Lonely Superpower,' *Foreign Affairs* 78, no. 2 (1999).

Ibbitson, John. 'Canada, Be Warned: A New Alliance Is Taking Shape.' *Globe and Mail* (Toronto), 24 February 2003, A13.

– 'So How Do We Get Bush to Like Canada?' *Globe and Mail* (Toronto), 14 April 2003, A13.

Ignatieff, Michael. 'Canada in the Age of Terror: Multilateralism Meets a Moment of Truth.' *Policy Options*, February 2003: 14–18.

Ikenberry, G. John, ed. *America Unrivaled: The Future of the Balance of Power*. Ithaca: Cornell University Press, 2003.

– *After Victory: Institutions, Strategic Restraint and the Rebuilding of Order after Major Wars*. Princeton: Princeton University Press, 2001.
– 'Institutions, Strategic Restraint, and the Persistence of Amercian Postwar Order.' *International Security* 23, no. 3 (1998).

Ikle, Fred, Albert Wohlstetter, et al., *Discriminate Deterrence: Report of the Commission on Integrated Long-Term Strategy*. Washington: US Government Printing Office for the Department of Defense, 1988.

Interfax News Agency, 'Georgia Voices Support for U.S. War against Iraq,' BBC Monitoring Former Soviet Union - Political, 3 February 2003.

IRNA news agency, Tehran, 'Senior Official Reaffirms Iran's Neutrality in US War against Iraq. ' BBC Monitoring Middle East – Political, 21 February 2003.

Joffe, Josef. 'Defying History and Theory: The United States as the "Last Remaining Superpower."' In John Ikenberry, ed. *America Unrivaled: The Future of the Balance of Power*. Ithaca: Cornell University Press, 2003.

Johnston, Douglas, and Sidney Weintraub. 'Altering U.S. Sanctions Policy.' Final Report of the CSIS Project on Unilateral Economic Sanctions. Washington: Center for Strategic & International Studies, 1999.

Jonas, George. 'The UN Is Its Own Worst Enemy.' *National Post* (Toronto), 19 March 2003. Available at http://www.wsws.org/articles/2002/feb2002/can-f19.shtml.

Jones, Keith. 'US War Plans Panic Canada's Elite.' *Weekly Standard* (Washington), 19 February 2002.

Kadish, Lt. Gen. Ronald T., USAF, Director, Ballistic Missile Defense Organization. Testimony before House Subcommittee on National Security, Veterans Affairs, and International Relations Committee on Government Reform, 8 September 2000. Available at Http://www.acq.osd.mil/ bmdo/ bmdolink/html/ kadish8sep00.html.

Kagan, Robert. 'The Benevolent Empire.' *Foreign Policy* 12, no. 111 (Summer 1998): 24.
– 'Multilateralism, American Style.' *Washington Post*, 13 September 2002.
– *Of Paradise and Power: America and Europe in the New World Order*. New York: Randcm House, 2003.
– 'Out to Torpedo Missile Defense.' *Washington Post*, 9 May 2001.
– 'Power and Weakness,' *Policy Review*, no. 113 (June–July 2002).
– 'A Real Case for Missile Defense.' *Washington Post*, 21 May 2001, B07.
–, and William Kristol. 'The Coalition Trap.' *Weekly Standard* 7, no. 5 (Washington), 15 October 2001.

Kaplan, Robert. 'The Coming Anarchy.' *Atlantic Monthly*, February 1994.
– *Warrior Politics: Why Leadership Demands a Pagan Ethos*. New York: Random House, 2001.

Kapralav, Yuriy. 'Effects of National Missile Defense on Arms Control and Strategic Stability.' Paper presented at the Missile Threat and Plans for Ballistic Missile Defense: Impact on Global Security, Rome, 18–19 January 2001.

Karatnycky, Adrian. 'Libya Unfit to Head UN Human Rights Group,' *Newsday*, 19 September 2002.

Katzman, Kenneth. 'Iraq: International Support for U.S. Policy.' Foreign Affairs and National Defense Division. Congressional Research Service Library of Congress Number 98-114 F, 19 February 1998.

Kaysen, Carl, et al. 'Nuclear Weapons after the Cold War.' *Foreign Affairs* 70, no. 4 (Fall 1991): 106–7.

Kelleher, Ann, and Laura Klein. *Global Perspectives: A Handbook for Understanding Global Issues*. New York: Prentice Hall, 1999.

Kelly, Michael. 'Immorality on the March,' *Washington Post*, 19 February 2003, A29.

Keohane, Robert. 'The Globalization of Informal Violence, Theories of World Politics, and the Liberalism of Fear.' Paper prepared for delivery at the annual meeeting of the American Political Science Association, Boston, 29 August–1 September 2002. Available at http://www.iyoco.org/911/911keohane.htm.

– *International Institutions and State Power: Essays in International Relations Theory*. Boulder: Westview, 1989.

– 'Institutionalist Theory and the Realist Challenge after the Cold War.' In *Neorealism and Neoliberalism: The Contemporary Debate*. David Baldwin, ed. New York: Columbia University Press, 1993.

Keohane, Robert, and Joseph S. Nye. 'Globalization: What's New? What's Not? (And So What?)' *Foreign Policy* no. 118 (Spring 2000).

– *Power and Interdependence*, 3d ed. New York: Longman, 2000.

Keohane, Robert, and Robert Axelrod. 'Achieving Cooperation Under Anarchy: Strategies and Institutions.' *World Politics* 38 (October 1985): 226–54.

Kerry, Senator John. 'New Era Not Here Yet, Say Critics.' *Washington Post*, 2 May 2001.

Kessler, Glenn, and Philip P. Pan. 'Missteps with Turkey Prove Costly: Diplomatic Debacle Denied U.S. A Strong Northern Thrust in Iraq.' *Washington Post*, 28 March 2003, A01.

Kingwell, Mark. 'What Distinguishes Us from the Americans?' *National Post* (Toronto), 5 March 2003.

Krauthammer, Charles. 'Bush Doctrine,' *Washington Post*, 24 January 2003.

– 'Lift the Sanctions Now,' *Washington Post*, 21 April 2003, A23.

– 'The Real New World Order: The American and the Islamic Challenge.' *Weekly Standard* 7, no. 9 (12 November 2001).

- 'The Unipolar Moment: America and the World.' *Foreign Affairs* 70, no. 1 (1990–91).

Krueger, Alan, and Jitka Maleckova. 'Education, Poverty, Political Violence and Terrorism: Is There a Causal Connection,' 2002. Unpublished manuscript.

Kupchan, Charles A. *The End of the Amercian Era: U.S. Foreign Policy and the Geopolitics of the Twenty-first Century.* New York: Alfred A. Knopf, 2003.

Lagnado, Alice. 'Moscow Defies US with Iran Arms Deal.' *Times* (London), 13 March 2001.

Laqueur, Walter. *The New Terrorism: Fanaticism and the Arms of Mass Destruction.* London: Oxford University Press, 2001.

Larson, E. *A New Methodology for Assessing Multi-Layer Missile Defense Options.* Washington: Rand Corporation, 1994.

Leblanc, Daniel. 'Frigates "Critical" to Guard against Terror, Former Admiral Says.' *Globe and Mail* (Toronto), 21 March 2003, A12.

- 'Canada Takes Afghan Mission.' *Globe and Mail* (Toronto), 13 February 2003, A1.

- 'Canadian Troops Headed to Afghanistan.' *Globe and Mail* (Toronto), 6 May 2003, A5.

LeDrew, Stephen. 'Think Strategically about Sovereignty.' *National Post* (Toronto), 23 January 2002.

Lee, Rensselaer W. *Smuggling Armageddon: The Nuclear Black Market in the Former Soviet Union and Europe.* New York: St. Martin's Press, 1998.

Lesser, Ian O., Bruce Hoffman, John Arquilla, David F. Ronfeldt, Michele Zanini, and Brian Michael Jenkins, eds. *Countering the New Terrorism.* Santa Monica: Rand, 1999.

Levy, Jack. 'An Introduction to Prospect Theory in International Relations.' *Political Psychology* 13, no. 2 (June 1992): 171–86.

Lieven, Anatol. 'The Roots of Terrorism, and a Strategy against It.' *Prospect Magazine* 68 (October 2001).

Limbaugh, David. 'The Phony Charge of "Unilateralism."' 16 July 2002. Available at www.worldnetdaily.com

Lindsay, James M., and Michael E. O'Hanlon. *Defending America: The Case for Limited National Missile Defense.* Washington: Brookings Institution Press, 2001.

Locke, Michelle. 'Computers Would Track Terror Threat.' Associated Press, 23 May 2002.

Loughlin, Sean. 'White House Touts International Support for Military Campaign.' CNN Washington Bureau, 20 March 2003. Available at www.cnn.com.

Mack, Andrew. 'North Korea's Matches – Our Powder Keg.' *Globe and Mail* (Toronto), 15 January 2003, A15.

– and Oliver Rohls. 'The Great Canadian Divide.' *Globe and Mail* (Toronto), 28 March 2003, A19.

MacKenzie, Lewis. 'Admit It, We're Engaged in Combat.' *National Post* (Toronto), 4 April 2003.

MacLeod, Ian. 'The "Coalition of the Willing."' *Ottawa Citizen*, 26 March 2003, 4.

Malone, David M., and Yuen Foong Khong, eds., *Unilateralism and U.S. Foreign Policy: International Perspectives*. London: Lynne Rienner Publishers, 2003.

Mandel, Robert. *Deadly Transfers and the Global Playground: Transnational Security Threats in a Disorderly World*. Westport, CT: Greenwood Publishing Group, 1999.

Mandelbaum, Michael. *The Ideas That Conquered the World: Peace, Democracy and Free Markets in the Twenty-first Century*. Washington: Public Affairs Press, 2002.

– 'The Inadequacy of American Power.' *Foreign Affairs* 81, no. 5 (September–October 2002).

Marshall, Tyler, and Kim Murphy. 'Arab Nations Work to Veil Cooperation with U.S.' *Los Angeles Times*, 20 March 2003.

Marzolini, Michael. 'Canada's Place Is with the U.S.' *National Post* (Toronto), 27 March 2003.

Mazarr, Michael J. *Missile Defenses and Asian-Pacific Security*. London: Palgrave, 1989.

McCallum, John. *DGPA Transcripts* - Defence Minister Statement in the House of Commons: Debating Opposition Motion on Support for NORAD, 29 May 2003, Reference: 03052901.

McCarthy, Shawn. 'PM Rejects War without UN,' *Globe and Mail* (Toronto), 18 March 2003.

McCarthy, Shawn and Daniel Leblanc. 'PM Scolds McCallum on Canada's Role in Iraq.' *Globe and Mail* (Toronto), 16 January 2003, A1.

McMahon, K. Scott. *Pursuit of the Shield: The U.S. Quest for Limited Ballistic Missile Defense*. New York: Rowman & Littlefield Publishing, 1997.

MacNamara, Don, BGEN (Ret'd). 'September 11, 2001 – September 11, 2002,' *On Track* 7, no. 3 (9 October 2002): 11–15.

Mearsheimer, John J. 'The Future of the American Pacifier.' *Foreign Affairs* 80, no. 5 (2001).

Michaels, Marguerite, and Karen Tumulty. 'Horse Trading on Iraq.' *Time*, 10 March 2003.

Micklethwait, John, and Adrian Wooldridge. *A Future Perfect: The Hidden Promise and Peril of Globalization*. New York: Crown Publishers, 1999.

Milbank, Dana, and Sharon LaFraniere. 'U.S., Russia Agree to Arms Pact: Nuclear Arsenals on Both Sides to Be Reduced by Two-Thirds.' *Washington Post*, 14 May 2002.

Milhollin, Gary. 'The Real Nuclear Gap.' *New York Times*, 16 June 2000, A33.

Missile Defence Debate: Guiding Canada's Role. A preliminary report on an international consultation on U.S. Missile Defence. The Liu Centre for the Study of Global Issues, University of British Columbia, 15–16 February 2001.

Mitchell, Gordon. *Strategic Deception: Rhetoric, Science, and Politics in Missile Defense Advocacy*. Ann Arbor: Michigan State University Press, 2000.

Mittleman, James H. *The Globalization Syndrome: Transformation and Resistance*. Princeton: Princeton University Press, 2000.

Mofina, Rick. 'Canadian Ships Won't Be Drawn into War: PM.' *Ottawa Citizen*, 20 March 2003.

Mohanty, Debra. 'Defence Industries in a Changing World: Trends and Options.' *Strategic Analysis*, January 2000.

Mousseau, Michael. 'Market Civilization and Its Clash with Terror.' *International Security* 27, no. 3 (2003), 5–29.

Mueller, John. *Retreat from Doomsday: The Obsolescence of Major War*. New York: Basic Books, 1989.

Murphy, Rex. 'Canada, Two-faced? It All Depends,' *Globe and Mail* (Toronto), 29 March 2003, A23.

Mutimer, David. 'Good Grief!: The Politics of Debating National Missile Defence.' *International Journal* 56, no. 2 (2001).

Myers, Steven Lee. 'Flight Tests by Iraq Show Progress of Missile Program.' *New York Times*, 1 July 2000, 1.

National Post (Toronto). 'Fidel Castro's Friends in Ottawa,' 19 April 2003.

– 'Graham Takes the Bait,' 12 August 2002.

– 'The Iraqi Files.' Editorial, 29 April 2003.

– 'North Korea's Nukes,' 28 April 2003.

– 'The War Canada Missed.' Editorial, 20 March 2003.

National Security Strategy of the United States. Washington: The White House, September 2002.

Natural Resources Defense Council Web page. Available at http: //www.nrdc .org/nuclear/nudb/datainx.asp.

New York Times. 'Misrepresenting the ABM Treaty,' 15 June 2001.

– 'No Frills Arms Control,' 14 May 2002.

Nichiporuk, Brian. *The Security Dynamics of Demographic Factors*. Santa Monica: Rand, 2000.

Nickson, Elizabeth. 'Don't Turn Iraq Over to the UN.' *National Post* (Toronto), 24 April 2003.

Nitze, Paul H., and J.H. McCall. 'Contemporary Strategic Deterrence and Precision-Guided Munitions.' In *Post Cold-War Conflict Deterrence*. Washington: National Academy Press, 1997.

Nunn, Sam, and Bruce Blair. 'From Nuclear Deterrence to Mutual Safety.' *Washington Post*, 30 June 1997, 22.

Nye, Joseph S. 'Divided We War.' *Globe and Mail* (Toronto), 24 March 2003, A15.

– 'The New Rome Meets the New Barbarians: How America Should Wield its Power.' *Economist*, 23 March 2002.

– *The Paradox of American Power: Why the World's Only Superpower Can't Go It Alone.* London: Oxford University Press, 2002.

Nye, Joseph S. and John D. Donahue, eds. *Governance in a Globalizing World.* Washington: Brookings Institution Press, 2000.

O'Hanlon, Michael E. 'Beyond Missile Defense: Countering Terrorism and Weapons of Mass Destruction.' Policy Brief 86. Washington: Brookings Institution Press, 2001.

O'Meara, Patrick, Howard D. Mehlinger, Mathew Krain, and Roxanna Ma Newman, eds. *Globalization and the Challenge of a New Century: A Reader.* Bloomington: Indiana University Press, 2000.

O'Neill, Brendan. 'Don't Mention the U-word.' 21 February 2003. Available at http://www.spiked-online.com/Articles/00000006DC77.htm.

O'Sullivan, Meghan. *Shrewd Sanctions: Economic Statecraft in an Age of Global Terrorism.* Washington: Brookings Institution Press, 2002.

O'Toole, Tara. 'Biological Weapons: National Security Threat and Public Health Emergency.' Presentation to the Centre for Security and International Studies, 22 August 2000.

Onwudiwe, Ihekwoaga. *The Globalization of Terrorism.* New York: Ashgate Publishing, 2000.

Pae, Peter. 'Kill Vehicle Scores a Hit with Proponents of Missile Defense Weapons: The Pentagon Says the Successful Tests May Restore Credibility to the Program.' *Los Angeles Times*, 26 March 2002.

Palmer, Randall. 'Canada to Push for Special Saddam Criminal Court.' Reuters, 27 March 2003.

Panofsky, K.H. 'The Remaining Unique Role of Nuclear Weapons in Post-Cold War Deterrence.' In *Post–Cold War Conflict Deterrence.* Washington: National Academy Press, 1997.

'The Paradoxes of Post–Cold War US Defense Policy: An Agenda for the 2001 Quadrennial Defense Review Project on Defense Alternatives.' Briefing Memo 18, 5 February 2001. Available at http://www.comw.org/pda/0102bmemo18.html.

Patrick, Stewart, and Shepard Forman eds., *Multilateralism and U.S. Foreign Policy: Ambivalent Engagement.* Washington: Center on International Cooperation Studies in Multilateralism, 2001.

Paul, James A. 'Oil in Iraq: The Heart of the Crisis.' Global Policy Forum, December 2002. Available at www.globalpolicy.org.

Payne, Keith B. *Deterrence in the Second Nuclear Age*. Lexington: University of Kentucky Press, 1996.

– *Missile Defense in the 21st Century: Protection against Limited Threats*. Washington: National Institute for Public Policy, 1991.

– 'Rational Requirements for U.S. Nuclear Forces and Arms Control: Executive Report.' *Comparative Strategy* 20 (2001): 105–28.

– and Lawrence Falk. 'Deterrence without Defence: Gambling on Perfection.' *Strategic Review* (Winter 1989): 25–40.

Payne, Keith B., Yuri Chkanikov, and Andrei Shoumikhin, 'A "Grand Compromise" with Russia on National Missile Defense?' *Defense News*, 8 May 2000.

Peterson, John. 'Europe, America and the War on Terrorism.' *Irish Studies in International Affairs* 13 (2002): 1–20.

Petersen, Scott. 'Iran's Nuclear Challenge: Deter, Not Antagonize.' *Christian Science Monitor*, 21 February 2002.

– 'US Grand Plan Proves Hard Sell Yesterday, Russia Rebuffed a Bush Proposal That Would Lift Barrier against US National Missile Shield,' *Christian Science Monitor*, 14 August 2001.

Pfaltzgraff, Robert L., ed. *Security Strategy and Missile Defense*. A special report. Washington: Institute for Foreign Policy Analysis, April 1998.

Pillar, Paul R. *Terrorism and U.S. Foreign Policy*. Washington: Brookings Institution Press, 2001.

Pipes, Daniel. 'God and Mammon: Does Poverty Cause Militant Islam?' *National Interest*, no. 66 (Winter 2001–2).

Pluchinsky, Dennis. 'Deadly Puzzle of Terrorism.' *Washington Times*, 11 September 2002.

Post-Cold War Nuclear Strategy Model - Introduction: Framing the Question.' Available at www.usafa.af.mil/inss/ocp20.htm.

Post, Jerrold, Kevin Ruby, and Eric Shaw. 'From Car Bombs to Logic Bombs: The Growing Threat of Information Terrorism.' *Terrorism and Political Violence* 12, no. 2 (2000): 97–122.

Potter, William C., and Harlan W. Jencks, eds. *The International Missile Bazaar: The New Suppliers Network*. Boulder, CO: Westview Press, 1994.

Preeg, Ernest H. 'Feeling Good or Doing Good with Sanctions: Unilateral Economic Sanctions and the U.S. National Interest.' *Significant Issues Series* 21, no. 3. Washington: Center for Strategic & International Studies, March 1999.

'Prime Minister's Position on the Iraq Crisis: 1998.' *National Network News* 10, no. 1 (2003): 6–7.

Pryce-Jones, David. 'Bananas Are the Beginning: The Looming War between America and Europe.' *National Review*, April 1999.

Rajagopalan, Rajesh. 'Nuclear Strategy and Small Nuclear Forces: the Conceptual.' *Strategic Analysis* 23 (October 1999).

Reeve, Simon. *The New Jackals*. London: Oxford University Press, 1999.

Reich, Walter. *Origins of Terrorism: Psychologies, Ideologies, Theologies, States of Mind*. Washington: Woodrow Wilson Center Press, 1998.

Rhinelander, John. 'ABM Treaty – Anachronism or Cornerstone?' Presentation to the U.S. Senate Staff hosted by Council for a Livable World, 1999. Available at http://www.clw.org/coalition/_briefv5n14.htm.

Rhodes, Edward. 'Nuclear Weapons and Credibility: Deterrence Theory beyond Rationality.' *Review of International Studies* 14, no. 1 (1988).

Richie, Alexandra. 'New Europeans Know Who Their Allies Are.' *National Post* (Toronto), 6 March 2003.

Richter, Burton. 'It Doesn't Take Rocket Science.' *Washington Post*, 23 July 2000, B02.

Rinkle, Benjamin. 'The ABM Treaty and Arms Race Myths.' *Washington Times*, 21 December 2001.

Roberts, Brad, Robert A. Manning, and Ronald N. Montaperto. 'China: The Forgotten Nuclear Power.' *Foreign Affairs* 79, no. 4 (2000).

Rose, William. 'U.S. Unilateral Arms Control Initiatives: When Do They Work? Contributions in *Military Studies* 82. New York: Greenwood Publishing Group, 1988.

Rosett, Claudia. 'Oil, Food and a Whole Lot of Questions.' *New York Times*, 18 April 2003.

Rossant, John. 'Europe Can't Afford to Stay Mad for Long.' *Business Week Online*, 20 March 2003.

Rubin, Barry. 'Don't Fight the Last War.' *Jerusalem Post*, 28 September 2001.

Ruggie, John Gerard, ed. *Multilateralism Matters: The Theory and Praxis of an Institutional Form*. New York: Columbia Univerity Press, 1993.

Rumsfeld, Donald, Defense Secretary. Transcript - Remarks on missile defense, U.S. State Department, 2 May 2001.

Russell, Charles, and Bowman Miller. 'Profile of a Terrorists.' Reprinted in *Perspectives on Terrorism*. Wilmington: Scholarly Resources Inc, 1983, 45–60.

Sadowski, Yahya. *The Myth of Global Chaos*. Washington: Brookings Institutiion Press, 1998.

Safire, William. 'Follow the Money.' *New York Times*, 21 April 2003.

Sagan, Scott, and Kenneth Waltz. *The Spread of Nuclear Weapons: A Debate Renewed*. 2d ed. New York: W.W. Norton and Co., 2003.

Sands, David R. 'Europe Warms to Missile Defense.' *Washington Times*, 6 February 2001.

Sanders, Richard. 'Who Says We're Not at War?' *Globe and Mail* (Toronto), 31 March 2003.

Sanders, Sol W. 'Cleaning the U.N. Trough.' *Washington Times*, 22 April 2003.

Sanger, David E. 'Administration Divided over North Korea.' *New York Times*, 20 April 2003.

Saracino, Peter. 'Canada Should Shun Missile System.' *Toronto Star*, 17 April 2000, A13.

Scholz, Christain M., and Frank Stahler. 'Unilateral Environmental Policy and International Competitiveness.' Kiel Institute of World Economics, January 2000.

Schorr, Daniel. 'Preempting Preemptive Action.' *Christian Science Monitor*, 3 January 2003.

Schmitt, Eric.'Back from Iraq, High-Tech Fighter Recounts Exploits.' *New York Times*, 23 April 2003.

Schweitzer, Glenn E. *A Faceless Enemy: The Origins of Modern Terrorism.* Oxford: Perseus Books Group, 2002.

Shadid, Anthony. 'Jordan to Allow Limited Stationing of U.S. Troops.' *Washington Post*, 30 January 2003, A10.

Shen, Dingli. 'China's Concern over National Missile Defence.' Paper presented at the International Conference on 'Challenges for Science and Engineering in the 21st Century,' Stockholm, Sweden, 14–18 June 2000, Session D3.

Simpson, Jeffrey. 'Choose Your Side: Puerile or Servile?' *Globe and Mail* (Toronto), 4 April 2003, A15.

– 'If Only Moral Superiority Counted as Foreign Aid.' *Globe and Mail* (Toronto), 21 January 2003, A17.

– 'Timing Is Everything for PM's New York Trip.' *Globe and Mail* (Toronto), 28 September 2001.

Skaalen, Lloyd, and Migs Turner. 'Put-up or Shut-up Canada!' *Journal of Homeland Security*, 22 March 2002.

Slocombe, Walter B. 'Preemptive Military Action and the Legitimate Use of Force: An American Perspective.' Prepared for meeting at the CEPS/IISS European Security Forum, Brussels, 13 January 2003.

Sly, Liz. 'Pakistan President Again Weighs Aiding U.S., Upsetting Populace.' Knight Ridder/Tribune News Service, 12 March 2003.

Smith, Iain. *The European Case for Missile Defense.* Washington: The Heritage Foundation, 2002.

Smith, Steve. 'The End of the Unipolar Moment: September 11 and the Future of World Order.' Essay prepared for the Social Science Research Council, 2001. Available at http://www.ssrc.org/sept11/essays/smith.htm.

Smith, Steve, and John Baylis. 'Introduction.' In *The Globalization of World*

Politics: An Introduction to International Relations. London: Oxford University Press, 2001.

Soros, George. *On Globalization.* Washington: Public Affairs Press, 2002.

Spector, Leonard, and Mark G. McDonough. *Tracking Nuclear Proliferation: A Guide in Maps and Charts.* Washington: Carnegie Endowment for International Peace, 1995.

Spencer, Jack. 'Debating the Dollars for Missile Defense.' *Washington Times,* 20 August 2001.

– 'Missile Defense: No Laughing Matter.' *Washington Times,* 27 August 2000.

Spiro, Peter J. 'The New Sovereigntists: American Exceptionalism and Its False Prophets.' *Foreign Affairs* 79, no. 6 (2000).

Standing Committee on National Defence and Veterans Affairs / Comité permanent de la défense nationale et des anciens combattants. Evidence recorded by electronic apparatus, 29 February 2000. Government of Canada. http://www.parl.gc.ca/InfoComDoc/36/2/NDVA/Meetings/Evidence/ndvaev19-e.htm.

Steinbruner, John. Testimony to Standing Committee on national Defence and Veterans Affairs / Comité Permanent de la Defense Nationale et des Anciens Combattants, Ottawa, 29 February 2000. Quoted at http://www.parl.gc.ca/infoComDoc/36/2/NDVA/Meegings/Evidence/ndvaev19-e.htm.

Stern, Jessica. *The Ultimate Terrorists.* Boston: Harvard University Press, 2000.

Steyn, Mark. 'Join America? Dream On.' *National Post* (Toronto), 20 January 2003.

– 'M. le Président's Imperiousness.' *National Post* (Toronto), 20 February 2003.

– 'The United Nations: Unfit to Govern.' *National Post* (Toronto), 28 April 2003.

Struzik, Ed. 'Biological Weapons Pose Threat to Canada, U.S., Scientist Says.' *Edmonton Journal,* 11 March 2001.

Sur, Serge, ed. 'Disarmament and Limitation of Armaments: Unilateral Measures and Policies.' New York: United Nations Publications, 1993.

Swaine, Michael, Rachael Swanger and Takashi Kawakami. *Japan and Ballistic Missile Defense.* Washington: Rand Corporation, 2001.

Telegraph (London), 'Last Chance for NATO.' Editorial, 25 April 2003.

'This Week with George Stephanopoulos.' ABC Interview with Prime Minister Jean Chrétien, 9 March 2003.

Thompson, Allan. 'Call to Arms.' *Toronto Star,* 15 February 2003.

Thompson, Mark, and Laura Brafford, 'Ready to Move In: U.S. Forces Could Be Primed to Start Fighting Iraq Again in Short Order.' *Time,* 2 December 2002: 40.

Thorne, Stephen. 'Our Ships Not in War, McCallum Says.' *Chronicle-Herald* (Halifax), 3 April 2003.

Tirenen, Walter. *White Paper for a Strategic Cyber Defense Concept: Deterrence through Attacker Identification.* University of Southern California: Information Science Institute, 2002. Available at http://www.isi.edu/gost/cctws/tirenen.html.

Toronto Star. 'Never Said Canada a Terrorist Haven: CSIS Boss.' Editorial, 29 April 2003.

Toughill, Kelly. 'Canadians Help U.S. Hunt in Gulf.' *Toronto Star*, 2 April 2003.

Traynor, Ian. 'West Scours Georgia for Nuclear Trash.' *Guardian* (Manchester), 27 March 2002.

Tupy, Marian L. 'South Africa Helps Libya Gain U.N. Human Rights Seat.' 2 February 2003. Available at www.cato.org.

Tyson, Ann Scott. 'Is More Terror in US Inevitable? "Matrix" behind Warnings: How the President Gauges Threats.' *Christian Science Monitor*, 23 May 2002.

– 'US, China Cautiously Rekindle Military Ties.' *Christian Science Monitor*, 20 February 2002.

Ungerer, Carl, and Marianne Hanson. *The Politics of Nuclear Non-Proliferation.* London: Allen & Unwin, 2002.

UN Security Council Resolution 955, UN SCOR, 49th year, 3453d mtg, at 1, UN Doc. S/RES/955, 1994.

'US Citizens Likely to Give Government More Power to Act Unilaterally.' Professionals for Cooperation 5. Moscow: Russian-American Academic Exchanges Alumni Association, 2002.

Utgoff, Victor. 'Extended Nuclear Deterrence and Coalitions for Defending against Regional Challengers Armed with Weapons of Mass Destruction.' Appendix C in *Post–Cold War Deterrence.* Washington: National Studies Board Research Council, 1997.

Vaisse, Justin. 'French Views on Missile Defense.' Washington: Brookings Institution Press, April 2001.

Van Crevald, Martin. *Nuclear Proliferation and the Future of Conflict.* New York: Free Press, 1993.

Vancouver Province. 'Hungary Open to U.S.' 25 February 2003, A6.

Vancouver Sun. 'Canadian Troops in Iraq Invisible to Chretien.' Editorial, 29 March 2003.

Von Hippel, David, Peter Hayes, et al. 'Modernizing the U.S. – DPRK Agreed Framework: The Energy Imperative.' Nautilus Institute for Security and Sustainable Development, 16 February 2001.

Vulliamy, Ed, Paul Webster, and Nick Paton. 'Scramble to Carve Up Iraqi Oil Reserves.' *Observer* (London), 6 October 2002.

Walker, Warren E. *Uncertainty: The Challenge for Policy Analysis in the 21st Century*. Santa Monica: Rand, 2001.

Wallace, Michael. 'Ballistic Missile Defense: The View from the Cheap Seats,' 2001. Available at http://www.wagingpeace.org/articles/bmd/Wallace_BMD_View_from_cheap_seats.htm.

Wallace, Richard. 'Qatar Base for War on Iraq.' *Daily Record* (U.K.), 13 September 2002, 2.

Walt, Stephen. 'Keeping the World "Off Balance": Self Restraint and U.S. Foreign Policy.' In John Ikenberry, ed. *America Unrivaled: The Future of the Balance of Power*. Ithaca: Cornell University Press, 2003.

– 'The Ties that Fray.' *National Interest* 54 (Winter 1998): 3–11.

Walt, Vivienne. 'U.S. Troops Keep Quiet on Iraq's Western Front.' *USA Today*, 17 March 2003, 05A.

Ward, John. 'First Gulf War Left Problems Unsolved,' 9 March 2003. Available at http://cnews.canoe.ca/CNEWS/Canada/2003/03/09/39880-cp.html.

Wark, Wesley. 'What's Going on Here?' *Globe and Mail* (Toronto), 5 August 2002, A11.

Warren, David. 'Shooting Down the Criticisms of Bush's Shield: Common Arguments against the U.S. Missile Defence Plan, and Why They'll Never Work.' *Ottawa Citizen*, 3 May 2001.

– 'Up with Your Missile Shield.' *Ottawa Citizen*, 22 February 2001, A18.

Washington Post. 'The Countermeasures Debate.' 9 August 2000.

Washington Times. 'A Pain in the Pyongyang.' 19 June 2001.

Watson, Roland, and Richard Beeston. 'America Pulls the Rug from under Turkey,' The Times Online, 7 March 2003. Available at http://www.timesonline.co.uk/article/0,,5470-602617,00.html.

Wax, Emily. 'France's Tentative Role in a Civil War: Troops in Ivory Coast Clash with Some Rebels While Staying Clear of Others.' *Washington Post*, 10 January 2003, A12.

Webster, William H. Arnaud de Borchgrave, Patrick R. Gallagher, Frank J. Cilluffo, and Bruce Berkowitz, eds. *Cybercrime, Cyberterrorism and Cyber warfare: Averting an Electronic Waterloo*. Washington: Center for Strategic and International Studies, 1998.

Wells, Paul. 'Germany, France: The Sum of All Fears.' *National Post* (Toronto), 30 January 2003, A13.

– 'Minister Lectured on Foreign Policy.' *National Post* (Toronto), 15 February 2003.

Wilen, Saul B. 'Counter-terrorism, Bit by Bit.' *Washington Times*, 3 April 2002.

Will, George F. 'Shrinking the U.N.' *Washington Post*, 20 February 2003, A39.

Williams, Shirley. 'An Offensive Defence.' *Guardian* (Manchester), 24 February 2001.

Wohlforth, William C. 'U.S. Strategy in a Unipolar World.' In John Ikenberry, ed. *America Unrivaled: The Future of the Balance of Power*. Ithaca: Cornell University Press, 2003.

Wolf, Jim. 'Hacking of Pentagon Computers Persists; Pace Undiminished This Year, Complicating Security Efforts.' *Washington Post*, 23 August 2000, A23.

Wolfsthal, Jon B. 'N. Korea: Hard Line Is Not the Best Line.' *Los Angeles Times*, 7 March 2001.

Worthington, Peter. 'Moral Causes and the UN Are Odd Companions,' *Toronto Sun*, 23 March 2003.

Xinhua News Agency, 'Gulf Countries Endorse UAE Call for Saddam's Exile.' 3 March 2003.

– 'First Batch of UAE Forces Arrrive in Kuwait.' 18 February 2003.

Yanarella, Ernest. *Missile Defense Controversy: Technology in Search of a Mission.* Lexington: University Press of Kentucky, 2002.

Young, Hugo. 'Secrets of Washington's Nuclear Madness Revealed.' *Guardian Weekly* (Manchester), 6 June 2000.

Zakaria, Fareed. 'Misapprehensions about Missile Defense.' *Washington Post*, 7 May 2001, A19.

Zangger Committee. Guidelines for Nuclear Transfers and Trigger List published by the IAEA as INFCIRC/209, September 1974.

Zoeller, Elisabeth. *Peacetime Unilateral Remedies: An Analysis of Countermeasures.* Ardsley, NY: Transnational Publishers, 1984.

Index